# Troubleshooting LC Systems

*Common sense is not so common.*

***—Voltaire***

# Troubleshooting LC Systems

*A Comprehensive Approach to Troubleshooting LC Equipment and Separations*

*John W. Dolan*

*and*

*Lloyd R. Snyder*

*LC Resources Inc., Walnut Creek, CA*

Humana Press • Totowa, New Jersey

**Library of Congress Cataloging in Publication Data**

Dolan, John W.
Troubleshooting LC Systems

Includes index.
1. High performance liquid chromatogrraphy—Equipment and supplies—Maintenance repair. I. Snyder, Lloyd R. II. Title.
QP519.9.H53D65 1989 543'.0894 88–34722
ISBN 0–89603–151–9


Suite 208, 999 Riverview Drive
Totowa, NJ 07512



Printed in the United States of America.

9 8 7 6 5 4

# Preface

Over the last 15 years, high-performance liquid chromatography (LC) has made the transition from an instrument used only by experts in research labs to a tool used for routine applications by relatively unskilled workers. With this transition have come major advances in instrumentation and column technology. In the past, the operator had to be a jack-of-all-trades, with a screwdriver, soldering iron, and various wrenches as constant companions in the LC lab. Today, many instruments contain microprocessors as powerful as those of mainframe computers of earlier days. With this technology has come a variety of self-diagnostic tools that allow the LC system to locate many of its own problems.

Traditionally, well-honed LC troubleshooting skills have been a result of years of work at the bench. Today the LC system itself often can do a better job of troubleshooting than the operator can. Yet many of the problems of the past are still the major problems of today: air bubbles, check valves, detector lamps, and, of course, problems with the separation. An added pressure on the operator of today's LC system is that of productivity—the lab often cannot afford unnecessary downtime. This means that the operator has to be a troubleshooting expert, or has to have that expertise at his or her fingertips. The present book was written to provide this expertise in an easy-to-use format for users at all levels of experience.

As a result of several years of teaching short courses on LC plus writing the LC Troubleshooting column for *LC/GC* magazine, we have become aware of two central themes necessary for successful LC troubleshooting. The first is an understanding of how the instrument operates. If the general principles of operation of each module in the LC system are understood, the operator is more likely to avoid problems that result from poor operational technique. The second area is preventive maintenance. For the most part, there is nothing magic about reliable LC operation. But like an automobile that needs regular oil changes and brake checks, the LC system will provide much better service when certain preventive maintenance procedures are practiced. These two needs are the reason for the strong emphasis on instrument operation and maintenance in this book.

Finally, a book cannot easily satisfy all the needs of all readers, so we have tried to increase its usefulness through its organization. For readers interested in gaining general knowledge, Chapters 6–13 cover the operation, maintenance, and troubleshooting of various modules. In general we have steered away from references to specific instrument brands and models because this would render the book out of date before it was printed. Instead, we have concentrated on generic LC system modules, so that once a technique is learned, it will be useful for years to come. Instrument-specific procedures are best found in operation and service manuals. Although some readers are interested in general information, others will come to this book with a specific problem that must be solved immediately. Study of the comprehensive index plus mastery of the troubleshooting tree of Table 2.3 should help these users find fast first-aid for the problem at hand.

We would be remiss if we did not express our sincere appreciation for the contribution of our colleagues in the review and criticism of the manuscript. Though their input has helped us identify missing topics and clarify several discussions, we accept any responsibility for inaccuracies herein. Participating in the review process were: Ben Buglio of Hoffmann-LaRoche, Ken Cohen of Boehringer-Ingelheim, Nelson Cooke of Beckman/Altex, Russell Gant of Perkin-Elmer, Joe Glajch of DuPont, Mack Harvey of Valco Instruments, Tom Jupille of LC Resources, Kerry Nugent of Michrom Bioresources, Herb Schwartz of Microphoretic Systems, Paul Upchurch of Upchurch Scientific, and Bob Weinberger of Applied Biosystems/Kratos. They greatly improved our book.

***John W. Dolan***
***Lloyd R. Snyder***

# Contents

## Section III
## Troubleshooting the Separation plus Other Problems

# Chapter 1

# INTRODUCTION



## If You Can't Wait

Most readers will be reading this chapter as a casual overview of a new book; if you fall into this category, continue. If, however, you are in a jam and have to get an LC (liquid chromatography) problem solved immediately, try one of these three approaches:

1. If you are not sure what the problem is (i.e., you only know the symptom), go to Chapter 2 and follow the flowchart until you find the cause of the problem.

2. If you have isolated the problem to a specific module, consult the appropriate chapter for that module (Chapters 6–13). There is a table of contents at the beginning of each chapter to guide you to the proper section. At the end of each chapter is a summary table giving solutions to the problems which were discussed.

3. If you know the cause of the problem, and it is not clearly classified as originating in a particular module, use the index at the end of the book to help find the proper information.

If you can't wait, read no further in this chapter.

## 1.1. An Expert at Your Fingertips

In many labs there is an expert who has the ability to fix an LC system faster and more surely than anyone else. This "guru" may be a PhD chemist with years of experience, or a technician with the proper "feel" for what's going on. It can be very frustrating to have spent several hours (or days) trying to fix an especially troublesome problem, only to have someone look over your shoulder and ask, "have you tried..." which then, of course, solves the problem.

Our goal in writing this book is to condense the expertise of many practical LC workers into a form that is useful for beginning and experienced chromatographers alike. We want this book to become your close companion and helper. There are details here on how to perform some pretty simple procedures, but—simple as they may be—they can be hard to learn unless you have someone to teach you. You should not be afraid to try any of the procedures discussed here, because there is little chance of causing irreparable or expensive damage to your LC system. We purposely have not included discussions of electronics problems, monochromator adjustments, and the like, which lie beyond the expertise of most workers.

We feel that you will be a better chromatographer and therefore be more fully equipped to solve problems quickly if you have an understanding of how the various LC modules work. For this reason, a discussion of the principles of operation of each module is included in the appropriate chapter. It is also important to have a good understanding of the chromatographic process. This is reviewed in Chapter 3.

As in the case of the expert discussed above, this book contains more details about the operation and troubleshooting of LC systems than most users care to know. When a problem arises, however, any of this information can be useful. Because our book is organized around the various system modules, there is considerable overlap in some places. We have often repeated the most useful information and cross-referenced the rest. For this reason, you may find that the index is the most expedient place to locate a specific topic.

## 1.2. LC vs HPLC

HPLC, HSLC, LC, high-performance liquid chromatography, high-pressure liquid chromatography, high-priced liquid chromatography, high-speed liquid chromatography, liquid chromatography, modern liquid chromatography.... No doubt you've heard all these terms (and more) to describe the same technique. What was high performance ten years ago is pretty low performance today; pressures are high only if there is a problem with the system. We're all talking about the same technique—separations on columns with small (3–10 μm) particles, in fairly short times (typically < 1 h), and at pressures up to 6000 psi. This is in contrast to traditional open-column or thin-layer (TLC) techniques. You probably have noticed by now that we've chosen to use just liquid chromatography (LC) to describe the technique. Although we (and many others) use it, you should note that the "LC" abbreviation is not used universally. For example, LC is not used in the description of HPLC-grade solvents (typically brand names) nor in other places where clarification is needed (e.g., to distinguish LC from another liquid chromatographic technique such as sample-cleanup columns).

## 1.3. How to Use This Book

The goal of any troubleshooting technique is to observe, isolate, and correct a problem with a minimum of time, money, and lost data. In order to quickly solve LC problems, it is necessary to understand how our book is organized and how best to use it. Then the right information can be readily accessed for a particular problem.

This book is divided into three main sections, each of which is written for a specific use. The first section, General Considerations, covers basic material that we feel all chromatographers should know in order to do their job well. Chapter 2 contains a comprehensive set of flow-charts to help you quickly isolate and correct LC problems. This is the chapter that you probably will use more than any other in the book, so you should understand how it is organized. You can read (or at least skim) the material in Chapters 3–5 as a review and/or to fill in those areas in which you need to be competent.

The second section of the book, Individual LC Modules, gives a detailed discussion of each of the major parts of the LC system. These chapters cover not only troubleshooting, but also the principles of operation and maintenance. Each chapter contains a list of spare parts that should be stocked for routine maintenance and emergency use. Most readers will not want to take the time to read all of these chapters at one sitting. You are encouraged to look them over so that you are aware of the kinds of information they contain. In this way, you'll have a headstart on finding information when you do need it.

The third section, Troubleshooting the Separation plus Other Problems, concentrates on problems related to the separation (Chapters 14–17) rather than equipment. If you work with method development or have to troubleshoot separation problems, you'll find reading Chapters 14 and 15 especially worthwhile.

We've taken extra care to make the index as comprehensive and useful as possible. Nothing is more frustrating than trying to find information in a book that has an inadequate index. With the index, Chapter 2, and the contents at the beginning of each chapter, you should be able to quickly find specific topics from a number of starting points.

The remainder of this chapter discusses how to use this book to help identify, isolate, and correct LC problems.

### *Observing That a Problem Exists*

Because it is first necessary to know that a problem exists, we stress a knowledge of the normal operation of the LC system. This includes keeping records of system history plus chromatographic performance under standard conditions, as well as using your five senses (and common sense) to know that the LC system is running properly. Written records of normal system operation can make the identification of problems easier. In order to help you decide whether an observation is normal or represents a problem, the chapters covering specific LC modules include a section describing normal operation.

### *Isolate the Problem*

When a problem exists, the problem must be classified and isolated. Chapter 4 (Principles of Troubleshooting) gives additional detail on how to classify a problem. The logic tree in Chapter 2 (Logical Approaches to Troubleshooting) helps you to quickly determine the problem area by giving Yes/No choices (or a limited number of selections). For example, you will observe the problem either (a) in the chromatogram itself, or (b) elsewhere in the system. Problems elsewhere in the system often are easier to fix, because they usually occur in the malfunctioning module. For example, a leaking fitting is easy to isolate, but problems with the chromatogram can be caused by many parts of the system (e.g., noise spikes in the chromatogram can be caused by air bubbles, a bad detector lamp, a poorly operating recorder, or something else). Further work is required to fully isolate chromatographic problems.

In this vein, it is useful to note that Chapter 2 is organized in a symptom–cause–solution manner, so all that you have to know is the symptom(s) that you see when you find a problem. The remaining chapters, on the other hand, are topical in nature and are organized in a cause–symptom–solution sequence. Thus, these chapters are most useful when you know (or suspect) the cause of the problem.

Several things should be kept in mind during problem isolation. *First,* know when to ask for help (see Chapter 5, Prevention of Problems). You should not be afraid to undertake procedures that you have not done before, but you should also know your limitations. For example, most LC users feel comfortable addressing hardware problems by themselves, but not electronic problems.

*Second,* keep notes on the effect of changes you make in the system. For example, if you suspect that an elevated system pressure is the result of a blocked injection-valve loop, record the pressure, mark the suspect loop so that you can identify it, replace the loop, and record the new pressure. Much troubleshooting time is wasted because of poor recordkeeping. This can result in (a) not being sure of what action corrected the problem, and (b, even worse) putting a faulty component back into the spare parts supply. Future troubleshooting should build on past experience, so that problems are solved more quickly.

*Third,* approach each problem in a logical, stepwise manner. This is discussed in more detail in Chapter 4 (Principles of Troubleshooting). Many times, simple Yes/No tests can be made to quickly isolate a problem. For example, when the pressure rises, you should not immediately assume that the column frit is blocked and therefore replace the frit or column. Perhaps the problem is caused by a worn injection valve or a blocked connecting tube. Follow a logical pressure-isolation sequence until the blockage is located.

Once the problem has been isolated to a particular module, you may or may not need further help in isolating the cause of the problem. Consult the "Problems and Solutions" section of the appropriate chapter for more help. Most chapters contain a final table summarizing the problems, symptoms, and solutions for the topics discussed in that chapter.

### *Correct the Problem*

Directions for the correction of many LC-system problems are given in chapters relevant to each module. These can be quickly found in one of three ways. *First,* if you are using the troubleshooting outline (Chapter 2), you will be directed to a specific section for directions. *Second,* you can use the index at the end of the book. *Third,* for added convenience, there is a table of contents at the beginning of each chapter to help you quickly locate information in that chapter. This may be more convenient to use than the index at the end of the book if you are interested in problems with just one module.

Several chapters contain exploded diagrams (Disassembly/Assembly Procedures) for some of the more common variations of the module under discussion. These, along with instructions for solving more general problems (e.g., mobile-phase degassing) will help you correct many LC problems. It should be understood, however, that this book is not a substitute for the User's Manual or Service Manual for the LC module or system. Manufacturer's manuals should always be consulted when detailed instructions are needed for the repair of a particular LC brand or model.

## 1.4. Other Sources of Troubleshooting Information

By its nature, this book cannot contain complete details on how to troubleshoot every brand and model of LC system. For this reason, you will have to rely on other sources of information, and we have listed some of these resources below.

### *Operation and Service Manuals*

Your primary source of specific details for repairing an LC module or system should be the operation and/or service manuals which came with the instrument when it was new. Though not all of these manuals are of equal quality, most contain exploded diagrams and step-by-step procedures for fixing the majority of hardware failures. If electronic diagnostics are available for your LC system, the manual should contain a discussion of the various error messages that you may encounter.

### *Manufacturer's Hot Lines*

Many LC manufacturers have toll-free (800-number) "hot" lines staffed by technicians ready to help you solve your LC problem. Even if the manufacturer does not publicly list this service, all manufacturers will provide some level of troubleshooting help —just call the company number and ask for "technical support".

### *LC/GC Troubleshooting Articles*

*LC/GC* magazine publishes a monthly column dealing specifically with LC Troubleshooting. Topics include (a) in-depth discussions of the operation and maintenance of the various LC modules, (b) case studies, and (c) responses to reader questions. This column can be a good source of up-to-date troubleshooting information. If you have a specific question, you can write to the editor for advice. Each year the December issue of LC/GC contains an index of the topics discussed for that year.

### *Troubleshooting Guides*

Several manufacturers (e.g., Phenomenex, Isco) have published short troubleshooting guides that can be obtained at no cost. Though these guides are not as comprehensive as this book, they contain compact troubleshooting trees that can help you to isolate and solve LC problems. An alternative to printed troubleshooting guides is *The HPLC Doctor,*[1] a computer-based expert system for LC troubleshooting; it uses logic trees similar to those in Chapter 2 to help you isolate and correct LC problems.

### *Troubleshooting Video*

A video course, *Troubleshooting HPLC Systems*[2] is also available. In three hours of video instruction, the course covers the highlights of this book. Watching a videotape is a good way to get started in LC troubleshooting; the live demonstrations (e.g., changing pump seals) are especially useful if you do not have someone in the lab to teach you such basic troubleshooting techniques.

### *Manufacturer's Literature*

A number of manufacturers publish short monographs[3,4] that contain valuable troubleshooting information. Though these are typically written for promotional purposes, they contain information that may not be available elsewhere. Contact the individual manufacturers for more information.

### *Contacting the Vendors*

Throughout this book we have provided illustrations of products available from various vendors. For more information, you should contact the manufacturer or distributor directly. The best source for addresses and phone numbers of these vendors is one of the buyer's guides published yearly by *LC / GC* magazine, *Analytical Chemistry, American Laboratory,* and other periodicals.

## 1.5. Summary

As with other tools that you use in the lab, you will get the most out of this troubleshooting book by using it properly. Though you may learn a great deal by reading it from cover to cover, the book was not written to be read in this way; it is more useful as a reference text. Because the use of the book often will occur under time pressure, we have tried to make its information available quickly and in a variety of formats. With the index, the logic trees of Chapter 2, the chapter contents, and the summary tables at the end of most chapters, you should be able to isolate and correct LC problems with a minimum of downtime.

## 1.6. References

[1]Jupille, T. H. and Buglio, B. (1986) *The HPLC Doctor,* LC Resources, Lafayette, CA.

[2]Dolan, J. W. and Snyder, L. R. (1986) *Troubleshooting HPLC Systems,* LC Resources, Lafayette, CA.

[3]Upchurch, P. (1988) *HPLC Fittings,* Upchurch Scientific, Oak Harbor, WA.

[4]Merrill, J. C. (1984) *A Laboratory Comparison of Popular HPLC Filters and Filter Devices for Extractables,* Gelman Sciences, Ann Arbor, MI.

# Chapter 2

# LOGICAL APPROACHES TO TROUBLESHOOTING

# HELP!

If you have a problem to solve and you don't have time to read anything else, use this chapter first. Read the following introduction to get an understanding of the organization of the chapter (this is summarized in Table 2.1). Then go directly to the troubleshooting trees (Tables 2.2 and 2.3).

## Introduction

This chapter may contain the most useful information in this book: flowcharts to guide you through LC-problem isolation and correction. The problem-solving logic is organized by the type of *symptom,* not the cause. For example, you may observe extra peaks in the chromatogram caused by an improperly set inject time on the system controller—the solution is found in the chromatogram discussion (extra peaks), because this is where the problem was first observed. If you already know what the problem is, it may be faster to use the information in later chapters (Chapters 6–17), which is organized by the type of *problem* (not the symptom).

Because different users will have different preferences on how to use this information, we have included two different sets of troubleshooting trees in this chapter (Tables 2.2 and 2.3). Each table is discussed in the following sections. A summary of how to use the various tables is contained in Table 2.1 and discussed in the following section. Tables 2.2 and 2.3 are grouped together at the end of the chapter.

## 2.1. Using the Troubleshooting Trees

This section covers how to most effectively use the troubleshooting trees of Tables 2.2 and 2.3. These steps are summarized in Table 2.1.

### *Step 1: Classify the Symptoms*

All LC problem symptoms have been classified into five (arbitrary) categories for this chapter:

Table 2.1
How to Use the Troubleshooting Trees

---

Step 1: Classify the symptoms:

Pressure (#1000)
Leaks (#2000)
Quantitation or data quality (#3000)
Hardware (#4000)
Chromatogram (#5000)

Step 2: Choose the level of detail desired

Overview (Table 2.2)
Problem area —> Primary symptom —>
Secondary symptom —> Likely cause

Detailed Information (Table 2.3)
Use Table 2.2 to find entry points for Table 2.3

Step 3: Identify the problem source (Table 2.3)

Step 4: Correct the problem

Use cross-references
Get help where necessary
Record all changes in system logbook

---

- Pressure (#1000)
- Leaks (#2000)
- Quantitation or data quality (#3000)
- Hardware (#4000)
- Chromatogram (#5000)

The number in parentheses (e.g., #2000 for leaks) indicates the entry-point line number for Table 2.3. That is, the troubleshooting tree containing information about leaks starts at line 2000. The first step in isolating the problem (Table 2.1, step 1) is to determine which of these five problem areas most closely fits the symptoms observed. Don't worry about making a mistake—you will be redirected to the proper section if you make an improper choice.

### *Step 2: Choose the Level of Detail Desired*

There are two different levels of detail in the troubleshooting trees. Experienced workers probably will want to look for the likely problem in Table 2.2, and then quit. For more complex problems, or for beginners, Table 2.2 can be used as a guide to find the proper entry point for Table 2.3.

*Overview of Likely Problems.* If you want a very simple classification of the problem symptoms plus a listing of the most likely causes of the problem, refer to Table 2.2. This problem listing is most useful for experienced workers as a reminder (checklist) of the types of things that can be responsible for a given symptom. In many cases, when you are given a list of the possible causes, you can quickly identify the most probable cause and correct the problem. For example, if the injector leaks at the syringe port, you follow the [leaks (2000A)] —> [normal, low, or erratic pressure] —> [leaks at injector] path. From the list of likely problems, you are reminded that the waste line could be siphoning. You can confirm and correct this problem quickly without further reference to the detailed flowcharts of Table 2.3. If you need more help, the entry point for Table 2.3 is given in the third column of Table 2.2 (2140B in this case).

*Entry-Point Directory for Troubleshooting Trees.* You can go directly to the detailed troubleshooting trees of Table 2.3, but it probably will be faster to find the best entry point from Table 2.2. For the syringe-port leak discussed above, you would follow the same [leaks] —> [normal, low, or erratic pressure] —> [leaks at injector] path. Then you would be directed (Table 2.2, column 3) to go to the entry point 2140B of Table 2.3 for further problem isolation. Note that you could have chosen the [hardware] —> [injector] —> [leaks] path and arrived at the same point. In both cases, you are directed to the Leaks category in the detailed problem isolation flowchart.

### *Step 3: Identify the Problem Source*

Once you enter Table 2.3, perform the tests and follow the flowchart until you find the cause of the problem. The organization and use of this table is discussed in Section 2.2 below. Two examples of using Table 2.3 are discussed in Section 2.3.

### *Step 4: Correct the Problem*

When you have identified the problem, you may be able to fix it without any further information. The troubleshooting tree often provides a cross-reference to a section or chapter in this book where you can find more information. You can also consult the contents listings at the beginning of the appropriate chapter or the index at the end of the book to help find more information. If you are directed to "get help," the problem generally is beyond the scope of this book. In this case, your primary information sources will be the operation and service manual for the LC system and/or technical support from the instrument manufacturer. When you have fixed the problem, don't forget to update the appropriate records in the system logbook.

## 2.2. Organization and Use of Table 2.3

Table 2.3 is a flowchart or logic tree based on a limited number of choices at each decision point. The table is taken in large part from *The HPLC Doctor,*[1] an LC-troubleshooting expert system that runs on a personal computer. In fact, this type of information is more easily accessed via the computer, so you may want to purchase a copy of this software to have in your lab.

### *Organization*

Table 2.3 is organized as a grid for easy cross-reference and location of specific information. Each line in the table is numbered along the left edge (the column with the number # sign). Each column is headed by a letter A–F. Thus, for example, if you are tracking a problem in the chromatogram section and a mixer problem is suspected, you are directed to 1810D (line 1810, column D) for more information on tracing mixer problems.

Each of the five major problem areas are found in a specific area of the table:

Pressure (lines #1000–1999)
Leaks (lines #2000–2999)
Quantitation or data quality (#3000–3999)
Hardware (#4000–4999)
Chromatogram (#5000–6999)

Column A contains only the major problem classification (e.g., pressure) and the primary symptom (e.g., high pressure). Column B contains the secondary symptom (if any), as well as other information. Thus, columns A and B of Table 2.3 closely parallel Table 2.2.

### *Decoding the Symbols*

The syntax of Table 2.3 is designed to be simple and easy to follow.

*Statements.* Statements (starting with no punctuation) ask questions, or give information or directions that will help you isolate the problem. These are self-explanatory: for example, (1050B) "loosen fitting at injector outlet," or (1080E) "probable pressure transducer... or pump problem."

*Observations.* The long dash (—) signifies a symptom that you observe. These symptoms usually are used as decision points (nodes) for the troubleshooting tree. For example, we are instructed (1050B) to [loosen fitting at injector outlet] in order to help trace a pressure problem. Two possible results are: the pressure is [—, still high] (1060B) or the pressure has returned to [—, normal] (1110B). Based on these observations, you are directed how to proceed to further isolate the problem (1060C or 1110C).

—> ***OK.*** When you have made a change that fixes the problem, you will see the —> **OK** statement (e.g., 1092F), indicating that you are done. All that remains is to record the problem and its fix in the system logbook.

—> ***GET HELP.*** The —> **GET HELP** message indicates that: (a) you have exhausted the ability of the troubleshooting tree to solve your problem, or (b) you need to get specific information from another source (e.g., detector flowcell rebuilding procedure from the service manual). If you feel that you have been following the wrong path to get to this point, you can try reentering Table 2.3 at another appropriate entry point.

*References.* Table 2.3 contains cross-references to other parts of the tree and to specific sections in the book where more information can be found relative to the problem at hand. For example, the statement at 1190F instructs you to reverse the column; the reference (Section 15.1) tells you where to look in the book for more information on column reversal. If you need more information than is given at the referenced section, use the index at the end of the book for additional information.

## 2.3. Two Examples

Here we discuss two examples of how the troubleshooting trees might be used in problem solving.

### *Case 1: Blocked Inlet-Line Frit*

Assume that the inlet-line frit in the reservoir is blocked. The likely symptom that we first observe is that the pump pressure fluctuates (because the pump is starving). Start with Table 2.2 with the [pressure] —> [cycling or erratic pressure] path. We are directed to start at 1590A in Table 2.3.

At 1590B we are instructed to check for leaks. There are no leaks, so we proceed to 1610C, where we need to indicate if we are using isocratic or gradient elution. Let's assume we're operating a gradient system. Now we are told (1620D) that long-term cycling is normal, but this doesn't fit our observations (short-term fluctuations), so we continue at 1630D. After helium sparging the mobile phase doesn't help (1650D), we check for restrictions in the inlet line (1660D). When we find that the solvent flow is restricted (1670D), we are instructed to replace parts until the restriction is removed (1676F). We replace the frit, record the changes in the logbook, and get back to work.

### *Case 2: One Broad Peak in the Chromatogram*

We're running an isocratic method with manual injections. The first two runs look fine, but there is a broad peak in the middle of the third and subsequent runs. The peak moves around in a window about 3 min wide from run to run, but it is always present. We start out in Table 2.2 with the [chromatogram] —> [interferences and/or extra peaks] —> [isocratic elution] path, and are directed to 6650B in Table 2.3.

In Table 2.3, we are asked whether the extra peaks are wider than their neighbors (6650C). Because the peak is wider, we are told that a late-eluting peak is likely (6660D). It turns out that the peak elutes two samples after it was injected, so it doesn't show up until the third sample. The retention time variation is the result of small differences in the time of each new injection relative to the last run. The easiest fix for the problem is to stagger the injections so that the extra peak comes out in an unimportant part of the chromatogram.

## 2.4. References

[1]Jupille, T. H. and Buglio, B. (1986) *The HPLC Doctor,* LC Resources, Lafayette, CA .

Table 2.2
Overview of Likely Problems

| Problem area | Primary symptom | Secondary symptom (Entry point)[a] | Likely causes |
|---|---|---|---|
| *Pressure (1000A)* | High pressure | (1020A) | Flow rate<br>Blockage<br>Column<br>Filter<br>Transducer |
| | Low pressure | (1250A) | Flow rate<br>Leaks<br>Transducer |
| | No pressure | No flow (1370B) | Power<br>Pump motor<br>Piston<br>Priming<br>Major leak<br>No solvent<br>Degassing |
| | | Flow observed (1510B) | Transducer |
| | Cycling or erratic pressure | (1590A) | Leaks<br>Degassing<br>Gradient<br>Blockage<br>Check valves<br>Mixer<br>Pump seals |

[a](Entry point) indicates location at which to enter Table 2.3.

*(continued)*

Table 2.2
Overview of Likely Problems *(continued)*

| Problem area | Primary symptom | Secondary symptom (Entry point) | Likely causes |
|---|---|---|---|
| *Leaks* (2000A) | High pressure | (1020A) | See Pressure |
| | Normal, low or erratic pressure | Leaks at fittings (2040B) | Fitting |
| | | Leaks at pump head (2070B) | Seals<br>Check valves |
| | | Leaks at injector (2140B) | Fittings<br>Seal<br>Syringe port<br>Waste-line siphoning |
| *Quantitation or Data-Quaity (3000)* | Poor linearity | Sudden onset (3030B) | Sample degradation<br>Wrong loop<br>Injection technique<br>Overload |
| | | Typical of this type of sample (3240B) | Overload<br>Non-linear detector |
| | Poor precision (reproducibility) | Inadequate resolution (5940A) | See chromatogram problems |
| | | Adequate resolution (3380B) | Retention time<br>Sample degradation<br>Detector problem |
| | Inaccurate results | Low levels less accurate than high levels (3510B) | Non-linear plot<br>Plot not through origin<br>Need more standards |
| | | High levels less accurate than low levels (3545B) | Overload |

*(continued)*

Table 2.2
Overview of Likely Problems *(continued)*

| Problem area | Primary symptom | Secondary symptom (Entry point) | Likely causes |
|---|---|---|---|
| *Quantiation or Data-Quality (3000) (continued)* | Inaccurate results | Accuracy of standard injections is OK, but spiked blanks are not (3550B) | Sample matrix effects |
| | | Sample peaks front, tail, or are doubled in some or all runs, but standard peaks are OK (3560B) | Interferences |
| | | Sample peaks and standards front, tail, or are doubled in some or all runs (3560B) | |
| *Hardware (4000 A)* | Pump problems | Pressure problems (1000A) | See Pressure problems |
| | | Leaks (2070B) | See Leak problems |
| | | No power (4050B) | Switch<br>Circuit<br>Fuse<br>Bad motor<br>Controller |
| | Injector problems | Leaks (2140B) | See Leak problems |
| | | Irreproducible results (3000A) | See Quantitation problems |
| | | Autosampler problem (Chapter 10) | See Chapter 10 |
| | | Doesn't turn (4180B) | Air supply<br>Binding valve<br>Bad actuator |
| | Detector problems | Noisy baseline (5020 A) | See chromatogram problems |

*(continued)*

Table 2.2
Overview of Likely Problems *(continued)*

| Problem area | Primary symptom | Secondary symptom (Entry point) | Likely causes |
|---|---|---|---|
| *Hardware (continued)* | Detector Problems | Power problems (4290B) | Power<br>Switch<br>Fuse |
| | | No signal (4370B) | Settings<br>Data system<br>Lamp |
| | | Response diminishes (6590C/6610C | Leaking cell gasket |
| *Chromatogram (5000A)* | Baseline noise or instability | Noise spikes (5030B) | Degassing<br>Detector lamp<br>Electrical interference |
| | | Short-term noise (5180B) | Detector setting<br>Electrical interference |
| | | Random noise (5240B) | Solvent impurities<br>Degassing<br>Equilibration<br>Temperature<br>Late peaks<br>Guard column<br>Column |
| | | Rhythmic noise, < 5-min cycle (5420B) | Also see pressure problems<br>Mixer |
| | | Baseline drift (5460B) | Solvent impurities<br>Equilibration<br>Temperature |
| | | Cycling baseline (> 5-min cycle) (5600B) | Electrical interference<br>Temperature<br>Mixer |

*(continued)*

Table 2.2
Overview of Likely Problems *(continued)*

| Problem Area | Primary Symptom | Secondary Symptom (Entry Point) | Likely Causes |
|---|---|---|---|
| *Chromatogram (continued)* | Low column efficiency and/or poor peak shape | Sudden onset (5640B) | Settings<br>Mixer<br>Guard column<br>Column<br>Frit<br>Extra-column effects |
| | | Typical of this type of sample (5800B) | Column<br>Guard-column |
| | | Extra-column | |
| | | | effects<br>Injection solvent<br>Overload<br>Secondary retention |
| | Insufficient resolution | Consistently inadequate for this type of sample (5940B) | See Chapters 14 and 15 |
| | | Adequate resolution previously observed (5950B) | Column<br>Guard column<br>Frit<br>Retention time |
| | Retention-time problems | $t_0$ has changed (5980B) | Flow rate<br>Leaks |
| | | $t_0$ has remained remained constant (6030B) | Equilibration<br>Temperature<br>Contaminated column<br>Guard column<br>Mobile phase<br>Overload |
| | Sensitivity changes | Retention changes (5980A) | See Retention-time problems |

*(continued)*

Table 2.2
Overview of Likely Problems *(continued)*

| Problem area | Primary symptom | Secondary symptom (Entry point) | Likely causes |
|---|---|---|---|
| *Chromatogram (continued)* | | Retention unchanged (6560B) | Settings<br>Detector<br>Data system<br>Sample decomposition<br>Column<br>Guard column |
| | Interferences and/or extra peaks | Gradient elution used (6620B) | Solvent<br>Frit<br>Column<br>Guard column<br>Injection solvent<br>Sample decomposition |
| | | Isocratic elution (6650B) | Late peaks<br>Frit<br>Column<br>Guard column<br>Injection solvent<br>Sample decomposition |

Table 2.3
Detailed Problem-Isolation Flowchart

**Notes**

1. It is most convenient to find the proper entry point for Table 2.3 by first using Table 2.2 for preliminary problem classification
2. General flow from general to specific problem classification is from left to right, and downward.
3. Table 2.3 is organized in a grid of numbered rows and lettered columns for cross-reference.
4. Major problem categories are listed below with the starting line numbers:
   - Lines #1000–1990: Pressure Problems
   - Lines #2000–2990: Leaks
   - Lines #3000–3990: Quantitation or Data-Quality
   - Lines #4000–4990: Hardware
   - Lines #5000 and above: Problems with the Chromatogram
5. See Section 2.2 for a discussion of how to use this table and Section 2.3 for examples.

| # | Column A | Column B | Column C | Column D | Column E | Column F |
|---|---|---|---|---|---|---|
| 1000 | **Pressure Problems** | | | | | |
| 1010 | Which best describes the problem? | | | | | |
| 1020 | ***—High pressure*** | Is flow-rate setting correct? | | | | |
| 1030 | | —No | Adjust —> **OK** | | | |
| 1040 | | —Setting is OK, continue below (1050B) | | | | |
| 1050 | | Loosen fitting at injector outlet | | | | |
| 1060 | | —Still high | Loosen fitting at injector inlet | | | |

| | | | | | | |
|---|---|---|---|---|---|---|
| 1070 | | | —Still high | Loosen fitting at pump outlet | | |
| 1080 | | | | —Still high | Probable pressure transducer, meter, controller, or pump problem —> **GET HELP** | |
| 1090 | | | | —Normal | If high-pressure mixer is used, loosen fittings at mixer inlet | |
| 1092 | | | | | —Pressure now is normal | Mixer is blocked; repair or replace mixer —> **OK** |
| 1094 | | | | | —Pressure still is high | Tubing blockage between pump and mixer; replace tubing —> **OK** |
| 1096 | | | | | If high-pressure mixer not used, blockage is between pump & injector; replace tubing, pre-column, and/or in-line filter —> **OK** | |
| 1100 | | | —Normal | Check for injector blockage | Refer to Table 10.9 | |
| 1110 | | —Normal | Is guard column or in-line filter in use? | | | |
| 1120 | | | —Yes | Loosen at outlet of filter or guard column | | |

*(continued)*

Table 2.3
Detailed Problem-Isolation Flowchart *(continued)*

| # | Column A | Column B | Column C | Column D | Column E | Column F |
|---|---|---|---|---|---|---|
| 1130 | ***—High pressure*** *(continued)* | | | —Still high | Loosen at inlet | |
| 1140 | | | | | —Still high | Blocked transfer line —> replace, **OK** |
| 1150 | | | | | Normal | Blocked frit, guard column, or filter —> replace, **OK** |
| 1155 | | | | —Normal | Continue at left (1160C) | |
| 1160 | | | —No | Loosen at column inlet | | |
| 1170 | | | | —Still high | Blocked transfer line —> replace, **OK** | |
| 1180 | | | | —Normal | Loosen at column outlet | |
| 1190 | | | | | —Still high and OK to reverse column | Reverse column (Sect. 15.1) |
| 1200 | | | | | | –Pressure normal–> **OK** (filter samples and/or add in-line filter upstream; see Sect. 17.2 or 11.3) |
| 1210 | | | | | | —No improvement, continue below (1220F, change frit) |

| | | |
|---|---|---|
| 1220 | —Still high and reversal not permitted | Change frit (Sect. 15.1) |
| | | —Pressure normal —> **OK** (filter samples and/or add in-line filter upstream; see Sect. 17.2 or 11.3) |
| 1230 | | —Still high; replace column —> **OK** |
| 1240 | —Normal | Loosen at detector inlet |
| 1242 | | —Still high, transfer line is blocked; replace —> **OK** |
| **1244** | | —Normal, continue below (1246F) |
| 1246 | | loosen fitting at detector outlet |
| 1248 | | —Pressure is normal, waste line is blocked; replace it —> **OK** |
| 1249 | | —Pressure is still high, detector flowcell or heat exchanger is blocked; rebuild or replace —> **GET HELP** |

*(continued)*

Table 2.3
Detailed Problem-Isolation Flowchart *(continued)*

| # | Column A | Column B | Column C | Column D | Column E | Column F |
|---|---|---|---|---|---|---|
| 1250 | ***—Low pressure*** | Check for leaks | | | | |
| 1260 | | —Leaks found | Go to Leaks (2020A) | | | |
| 1270 | | —No leaks | Check flow setting | | | |
| 1280 | | | —Incorrect | Adjust —> **OK** | | |
| 1290 | | | —Setting is OK, continue below (1292C) | | | |
| 1292 | | | Verify that proper mobile phase is being used | | | |
| 1294 | | | —Wrong mobile phase | Replace with proper formulation —> **OK** | | |
| 1296 | | | —Correct mobile phase is used | Make volumetric flow check | | |
| 1300 | | | | —Incorrect | Possible bubbles in pump; degas mobile phase, bleed pump and re-equilibrate | |
| 1302 | | | | | —> **OK** | |
| 1304 | | | | | —Flow still off, continue below (1306E) | |
| 1306 | | | | | recalibrate pump per manufacturer's procedure | |
| 1310 | | | | | —> **OK** | |

| | | | | | | |
|---|---|---|---|---|---|---|
| 1320 | | | | | —Still incorrect go to 1480C (replace check valves) | |
| 1330 | | | | —Flow OK | Check retention times | |
| 1340 | | | | | —Retention OK | Probable transducer failure —> **GET HELP** |
| 1350 | | | | | —Retention not OK | Go to Chromatogram problems, retention (5980A) |
| 1360 | —*No pressure* | Check flow at system outlet | | | | |
| 1370 | | —No flow | Verify pump operation (power on, piston indicators, noise) | | | |
| 1380 | | | —Pump not working | —> Go to Pump hardware problems (4050C) | | |
| 1390 | | | —Pump is working, but no flow is observed, continue below (1400C) | | | |
| 1400 | | | Reprime pump | | | |
| 1410 | | | —> **OK** | Take precautions to prevent loss of prime (Sect. 6.3): (a) use degassed mobile phase, (b) mount reservoir higher than pump, (c) use 10-μm inlet-line frit, etc. | | |

*(continued)*

Table 2.3
Detailed Problem-Isolation Flowchart *(continued)*

| # | Column A | Column B | Column C | Column D | Column E | Column F |
|---|---|---|---|---|---|---|
| 1420 | ***Low pressure*** *(continued)* | | —Still no flow, continue below (1430C) | | | |
| 1430 | | | Verify sufficient mobile phase in reservoir; disconnect inlet line at inlet check valve | | | |
| 1440 | | | —No solvent runs from line | Probable inlet-line blockage: (a) replace inlet-line frit and/or inlet-line, (b) elevate reservoir, (c) replace solvent proportioning valves, and/or (d) verify that reservoir is vented | | |
| 1450 | | | | —> **OK** | | |
| 1460 | | | | — No improvement —> **GET HELP** | | |
| 1470 | | | —Solvent flows freely from inlet line, continue below (1480C) | | | |
| 1480 | | | Replace check valves and reprime | | | |
| 1490 | | | —> **OK** | | | |
| 1500 | | | —No better | Probable broken piston or other | | |

| | | | | | |
|---|---|---|---|---|---|
| | | | | pump problem —> go to Pump hardware problems (4050C) | |
| 1510 | | —Flow observed | Perform volumetric flow-rate check | | |
| 1520 | | | —Incorrect | Recalibrate pump per manufacturer's procedure | |
| 1530 | | | | —> **OK** | |
| 1540 | | | | —Still no pressure | Probable transducer failure —> **GET HELP** |
| 1550 | | | —Flow OK | Check retention times | |
| 1560 | | | | | |
| 1570 | | | | —Retention OK | Probable transducer failure —> **GET HELP** |
| 1580 | | | | — retention not OK | Go to Chromatogram problems, retention (5980A) |
| 1590 | ***—Cycling or erratic pressure*** | Check for leaks | | | |
| 1600 | | —Leaks found | Go to Leaks flowchart (2020A) | | |
| 1610 | | —No leaks | Are you using isocratic or gradient-elution? | | |

*(continued)*

Table 2.3
Detailed Problem-Isolation Flowchart *(continued)*

| # | Column A | Column B | Column C | Column D | Column E | Column F |
|---|---|---|---|---|---|---|
| 1620 | —***Cycling or erratic pressure*** *(continued)* | | —Gradient | Pressure cycling at one cycle per gradient run is normal with gradients, due to viscosity changes —> **OK**; otherwise continue with isocratic option at left (1630C) | | |
| 1630 | | | —Isocratic | Helium-degas mobile phase (Sect. 6.1) and re-equilibrate | | |
| 1640 | | | | —> **OK** | Probable bubble problems in pump heads, continue to operate with de-gassed mobile phase | |
| 1650 | | | | —No better, continue below (1660D) | | |
| 1660 | | | | Check for blockage by disconnecting tubing at inlet check-valve(s) | | |
| 1670 | | | | —Solvent does not siphon freely from reservoir when disconnected | If low-pressure mixer is used, disconnect line(s) at inlet to mixer | |
| 1674 | | | | | —Siphons freely | Mixer or connecting tubing to pump blocked; repair or |

| | | | |
|---|---|---|---|
| | | | replace —> **OK** (note: some mixers may not allow siphoning when they are working normally) |
| 1676 | | —Still doesn't siphon | Probable solvent inlet-line frit or line blockage; replace frit and/or line —> **OK** |
| 1680 | —Solvent siphons freely, continue below (1690D) | | |
| 1690 | Purge check valves (Sect 7.6) | | |
| 1700 | —Better | Probable bubbles or dirt in check valves; continue operation with degassed and filtered mobile phase (Sect. 6.3) —> **OK** | |
| 1710 | —No better, continue below (1720D) | | |
| 1720 | Check low-pressure fittings, tighten carefully (Section 9.1) | | |
| 1730 | —> **OK** | | |
| 1740 | —No better, continue below (1750D) | | |
| 1750 | Replace check valves one at a time | | |

*(continued)*

Table 2.3
Detailed Problem-Isolation Flowchart *(continued)*

| # | Column A | Column B | Column C | Column D | Column E | Column F |
|---|---|---|---|---|---|---|
| 1760 | ***—Cycling or erratic pressure*** *(continued)* | | | —Better | Dirty or damaged check valve; discard or rebuild old valves | |
| 1770 | | | | —No better, continue below (1780D) | | |
| 1780 | | | | Replace pump seals (check for broken pistons) | | |
| 1790 | | | | —> **OK** | Place pump-seal replacement on periodic maintenance schedule (Sect. 7.6) | |
| 1800 | | | | —No better, continue below (1810D) | | |
| 1810 | | | | If low-pressure mixer is in use, bypass mixer and proportioning valves; place all solvent inlet lines in same reservoir | | |
| 1820 | | | | —Better | Probable defective proportioning valve and/or mixer (Sect. 7.3); replace unit —> **OK** | |
| 1830 | | | | —No better —> **GET HELP** | | |

| | | | | | |
|---|---|---|---|---|---|
| 2000 | **Leaks Observed** | | | | |
| 2010 | ***—Pressure is high*** | Go to Pressure (1020A) | | | |
| 2020 | ***—Pressure normal, low, or erratic*** | Where is the leak? | | | |
| 2040 | | —Leaks at one or more of the fittings | Tighten leaking fittings (do not over-tighten) (Sect. 9.3) | | |
| 2050 | | | —Still leaking | Replace fitting —> **OK** | |
| 2060 | | | —Leak fixed | —> **OK,** return to operation | |
| 2070 | | —Leaks at pump head | Where is the leak on the pump? | | |
| 2080 | | | —Leaks at check valve | Tighten check valve (do not overtighten) | |
| 2090 | | | | —Still leaking | Replace check valve with new or rebuilt part (Sect. 7.6) —> **OK** |
| 2100 | | | | —> **OK,** return to operation | |
| 2110 | | | —Leaks at pump seals (generally drips behind head) | Replace seals (Sect. 7.6) —> OK | Flush buffers from system daily to prolong seal life (Sect. 7.5) |
| 2120 | | | —Leaks at other Fitting on pump | Tighten or replace fitting —> **OK** | |
| 2130 | | | — leaks elsewhere on pump | —> **GET HELP** (check service manual) | |

*(continued)*

Table 2.3
Detailed Problem-Isolation Flowchart *(continued)*

| # | Column A | Column B | Column C | Column D | Column E | Column F |
|---|---|---|---|---|---|---|
| 2140 | ***—Pressure normal low or erratic*** *(continued)* | —Leaks at injector | Where does the injector leak? | | | |
| 2150 | | | —At one of the fittings | Tighten fitting | | |
| 2160 | | | | —Still leaks | Replace tube fitting,rinse valve body, and retighten | |
| 2170 | | | | | —Still leaks | Replace valve with new or rebuilt one —> **OK** |
| 2180 | | | | | —> **OK,** return to operation | |
| 2190 | | | | —Leaks out fill- or waste-port | Check for siphoning from valve or waste reservoir | |
| 2200 | | | | | —Siphoning observed | Reroute waste line to break siphon action and/or crimp line to prevent back-flow (Sect. 10.5) —> **OK** |
| 2210 | | | | | —No siphoning | Valve damaged or out of adjustment, see valve-body leaks below (2220C) |

| | | | | | |
|---|---|---|---|---|---|
| 2220 | | —Valve body leaks (generally around rotor seal) | Check that valve is adjusted to operate at least 1000 psi above system pressure | | |
| 2230 | | | —Out of adjustment | Adjust valve per manufacturer's instructions | |
| 2240 | | | | —> **OK** | Continue with operation |
| 2250 | | | | —Still leaks | Replace with new or rebuilt valve —> **OK** |
| 2260 | | | —Adjustment is OK | Replace with new or rebuilt valve —> **OK** | |
| 2270 | —Leaks at detector | Where is the leak? | | | |
| 2280 | | —Fittings | Tighten fittings (carefully) | | |
| 2290 | | | —> **OK** | | |
| 2300 | | | —Still leaks | Replace fitting —> **OK** | |
| 2310 | | —Leaks at flowcell | Is a back-pressure restrictor in use? | | |
| 2320 | | | —Yes | Remove the restrictor | |
| 2330 | | | | —Now OK | Operate with no restrictor, or adjust to lower pressure limit —> **OK** |

*(continued)*

Table 2.3
Detailed Problem-Isolation Flowchart *(continued)*

| # | Column A | Column B | Column C | Column D | Column E | Column F |
|---|---|---|---|---|---|---|
| 2340 | ***—Pressure normal, low or erratic*** *(continued)* | | | | —Still leaks | Blocked flowcell or heat exchanger; rebuild or replace —> **OK** |
| 2350 | | | | —No | Blocked flowcell or heat exchanger; rebuild or replace —> **OK** | |
| 3000 | **QUANTITATION or DATA-QUALITY PROBLEMS** | | | | | |
| 3010 | Which best describes your problem? | | | | | |
| 3020 | ***—Poor linearity*** | How did the problem arise? | | | | |
| 3030 | | —Sudden onset | Check all system parameters (especially auto-sampler settings) | | | |
| 3040 | | | —Problem located | Fix problem —> **OK** | | |
| 3050 | | | —Everything looks OK, continue below (3060C) | | | |
| 3060 | | | Prepare and inject a fresh series of standards | | | |

| | | | |
|---|---|---|---|
| 3070 | | —> **OK** | Probable standard degradation; make fresh standards more often |
| 3080 | | —No better, continue below (3090C) | |
| 3090 | | Have you recently changed the sample loop? | |
| 3100 | | —Yes, new loop | Probable error in loop size; replace loop |
| 3110 | | | —> **OK** |
| 3120 | | | — No better, continue at left (3130C) |
| 3130 | | —No, same loop; continue below (3140C) | |
| 3140 | | What type of injection are you making? | |
| 3150 | | —Partial-loop injection | Switch to filled loop injection, or inject <50% of loop volume (Section 10.1) |
| 3160 | | | —> **OK** |
| 3170 | | | —No better, continue at left (3180C) |
| 3180 | | —Filled-loop injection | Push at least 300% of loop volume through loop to assure thorough filling (Section 10.1) |

*(continued)*

Table 2.3
Detailed Problem-Isolation Flowchart *(continued)*

| # | Column A | Column B | Column C | Column D | Column E | Column F |
|---|---|---|---|---|---|---|
| 3190 | ***—Poor linearity*** *(continued)* | | | —> **OK** | | |
| 3200 | | | | —Still bad | check chromatogram for different peak shapes | —Different peaks shapes found. Possible effect on integration; adjust integration parameters —> **OK** |
| 3210 | | | | | | —Similar peak shapes or integration adjustment didn't help; check Chap. 16 and Table 16.4 for more ideas —> **GET HELP** |
| 3240 | | —Problem is typical of this type of sample | Check data system and detector settings | (Just to be sure you didn't make a mistake) | | |
| 3250 | | | —Problem found | Correct problem —> **OK** | | |
| 3260 | | | —Everything looks OK, continue below (3270C) | | | |
| 3270 | | | Are you using peak height or area? | | | |
| 3280 | | | —Using peak height | Switch to using peak area | | |
| 3290 | | | | —> **OK** | | |

| | | | | | | |
|---|---|---|---|---|---|---|
| 3300 | | | | —No better, continue at left | | |
| 3310 | | | —Using peak area | Does peak shape or retention change with sample loading? | | |
| 3320 | | | | —Yes, peaks broaden or retention changes noticeably (>5%) | Probable sample/-detector-overload problem; dilute sample 4x, increase detector sensitivity 4x and reinject (Sections 14.2, 16.2) | |
| 3330 | | | | | —Better | Adjust method for smaller sample-mass injections —> **OK** |
| 3340 | | | | | —No better, continue at left (3350D) | |
| 3350 | | | | —No changes observed | Something else seems to be wrong with your method; consult Chapter 16 and Table 16.4 for more ideas —> **GET HELP** | |
| 3360 | ***—Poor precision (reproducibility)*** | Is the resolution adequate? | | | | |
| 3370 | | —No, Rs <1.5 | —> Go to Chromatogram problems, inadequate resolution (5940A) | | | |

*(continued)*

Table 2.3
Detailed Problem-Isolation Flowchart *(continued)*

| # | Column A | Column B | Column C | Column D | Column E | Column F |
|---|---|---|---|---|---|---|
| 3380 | ***—Poor precision reproducibility*** *(continued)* | —Yes, Rs >1.5 | Are retention times constant from run to run? | | | |
| 3390 | | | —No, variations of >1-2% are seen | —> Go to Chromatogram problems, retention-time problems (5980A) | | |
| 3400 | | | — Retention times are constant, continue below | | | |
| 3410 | | | Make several injections of a well-characterized ("known") sample (e.g., column-test solution) | | | |
| 3420 | | | —Reproducibility is still poor | Likely problem with injector or injection technique; trace this in linearity section above (start at 3090C) | | |
| 3430 | | | —Now OK, continue below (3440C) | | | |
| 3440 | | | Make several injections of fresh standards | | | |

| | | | | |
|---|---|---|---|---|
| 3450 | | | —Now OK | Probable degradation of standards and/or samples; make fresh more often and/or refrigerate before use (Chapter 16) —> **OK** |
| 3460 | | | —Still bad | Probable detector problem; check the possibilities below (3470D), or —> **GET HELP** |
| 3470 | | | | (a) Outside linear range (see detector manual) |
| 3480 | | | | (b) Dirty flowcell (see manual or Section 12.2 for cleaning instructions) |
| 3490 | | | | (c) Weak lamp (replace lamp, Section 12.2) |
| 3500 | ***—Inaccurate results*** | Which best describes the problem? | | |
| 3510 | | —Low levels less accurate than high levels | Which best defines the calibration plot? | |
| 3520 | | | —Plot is linear and goes through the origin (0,0) | To a certain extent, less precision is expected with low levels due to measurement errors, but this may be construed as poor accuracy; |

*(continued)*

Table 2.3
Detailed Problem-Isolation Flowchart *(continued)*

| # | Column A | Column B | Column C | Column D | Column E | Column F |
|---|---|---|---|---|---|---|
| 3520 | ***—Inaccurate results*** *(continued)* | | | you may have to live with this, otherwise consider the other possibilities at left (3530C–3560C) | | |
| 3530 | | | —Plot is linear, but does not go through the origin | Modify method so that plot goes through the origin (see Section 16.3 for information); or see 3540D | | |
| 3540 | | | —Plot is nonlinear | Non-linear plots tend to be less accurate than linear ones; you might improve things by using more standards over your calibration range (Section 16.3) | | |
| 3545 | | —High levels less accurate than low levels | Check for sample/-detector overload (Section 16.2) —> go to 3240B | | | |
| 3550 | | —Accuracy of standard injections is OK, but spiked blanks are not | Possible sample matrix effects, see Section 16.3 for help | | | |
| 3560 | | —Sample peaks front, tail, or are | Possible interferences check peak purity | | | |

| | | | | |
|---|---|---|---|---|
| | | doubled in some or all runs | by (a) another LC method, (b) wavelength ratioing, or (c) LC/mass spec (see Section 16.3 for help) | |
| 3570 | | | Possible on-column degradation; check by trapping peak and reinjecting at higher (e.g., 10x) sensitivity | |
| 4000 | **Hardware Problems** | | | |
| 4010 | Where does the problem seem to be located? | | | |
| 4012 | ***—Buzzers, error messages, or other indicators*** | See operation or service manuals —> **GET HELP** | | |
| 4014 | ***—Reservoir or inlet lines*** | Bubbles seen in inlet line | Do the bubbles move? | |
| 4016 | | | —Yes, bubbles are drawn into pump | Poor degassing, blocked frit, loose fittings, or unvented reservoir cap likely; correct in stepwise manner —> **OK** |
| 4018 | | | —Bubbles are stationary | Generally this is not a problem; remove bubbles by opening purge valve or siphoning solvent; additional solvent degassing may help prevent this problem —> **OK** |

*(continued)*

Table 2.3
Detailed Problem-Isolation Flowchart *(continued)*

| # | Column A | Column B | Column C | Column D | Column E | Column F |
|---|---|---|---|---|---|---|
| 4020 | ***—Pump Problems*** | Which best describes the nature of the pump problem? | | | | |
| 4030 | | —Pressure problems (low pressure, high pressure, erratic pressure readings, etc.) | —> Go to Pressure to trace this problem (1000A) | | | |
| 4040 | | — leaks (behind heads, at fittings, etc.) | —> go to Leaks for this problem (2070B) | | | |
| 4050 | | —No power (display and indicator lights do not operate, no sound, etc.) | Is the pump plugged into a working power line and switched on? | | | |
| 4060 | | | —No | Remedy problem (oops!) —> **OK** | | |
| 4070 | | | —Yes, continue below (4080C) | | | |
| 4080 | | | Check fuse for continuity (look on back panel or consult operator's manual) | | | |
| 4090 | | | —Fuse is bad | Replace fuse with one of the correct rating | | |

| | | | | |
|---|---|---|---|---|
| | | | | —> **OK** |
| | | | | —Fuse blows again —> **GET HELP** |
| 4100 | | | —Fuse is OK, continue below (4110C) | |
| 4110 | | | If pump is operated by a controller, run pump in stand-alone mode or replace controller with good one | |
| 4120 | | | —Now OK | Probable controller problem —> **GET HELP** |
| 4130 | | | —Still bad | Probable major pump problem —> **GET HELP** |
| 4140 | ***—Injector problem*** | Which best describes the problem? | | |
| 4150 | | —Leaks at injector body or fittings | —> Go to leaks (2140B) | |
| 4160 | | —Poor quantitation (irreproducible results) | —> Go to quantitation (3000A) | |
| 4170 | | —Autosampler problem | This chart doesn't cover autosampler problems, see Chapter 10 and Table 10.11 for more help | |

*(continued)*

Table 2.3
Detailed Problem-Isolation Flowchart *(continued)*

| # | Column A | Column B | Column C | Column D | Column E | Column F |
|---|---|---|---|---|---|---|
| 4180 | —***Injector problem*** *(continued)* | —Injector binds or doesn't turn | Which type of valve are you using? | | | |
| 4182 | | | —Manually operated valve | Adjust rotor-seal pressure | | |
| 4184 | | | | —> **OK** | | |
| 4186 | | | | —Still binds, or leaks now | Rebuild or replace valve —> **OK** | |
| 4190 | | | —Automated valve | Is actuation electric or pneumatic? | | |
| 4200 | | | | —Electric | Verify that valve is plugged in and proper connections have been made; remove valve body or loosen rotor seal to remove load | |
| 4210 | | | | | —Actuator works OK without load | Probable worn or misadjusted valve; adjust or replace valve |
| 4220 | | | | | | —> **OK** |
| | | | | | | —Still bad, continue below |
| | | | | | | Weak actuator, replace |
| | | | | | | —> **OK** |
| | | | | | | — still bad —> **GET HELP** |

| | | | | | |
|---|---|---|---|---|---|
| 4230 | | | | —Actuator doesn't turn without load | Replace actuator |
| | | | | | —> **OK** |
| | | | | | —Still bad —> **GET HELP** |
| 4240 | | | —Pneumatic | Verify that you have adequate air pressure for valve operation (generally about 60 psi); adjust if necessary | |
| 4250 | | | | —> **OK** | |
| 4260 | | | | —Still doesn't turn, isolate valve or actuator problem using steps for electric valve above (start at 4200E) | |
| 4272 | —Difficult to fill loop (high backpressure) | Probable blocked loop; replace | | | |
| 4273 | | —> **OK** | | | |
| 4274 | | —Still hard to fill | Try a new syringe | | |
| 4275 | | | —> **OK** | | |
| 4276 | | | —Still hard to fill | Blocked port or rotor passage; rebuild, replace, or see Table 10.9 for problem isolation —> **OK** | |

*(continued)*

Table 2.3
Detailed Problem-Isolation Flowchart *(continued)*

| # | Column A | Column B | Column C | Column D | Column E | Column F |
|---|---|---|---|---|---|---|
| 4277 | ***—Detector problems*** | Which best describes the detector problem? | | | | |
| 4280 | | —Noisy baseline (peaks are seen, but baseline is unsatisfactory) | —> Go to chromatogram, baseline problems (5020A) | | | |
| 4290 | | —Power problems (indicator lamps, display, lamp, fan, etc.) | Check that detector is plugged into a working power line and the power switch is on | | | |
| 4300 | | | —Problem located | Correct problem (oops!) —> **OK** | | |
| 4310 | | | —Everything looks fine, continue below (4320C) | | | |
| 4320 | | | Check the fuse for continuity (on back panel, or check manual for location) | | | |
| 4330 | | | —Fuse is bad | Replace with fuse of proper rating | | |
| 4340 | | | | —> **OK** | | |
| 4350 | | | | —Fuse blows again or no improvement —> **GET HELP** | | |

| | | | | | | |
|---|---|---|---|---|---|---|
| 4360 | | | —Fuse is OK | Probable internal electronic failure, check internal diagnosics and/or manual —> **GET HELP** | | |
| 4370 | | — No signal is observed (detector appears to work OK, but no peaks are seen, only flat, unresponsive baseline) | Check all detector settings (especially "short" or "zero" switch) | | | |
| 4380 | | | —Problem found | Remedy problem —> **OK,** otherwise continue at left (4390C) | | |
| 4390 | | | —Everything looks OK, continue below (4395C) | | | |
| 4395 | | | Check lamp energy level on meter (if available); if no meter, inject uracil (reversed-phase) or toluene (normal phase) and look for signal at $t_0$ | —Energy too low, or no $t_0$ signal on injection | Probable lamp failure, replace lamp (Sect. 12.2) | —> **OK**<br>— problem persists, continue below at left (4400C) |
| 4400 | | | Check data-system settings and proper signal cable connections | | | |
| 4410 | | | —Problem found | Remedy problem —> **OK,** otherwise continue at left (4420C) | | |

*(continued)*

Table 2.3
Detailed Problem-Isolation Flowchart *(continued)*

| # | Column A | Column B | Column C | Column D | Column E | Column F |
|---|---|---|---|---|---|---|
| 4420 | ***—Detector problems*** *(continued)* | | —Everything looks OK, continue below (4430C) | | | |
| 4430 | | | Confirm lamp is on (look for purple glow, but don't look directly at lamp) | | | |
| 4440 | | | —Lamp is not on | Probable lamp failure, replace lamp (Sect. 12.2) | | |
| 4450 | | | | —> **OK** | | |
| 4460 | | | | — Still no signal probable electronic problem —> **GET HELP** | | |
| 4470 | | | —Lamp is on | This doesn't make sense; verify all settings and connections one more time, then —> **GET HELP** | | |
| 5000 | **Chromatogram Problems** | (See Chapters. 14, 15, and 17 for more information on chromatogram problems) | | | | |
| 5010 | Which best describes the problem? | | | | | |

| | | | | | | |
|---|---|---|---|---|---|---|
| 5020 | ***—Baseline noise or instability*** | Which best describes the problem? | | | | |
| 5030 | | —Noise spikes | Shut off the pump; is spiking better or about the same? | | | |
| 5040 | | | —Better | Helium-degas mobile phase and re-equilibrate (Section 6.4) | | |
| 5050 | | | | —> **OK,** continue to operate with degassed mobile phase | | |
| 5060 | | | | —No better | Add post-detector restrictor (don't exceed flowcell pressure limits) (Section 12.2) | |
| 5070 | | | | | —> **OK,** continue to operate with restrictor | |
| 5080 | | | | | —No better, continue below (5090D) | |
| 5090 | | | | | Onset of spiking coincided with a recent column or guard column change | Remove trapped air by flushing with 20 column-volumes of degassed mobile phase |
| | | | | | | —> **OK,** continue with operation; othererwise, continue at left (5100D) |

*(continued)*

Table 2.3
Detailed Problem-Isolation Flowchart *(continued)*

| # | Column A | Column B | Column C | Column D | Column E | Column F |
|---|---|---|---|---|---|---|
| 5100 | ***—Baseline noise or instability*** *(continued)* | | | | Recently changed from an immiscible mobile phase | Remove immiscible mobile phase by flushing with mutually-miscible solvent (e.g., isopropanol)<br>—> **OK,** continue with operation<br>—No better —> **GET HELP** |
| 5110 | | | —Spiking is about the same with pump off | Detector lamp is suspect, replace lamp (Section 12.2) | | |
| 5120 | | | | —> **OK** | | |
| 5130 | | | | —No better | Check all electrical connections; disconnect and reconnect all cables | |
| 5140 | | | | | —> **OK** | |
| 5150 | | | | | If no better, disconnect other electrical devices from same line as LC | |
| 5160 | | | | | —> **OK** | Install isolation transformer or move to different circuit |

| | | | | | |
|---|---|---|---|---|---|
| 5170 | | | | — Still bad | Substitute known good detector into system<br>—If better, you have a detector problem —> **GET HELP**<br>—If not better, you probably have data-system or chart-recorder problem—> isolate by substitution or **GET HELP** |
| 5180 | —Short-term baseline noise (buzz) | Detector has been set to a more sensitive setting recently | It is normal to have increased noise when sensitivity is increased —> **OK** | | |
| 5190 | | Detector sensitivity has not been changed | Disconnect and reconnect all electrical connections; check carefully between detector and data system | | |
| 5200 | | | —> **OK** | | |
| 5210 | | | —Still bad | Disconnect other electrical devices from same line as LC | |
| 5220 | | | | —> **OK** | Install isolation transformer or move to different circuit |

*(continued)*

Table 2.3
Detailed Problem-Isolation Flowchart *(continued)*

| # | Column A | Column B | Column C | Column D | Column E | Column F |
|---|---|---|---|---|---|---|
| 5230 | ***—Baseline noise or instability (continued)*** | | | | —Still bad | Shut detector off — if better, you have a detector problem —> **GET HELP**<br>—If not better, you probably have a data system problem —> **GET HELP** |
| 5240 | | —Noise is random | Gradient elution is being used | Noise pattern is the same from run to run | Probable impurities in mobile-phase solvents; use HPLC-grade solvents (Section 6.3) —> **OK**<br>—Still bad, see Section 17.1 (gradient elution) | |
| 5250 | | | If isocratic elution is being used, shut off the pump | | | |
| 5260 | | | —Noise is diminished or gone | Helium-degas the mobile phase and re-equilibrate (Section 6.1) | —> **OK,** continue to operate with degassed mobile phase | |
| 5280 | | | | | —No improvement, continue at left (5290D) | |

| | | |
|---|---|---|
| 5290 | Make a new batch of mobile phase and equilibrate for 20 column volumes | —> **OK**<br><br>— No improvement, continue at left (5310D) |
| 5310 | Insulate and/or thermostat column and connecting tubing (Section 11.3) | —> **OK**<br><br>—No improvement, continue at left (5320D) |
| 5320 | Equilibrate system without making injections for about an hour | |
| 5330 | —Better | Probable cause is late eluting peaks; change method to remove these peaks or elute them earlier; or use a step-gradient at the end of each run to clear late eluters —> **OK** |
| 5340 | —No better, continue below (5350D) | |
| 5350 | Change guard column if one is used | |
| 5360 | —Better | Probable cause is bleed-through from guard column; replace guard column more often —> **OK** |

*(continued)*

Table 2.3
Detailed Problem-Isolation Flowchart *(continued)*

| # | Column A | Column B | Column C | Column D | Column E | Column F |
|---|---|---|---|---|---|---|
| 5370 | —***Baseline noise or instability*** *(continued)* | | | —No better, or guard column not used, continue below (5380D) | | |
| 5380 | | | | Replace analytical column | | |
| 5390 | | | | —Better | Probable cause is old or contaminated column; use guard column and/or flush column regularly (Section 11.3) to minimize this problem —> **OK** | |
| 5400 | | | | —No better —> **GET HELP** | | |
| 5410 | | | —Problem remains, —> follow electrical interference isolation procedure in Short-term noise above (start at 5190D) | | | |
| 5420 | | —Noise is rhythmic (regular pattern, cycle <5 min) | Pressure fluctuates —> go to cycling Pressure (start at 1590A) | | | |
| 5430 | | | Pressure is steady | Switch pump off and allow system to equilibrate | | |

| | | | | | |
|---|---|---|---|---|---|
| 5440 | | | | —Better | Probable mixer problem, continue at 1810D for problem isolation |
| 5450 | | | | —No better | Probable electrical interference, detector, or data system problem, continue at Short-term noise (start at 5190D) |
| 5460 | | —Baseline drift is observed | Gradient elution is being used | Change to fresh, degassed, HPLC-grade solvents | |
| 5470 | | | | —> **OK** | Gradient elution requires the best quality solvents available (Section 17.1) |
| 5480 | | | | —No better | Some drift is Normal with gradient elution due to refractive index, absorbance, etc. (Section 17.1); try to live with this, or look for other problems under isocratic option to left (5490C) |
| 5490 | | | Isocratic elution is being used | Did onset of problem coincide with a recent change of solvent, guard column, or column? | |

*(continued)*

Table 2.3
Detailed Problem-Isolation Flowchart *(continued)*

| # | Column A | Column B | Column C | Column D | Column E | Column F |
|---|---|---|---|---|---|---|
| 5500 | —***Baseline noise or instability*** *(continued)* | | | —Yes | Allow further equilibration (at least 20 column volumes of solvent) | (Longer equilibration can be required when switching to/from ion-pairing or between LC modes) |
| 5510 | | | | | —> **OK** | |
| 5520 | | | | | —No better, continue at left (5530D) | |
| 5530 | | | | —No better (system fully equilibrated); continue below (5540D) | | |
| 5540 | | | | Improve temperature control of column, transfer lines, and/or detector by insulating or using column oven, detector heater, and so on | —> **OK**<br>—No better, continue at left (5570D) | |
| 5570 | | | | Are retention times changing from run to run | | |
| 5580 | | | | —Yes | Go to Retention-time variation (5980A) | |
| 5590 | | | | —No | Possible detector problem —> **GET HELP** | |

| | | | | |
|---|---|---|---|---|
| 5600 | | —Cycling baseline is observed (cycle >5 min) | Likely causes are: | |
| 5610 | | | Electrical interference | Go to 5190D for problem isolation |
| 5620 | | | Temperature problems | Improve temperature control of column, transfer lines, and/or detector by insulating or using column oven, detector heater, and so on; if no better —> **GET HELP** |
| 5630 | | | Mixer problems | Go to 1810D |
| 5640 | ***—Low column efficiency and/or poor peak shape*** | Sudden (unexpected) onset of problem | Check all system parameters (including detector time constant); reset if necessary | |
| 5650 | | | —> **OK** | |
| 5660 | | | —No better, continue below (5670C) | |
| 5670 | | | Are retention times OK? | |
| 5680 | | | —No | Probable mobile-phase, proportioning valve, or mixer problem —> go to Retention-time problems (5980A) |

*(continued)*

Table 2.3
Detailed Problem-Isolation Flowchart *(continued)*

| # | Column A | Column B | Column C | Column D | Column E | Column F |
|---|---|---|---|---|---|---|
| 5685 | ***—Low column efficiency and/or poor peak shape*** *(continued)* | | —Yes, continue below (5690C) | | | |
| 5690 | | | Run reference standard (column test) | | | |
| 5700 | | | —Now OK | Probable sample preparation problem, see Section 17.2 for help | | |
| 5710 | | | —No better, continue below (5720C) | | | |
| 5720 | | | Remove guard column | —Better | Probable bad guard column or extra-column effects; install new guard column and/or fix extra-column effects —> **OK** | |
| 5730 | | | | —No better, continue below (5740D) | | |
| 5740 | | | | Replace analytical column | | |
| 5750 | | | | —> **OK** | (See Section 11.3 and Chapters 14, 15 and 17 for ideas on extending column life) | |

| | | | | | |
|---|---|---|---|---|---|
| 5760 | | | | —No better, continue below (5770D) | |
| 5770 | | | | Are earlier bands broadened more or less than later ones? | |
| 5780 | | | | —Early bands are broadened more | Probable extra-column effects, check connecting tubing dimensions and fittings for proper use and assembly —> **OK** |
| 5790 | | | | —No, —> **GET HELP** | |
| 5800 | | Symptoms are typical of this type of sample | Run reference (column test) sample (Section 4.1) | | |
| 5810 | | | —No better | Probable column or plumbing problem, trace starting at sudden onset (5640C) | |
| 5820 | | | —Now OK, continue below (5830C) | | |
| 5830 | | | Check for symptoms below: | | |
| 5840 | | | —Early peaks tail more than later peaks | If possible, use weaker injection solvent (50% or less of mobile-phase strength) | |

*(continued)*

Table 2.3
Detailed Problem-Isolation Flowchart *(continued)*

| # | Column A | Column B | Column C | Column D | Column E | Column F |
|---|---|---|---|---|---|---|
| 5850 | ***—Low column efficiency and/or poor peak shape*** *(continued)* | | | —Now OK | Modify method to use weaker injection solvent so that solvent effects are minimized (Section 14.3) —> **OK** | |
| 5860 | | | | —No improvement, continue below | | |
| 5870 | | | | Use smaller injection volume (<25 uL) | | |
| 5880 | | | | —Now OK | Modify method to use smaller injection volume or use larger volume with weaker solvent (Section 14.3) —> **OK** | |
| 5890 | | | | —No improvement, continue at left (5900C) | | |
| 5900 | | | —All bands broadened or tail | Dilute sample 4x, increase detector sensitivity 4x, and reinject | | |
| 5910 | | | | —Better | Probable sample overload; modify method for smaller mass injections (Section 14.2) —> **OK** | |

| | | | | | |
|---|---|---|---|---|---|
| 5920 | | | | —No improvement, continue at left (5930C) | |
| 5930 | | | —Some bands broadened, others OK | Secondary retention likely; add triethylamine acetate, and so on —> see Section 14.7 for guidelines | |
| 5932 | | | —Angular tips of peaks | Check for proper detector and data-system settings (sensitivity, wavelength, and so on) | |
| 5934 | | | | —Wrong settings found | Correct the error —> **OK** |
| 5936 | | | | —Settings are correct | Probable sample overload —> go to 5900D to verify |
| 5938 | | | —Peaks are angu-ar at the baseline | Baseline zero is too low; reset —> **OK** | |
| 5940 | ***—Inadequate resolution*** | Resolution consistently inadequate with this type of sample | This is a method-development problem, not a troubleshooting problem; see Chapters 14 or 15, ref. 11 of Chapter 2, or DryLab software for help | | |
| 5950 | | This type of sample has previously had adequate resolution | Which best describes the loss of resolution? | | |

*(continued)*

Table 2.3
Detailed Problem-Isolation Flowchart *(continued)*

| # | Column A | Column B | Column C | Column D | Column E | Column F |
|---|---|---|---|---|---|---|
| 5955 | ***—Inadequate resolution*** *(continued)* | | —Change in band width | Go to low column efficiency (start at 5800C) | | |
| 5960 | | | —Change in retention times below (5980A) | Go to retention-time problems | | |
| 5962 | | | —Peak spacing has changed for only some of the bands | Has $t_0$ changed? | | |
| 5964 | | | | —No | Try a fresh sample | |
| 5966 | | | | | —> **OK** | Sample degradation likely |
| 5968 | | | | | —No improvement, continue below (5970E) | |
| 5970 | | | | | Try a new batch of mobile phase | |
| 5972 | | | | | —> **OK** | |
| 5973 | | | | | —No better, continue below (5974E) | |
| 5974 | | | | | Try a new column | |
| 5975 | | | | | —> **OK** | |
| 5976 | | | | | —no better, or changed selec-tivity | Secondary retention is causing trouble; use mobile-phase |

| | | | | | | |
|---|---|---|---|---|---|---|
| | | | | | | additives, different column, or modify method (see Chapters 14, 15, and 17) |
| 5977 | | | | — Yes, $t_0$ has changed | Flow rate or column dimensions have changed; correct the error —> **OK** | |
| 5980 | ***—Retention time problems*** | $t_0$ has changed | Probable flow-rate problem, check flow-rate setting on pump | | | |
| 5990 | | | —Setting is wrong | Reset —> **OK** | | |
| 6000 | | | —Setting is OK | Check pressure | | |
| 6010 | | | | —Pressure is abnormal | —> Go to Pressure problems (1010A) | |
| 6020 | | | | —Pressure is normal | This doesn't make sense; check for other symptoms (e.g., leaks in combination with high column pressure) —> **GET HELP** | |
| 6030 | | $t_0$ has remained constant | What is the nature of the retention change? | | | |
| 6040 | | | —Systematic; changes in the same direction from one run to the next | Make sure column has been equilibrated with at least 20 column-volumes of mobile phase before injection | | |
| 6050 | | | | —> Now **OK** | Allow sufficient equilibration before injection | |

*(continued)*

Table 2.3
Detailed Problem-Isolation Flowchart *(continued)*

| # | Column A | Column B | Column C | Column D | Column E | Column F |
|---|---|---|---|---|---|---|
| 6060 | —***Retention time problems*** *(continued)* | | | — No change, continue below (6070D) | | |
| 6070 | | | | Control temperature of column (insulate or use column oven) | | |
| 6080 | | | | —> Now **OK** | Protect column from temperature changes | |
| 6090 | | | | — No better, continue below (6100D) | | |
| 6100 | | | | Reactivate column per manufacturer's instructions or strong-solvent flush (Section 11.3) | | |
| 6110 | | | | —> Now **OK** | Buildup of contaminants on column should be minimized by (a) better sample cleanup, (b) use of guard column, and/or (c) regular column flushing (Section 11.3) | |
| 6120 | | | | —Little or no improvement, continue below (6130D) | | |

| | | | |
|---|---|---|---|
| 6130 | | Change guard column and/or analytical column | |
| 6140 | | —> **OK** | Change guard column regularly to protect analytical column |
| 6150 | | — No change, continue below (6160D) | |
| 6160 | | Try a new batch of mobile phase; cap reservoir and reduce helium sparge to trickle | |
| 6170 | | —> **OK** | Probable evaporative loss of one or more mobile-phase components; use setup that minimizes evaporative loss (Section 6.4) |
| 6180 | | — Still have problems —> **GET HELP** | |
| 6190 | — Retention changes randomly (unpredictable from one run to the next) | Does problem correlate with the mass of sample injected? | |
| 6200 | | —Yes, higher-mass samples are worse | Probable column overload, check by diluting sample 4x, increasing detector sensitivity 4x and reinjecting |

*(continued)*

Table 2.3
Detailed Problem-Isolation Flowchart *(continued)*

| # | Column A | Column B | Column C | Column D | Column E | Column F |
|---|---|---|---|---|---|---|
| 6210 | —***Retention time problems*** *(continued)* | | | | —Improved | Adjust method so that smaller sample mass is injected (Section 14.2) —> **OK** |
| 6220 | | | | | —No better, continue at left (6230D) | |
| 6230 | | | | —Not mass-dependent, continue below | | |
| 6240 | | | | Are you using gradient or isocratic elution? | | |
| 6250 | | | | —Gradient | Allow longer equilibration time between runs | |
| 6260 | | | | | —Improvement noted | Adjust method to allow for better re-equilibration between runs |
| 6270 | | | | | —No change, continue with isocratic option at left (6280D) | |
| 6280 | | | | —Isocratic, continue below (6290D) | | |

| | | | | |
|---|---|---|---|---|
| 6290 | | How is the mobile phase being prepared? | | |
| 6300 | | —On-line automatic mixing | Premix mobile phase and run with all solvent inlet lines in the same reservoir | |
| 6310 | | | —Now OK | Probable mixer problem, repair or replace mixer —>**OK** |
| 6320 | | | —No difference, continue at left (6330D) | |
| 6330 | | —Manual mixing, continue below (6340D) | | |
| 6340 | | Control temperature of column (insulate or use column oven) | | |
| 6350 | | —> Now **OK** | Protect column from temperature changes | |
| 6360 | | —No better —> **GET HELP** | | |
| 6370 | —Retention changes abruptly, then stays constant for a number of runs | Did retention change correlate with a change in mobile phase? | | |

*(continued)*

Table 2.3
Detailed Problem-Isolation Flowchart *(continued)*

| # | Column A | Column B | Column C | Column D | Column E | Column F |
|---|---|---|---|---|---|---|
| 6410 | ***—Retention time problems*** *(continued)* | | | —Yes | Check for use of improper mobile phase, correct; and/or allow longer equilibration before injecting —> **OK** | |
| 6420 | | | | —No, continue below (6430D) | | |
| 6430 | | | | Did retention change correlate with a column and/or guard-column change? | | |
| 6440 | | | | —Yes | Make sure equivalent columns are used (same manufacturer and part number); method may need adjustment to allow for column-to-column variation(Section 11.4) —> **OK** | |
| 6450 | | | | —No, continue below (6460D) | | |
| 6460 | | | | Did retention change correlate with a change in ambient temperature? | | |

| | | | | |
|---|---|---|---|---|
| 6470 | | | —Yes | Protect column from temperature changes (insulate or use column oven) —> **OK** |
| 6480 | | | —No, continue below (6490D) | |
| 6490 | | | Flush and/or reactivate column (see manufacturer's instructions and/or Section 11.3) | |
| 6500 | | | —Retention returned to correct valve | Probable column contamination; minimize this problem by (a) changing sample pretreatment, (b) using guardcolumn, and/or (c) flushing column regularly (Section 11.3) —> **OK** |
| 6510 | | | —No change, continue below (6520D) | |
| 6520 | | | Replace column and/or guard column | |
| 6530 | | | —Retention now OK | Probable column contamination, see recommendations directly above (6500E) —> **OK** |
| 6540 | | | —No improvement —> **GET HELP** | |

*(continued)*

Table 2.3
Detailed Problem-Isolation Flowchart *(continued)*

| # | Column A | Column B | Column C | Column D | Column E | Column F |
|---|---|---|---|---|---|---|
| 6550 | ***—Changes sensitivity*** | Change in sensitivity coincides with changes in retention | Go to Retention time problems (5980A) | | | |
| 6560 | | Check all system parameters (especially detector wavelength and sensitivity, data-system sensitivity and threshold, and sample volume and concentration) | | | | |
| 6570 | | —Settings are wrong | reset —> **OK** | | | |
| 6580 | | —Settings are OK | Which best describes the plate number for all peaks in the chromatogram? | | | |
| 6590 | | | —All peaks show good plate numbers | Detector or data-system problem suspected —> **GET HELP** | | |
| 6600 | | | —Some peaks are OK, others show low plate numbers | Probable sample decomposition; rerun a fresh sample —> **OK;** | | |

| | | | | |
|---|---|---|---|---|
| | | | | otherwise check Chapters 14 and 15 for ideas |
| 6610 | | | —All peaks show low plate numbers | Go to Low column-efficiency section above (5640A) |
| 6620 | ***—Interferences and/or extra peaks*** | Gradient elution is being used | Run a blank gradient | |
| 6630 | | | —The interfering peaks appear in the blank gradient | Extra peaks are solvent impurities use better quality solvents or adjust method to move interfering peaks relative to peaks of interest (Section 17.1) —> **OK** |
| 6640 | | | —No interfering peaks are seen, continue below (6650C) | |
| 6650 | | Isocratic elution is being used | Are extra peaks about the same width or wider than neighboring peaks? | |
| 6660 | | | —Extra peaks are wider | Probable late eluting peaks; (a) allow more time between injections, (b) adjust injection time so extra peaks are an unimportant part of run, (c) use a strong solvent step at the end |

*(continued)*

Table 2.3
Detailed Problem-Isolation Flowchart *(continued)*

| # | Column A | Column B | Column C | Column D | Column E | Column F |
|---|---|---|---|---|---|---|
| 6660 | ***—Interferences and/or extra peaks*** *(continued)* | | | of the run to strip late eluters from the column, and/or (d) improve sample cleanup —> **OK** | | |
| 6670 | | | —Extra peaks are about same width, continue below (6680C) | | | |
| 6680 | | | Are extra peaks paired (double peaks with sample peaks? | | | |
| 6690 | | | —Double (distorted) peaks are observed | Which peaks are doubled? | | |
| 6700 | | | | —All peaks are doubled | Column void or blocked frit suspected; back-flush, replace frit, or replace column (Sections 11.3, 11.4) —> **OK;** otherwise continue at left (6710E) | |
| 6710 | | | | —Only early peaks are doubled | Is injection solvent stronger than mobile phase? | |
| 6720 | | | | | —Yes | Use injection solvent no stronger than mobile phase (Sect. 14.12) —> **OK;** otherwise continue at left (6725E) |

| | | | | |
|---|---|---|---|---|
| 6725 | | | —No | Possible column void or blocked frit; see 6700E for corrective measures —> **OK;** otherwise continue at left (6730E) |
| 6730 | | —Only late peaks are doubled | Suspected sample contaminant or degradation | (a) use fresh sample, (b) modify method to resolve interferences, and/or (c) use better cleanup —> **OK;** otherwise —> **GET HELP** (see Chapters 14 and 15) |
| 6740 | —No peak doubling is observed | Inject a freshly-made sample | | |
| 6750 | | —> **OK** | Sample degradation likely | |
| 6760 | | —Still extra peaks | Sample interferences likely; confirm by increasing *N*, modifying mobile phase, or changing method; change sample cleanup if necessary (Chapter 17) | |

# Chapter 3

# SEPARATION BASICS

## Introduction

Effective troubleshooting requires an understanding of how the LC system works. It is assumed that the reader of this book has already been exposed to the fundamentals of LC separation. In this chapter we will review some of these basics, particularly those that relate directly to troubleshooting. For further information, consult Snyder and Kirkland.[1]

## 3.1. The Separation Process

A simple LC system is illustrated in Fig. 3.1. A bottle or mobile-phase *Reservoir* holds a liquid that will be delivered by a high-pressure *Pump* to the remainder of the system. The mobile-phase leaving the pump next flows through a sample-injection system or *Sample Injector.* This module allows a precise volume of sample solution to be inserted into the flowing mobile-phase stream. The mobile phase plus sample enters the *Column,* where the sample will be separated into its components. The sample components eventually leave the column and are measured by the *Detector.* The response of the detector is recorded by an analog or digital device to provide a record of the separation. This can in turn be converted into an analytical result by either manual or electronic means. In some cases an analog *Recorder* is used, in other cases a *Data Processor* provides for the storage and interpretation of raw data.

The separation of a three-component sample within the LC column is illustrated in Fig. 3.2. A hypothetical mixture represented by circles, squares, and triangles is shown at the time of sample injection (Fig. 3.2A), and at later times during the separation (Figs. 3.2B,C). It is seen that as the sample is carried

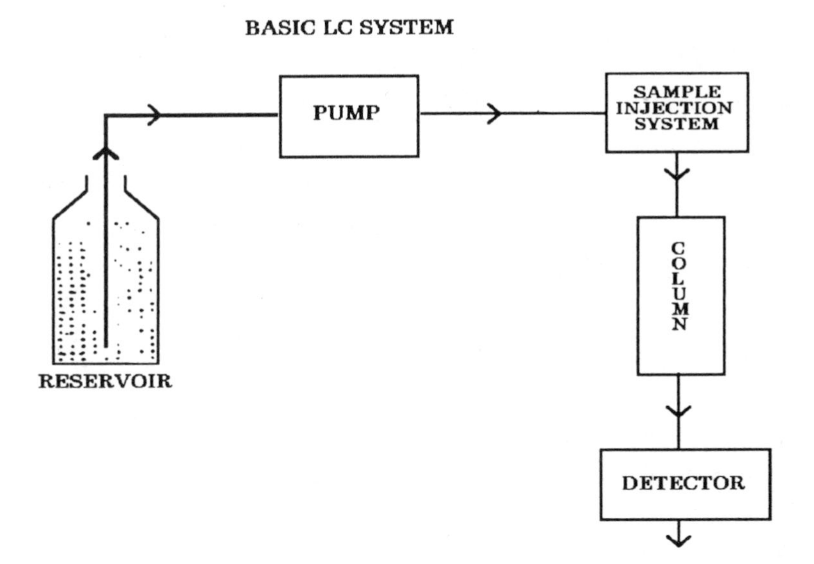


**Fig. 3.1.** Schematic of a simple LC system.

through the column by the flow of mobile phase, the different compounds move at different average speeds (differential migration) and eventually separate. Also seen in Fig. 3.2C is the spreading of molecules of the same kind (e.g., triangles) so as to occupy a larger volume of the column than at the time of injection. Eventually the sample components leave the column and are measured by the LC detector—as seen in Fig. 3.2D.

These two processes—differential migration of different compounds and molecular spreading of the same compound—determine the final separation. If these two phenomena are controlled, the desired separation can be obtained (as in Fig. 3.2D). Four characteristic features of the chromatogram can be noted:

1. Each compound leaves the column as a symmetrical *Peak* or *Band* (Gaussian curve).

2. Each compound appears in the chromatogram at a characteristic time—the *Retention Time* $t_R$ of the band (illustrated for band B of Fig. 3D).

3. Adjacent bands are separated by a certain time, equal to the difference in the $t_R$-values; larger retention time differences mean better separated bands.

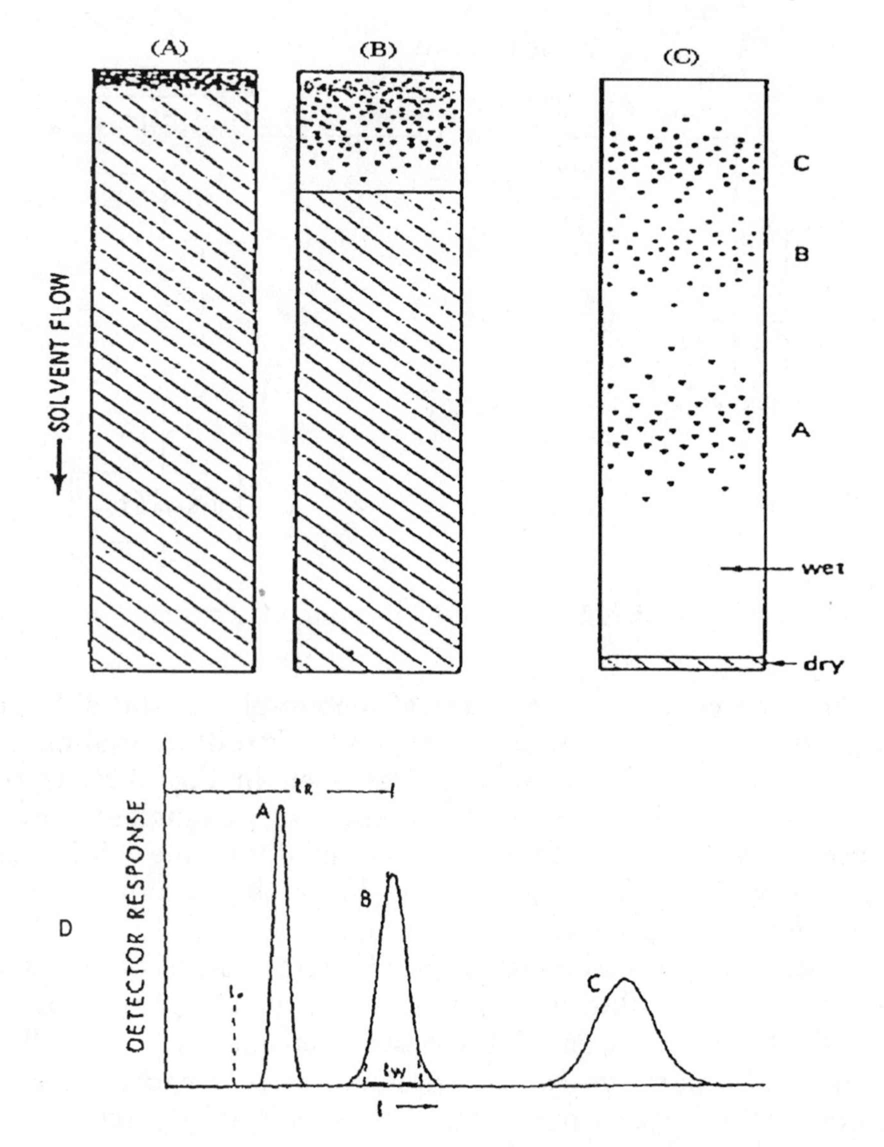


**Fig. 3.2.** Separation of a hypothetical three-component sample by LC. (A-C) Separation inside the column; (D) the resulting chromatogram.

4. Each band has a certain *bandwidth*, which is seen to increase from band A to B to C in Fig. 3.2D; wider bands are more poorly separated from neighboring bands than are narrower bands.

Table 3.1. Different LC Methods

| Method | Description |
|---|---|
| Reversed-phase chromatography | Uses nonpolar column packings and water/organic mobile phases; e.g., 50% methanol/water by volume; the most commonly used method |
| Ion-pair chromatography | Similar to reversed-phase LC, except an "ion-pairing" reagent plus buffer is added to the mobile phase; used for separating acids and bases |
| Normal-phase chromatography | Uses polar column packings and organic mobile phases; e.g., 20% ethyl acetate/hexane by volume |
| Size-exclusion chromatography | Uses inert packings with pores of a defined diameter; mobile phases can be aqueous or organic; used for separating compounds by molecular weight, especially macromolecules |
| Ion-exchange chromatography | Uses columns with charged functional groups that bind ions of opposite charge; mobile phase usually a mixture of buffer, salt, and water |

LC separation can be carried out using different kinds of columns, leading to different LC methods. The presently used LC methods are summarized in Table 3.1.

### *Pressure and Flow Relationships*

The flow of mobile phase through the column can be set to a particular value, the *Flow Rate,* which usually is measured in mL/min. Depending on the flow rate and other separation conditions a certain *Pressure Drop* occurs across the column. This pressure accounts for most of the pressure reading measured at the head of the column (and indicated by the LC controller or pressure gauge). Although LC operation typically involves high pressures (e.g., 1000–5000 psi), the system pressure must be maintained below some maximum value. Therefore it is important to be able to predict how pressure $P$ will change as conditions are varied. We can express this relationship in the following way:

$$P = 250\, L\, \eta\, F/d_p^{\,2}\, d_c^{\,2}$$

$$= 1200\, L\, \eta\, F/d_p^{\,2} \quad \text{(0.46-cm id columns)} \qquad (3.1)$$

Here $L$ is the length of the column (cm), $\eta$ is mobile-phase viscosity in cPoise, $F$ is the flow rate (mL/min), $d_c$ is the column internal-diameter (cm), and $d_p$ is the diameter (µm) of particles of column-packing within the column. In reversed-phase LC (the most common LC method), values of $\eta$ range between 0.5 and 1.5, the particle size is usually 5 µm, and columns of 15–25 cm in length are common. For a flow rate of 1 mL/min, the column pressure therefore usually will be between 500 and 2000 psi.

The pressure (with flow rate constant) can vary for any of the following reasons:

1. Change in the mobile phase composition or temperature (increase in temperature lowers pressure)
2. Change in the column (length or particle size)
3. Blocking of the column by particulate matter (a normal process that usually causes a gradual increase in pressure, other conditions the same)

This assumes that the column is well packed with particles of a narrow size distribution, which is the case for most commercially available columns today.

Table 3.2 summarizes viscosity values for the common mobile phases used for reversed-phase LC. Viscosities and other solvent properties of interest in LC are given in refs. (2,3). A copy of one of the latter books should be in every LC laboratory.

## 3.2. Retention in LC

When various separation conditions (column, mobile phase, temperature) are held constant, a given compound leaves the column with a particular retention time $t_R$. Because the retention time does not change from sample to sample, values of $t_R$ can be used to identify a particular band within the chromatogram. Let's

Table 3.2. Viscosity of Mobile Phases Used in Reversed-phase Chromatography (Water/Organic Mixtures at 25°C)

| Organic, % volume | Viscosity, cpoise | | |
|---|---|---|---|
| | Methanol | Acetonitrile | Tetrahydrofuran |
| 0 | 0.89 | 0.89 | 0.89 |
| 20 | 1.40 | 0.98 | 1.22 |
| 40 | 1.62 | 0.89 | 1.38 |
| 60 | 1.54 | 0.72 | 1.21 |
| 80 | 1.12 | 0.52 | 0.85 |
| 100 | 0.56 | 0.35 | 0.46 |

assume that in our LC procedure the compound anthracene has a retention time of 11.0 min (as measured for a pure sample of anthracene). Now further assume that we inject a sample suspected to contain anthracene. If we see a band with $t_R = 11.0$ min, we can tentatively assume that this band corresponds to anthracene. Of course, it is always possible that some other compound also has the same retention time—in which case we would be mistaken in our identification of the 11.0 min band. However, this is not very common.

Besides using retention times to identify different compounds in the sample, we can change retention times in order to improve the separation (by changing separation conditions). We will discuss this in the next section. In either case, we need to know how retention varies with certain conditions. Retention time $t_R$ can be expressed as:

$$t_R = t_0 (1 + k') \qquad (3.2)$$

Here $t_0$ is the *column Dead-Time,* equal to the retention time for solvent or mobile-phase molecules. The quantity k' is the *Capacity Factor* of the band. The column dead-time can be measured or estimated in various ways that are summarized in Fig. 3.3. Sometimes a characteristic, odd-shaped "blip" is seen at the beginning of the chromatogram (Fig. 3.3A). In other cases, the first disturbance in the baseline resembles that in Fig. 3.3B. In

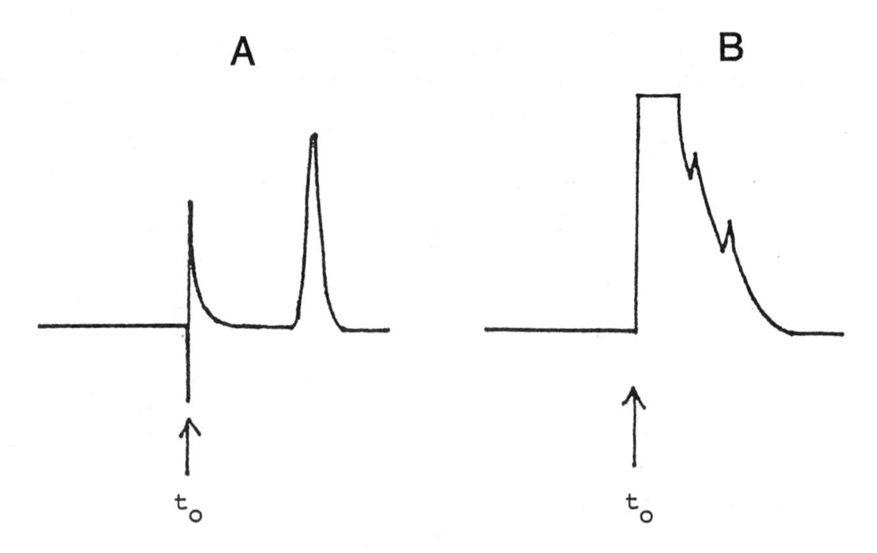


**Fig. 3.3.** The measurement of the column dead-time $t_0$.

either case, the arrows in Fig. 3.3 indicate the value of $t_0$ (the *x*-axis of these and other chromatograms is always in time units).

We can also approximate $t_0$ from the equation

$$t_0 \cong 0.5\, L\, d_c^{\,2}/F \tag{3.3}$$

Rough values of $t_0$ from Eq. 3.3 allow us to confirm that a $t_0$-value determined as in Fig. 3.3 is correct. Sometimes it is not obvious that the first baseline disturbance is actually at $t_0$.

Once we have a value of $t_0$ for a given column, we can calculate its *Dead Volume,* $V_m$:

$$V_m = t_0 F \tag{3.4}$$

Knowing $V_m$, we always can calculate $t_0$ for that column from $V_m$ and the flow rate used for a particular LC run. Measurements of $t_0$ for a particular chromatogram will often prove useful in troubleshooting.

The band capacity-factor $k'$ remains the same as flow rate is changed. This parameter ($k'$) is important both for understanding and controlling separation, and also in troubleshooting. We can obtain $k'$ from Eq. 3.2:

$$k' = (t_R - t_0)/t_0 \tag{3.5}$$

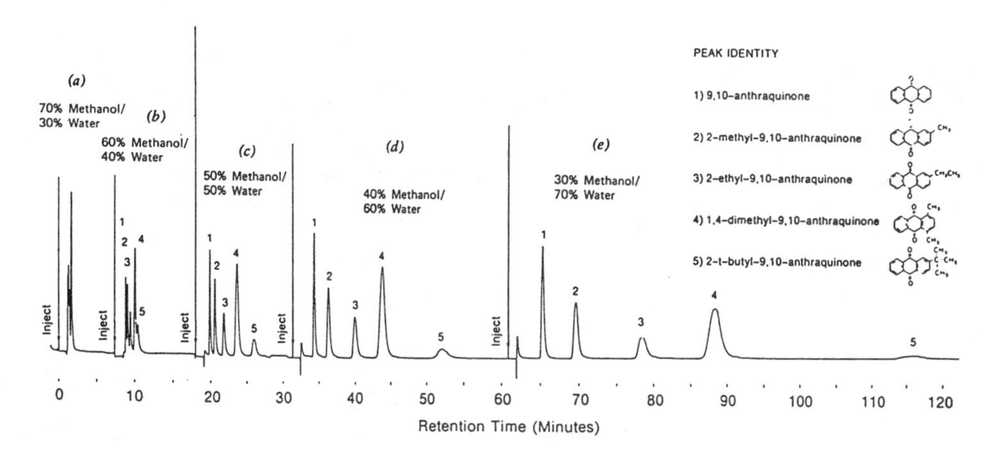


**Fig. 3.4.** The effect of solvent composition (solvent strength) on LC separation. Sample, a mixture of five alkyl anthraquinones; reversed-phase C18 column, mobile phase, methanol/water mixtures as indicated. Reprinted from ref. (1) with permission.

## *Solvent Strength*

In LC separation it is important to be able to vary sample retention; i.e., to increase or decrease values of $k'$ and $t_R$. This is almost always done by varying mobile phase composition or *Solvent Strength.* An example is shown in Fig. 3.4. Here the same sample is repetitively injected, with a change in the composition of the methanol/water mobile phase for each run. The first injection is with 70% methanol/water by volume and the last injection is with 30% methanol/water by volume. As the percent methanol volume decreases, it is seen that sample retention increases. The first run takes about 4 min for the sample to clear the column, while the last run requires about an hour. We refer to solvents that give short run times or reduced retention (small $t_R$ and $k'$ values) as *Strong Solvents.* Likewise, solvents that give longer run times and increased $k'$ or $t_R$ values are called *Weak Solvents.* In reversed-phase LC, water is a weak solvent and methanol is a strong solvent. As we increase the percent methanol in a methanol/water mobile phase, the mobile phase becomes stronger.

Table 3.3 summarizes some useful information on solvent strength in LC. In part A of the table, the strength of common reversed-phase solvents is tabulated with methanol/water as a

Table 3.3. Solvent Strength in LC

(A) *Reversed-phase chromatography—equal-solvent-strength mixtures (data from ref. 4).*

| Methanol/ water* | Acetonitrile/water* | Tetrahydrofuran/water* | Relative $k'$ |
|---|---|---|---|
| 0 | 0 | 0 | 100 |
| 10 | 6 | 4 | 40 |
| 20 | 14 | 10 | 16 |
| 30 | 22 | 17 | 6 |
| 40 | 32 | 23 | 2.5 |
| 50 | 40 | 30 | 1 |
| 60 | 50 | 37 | 0.4 |
| 70 | 60 | 45 | 0.2 |
| 80 | 73 | 53 | 0.06 |
| 90 | 86 | 63 | 0.03 |
| 100 | 100 | 72 | 0.01 |

B) Normal-phase chromatography—increasing solvent strength of pure solvents

1. Hexane (FC113†) (weak)
2. Chloroform
3. Methylene chloride
4. MTBE†
5. Ethyl ether
6. Tetrahydrofuran
7. Ethyl acetate
8. Acetonitrile
9. Propanol
10. Methanol (strong)

*10%v decrease in the organic component increases $k'$ by 2- to 3-fold

†FC-113 is 1,1,2-trifluorotrichloroethane; MTBE is methyl-*t*-butyl ether.

reference. Thus 30% methanol/water by volume gives about the same run times (same solvent strength) as 22% acetonitrile/water by volume or 17% THF/water by volume. Increasing the organic solvent concentration in the mobile phase by 10%v will generally decrease $k'$ or $t_R$ by about two- to threefold. In Table 3.3B the relative solvent strength of different commonly-used solvents is shown for normal-phase LC (e.g., silica or polar-bonded-phase column). Nonpolar solvents such as hexane are weak, and polar solvents such as methanol are strong. Mixtures of a weak and strong solvent will have intermediate strength.

### *Band Spacing*

A change in solvent strength (change in %v of the stronger solvent) generally increases the retention times of all the bands in

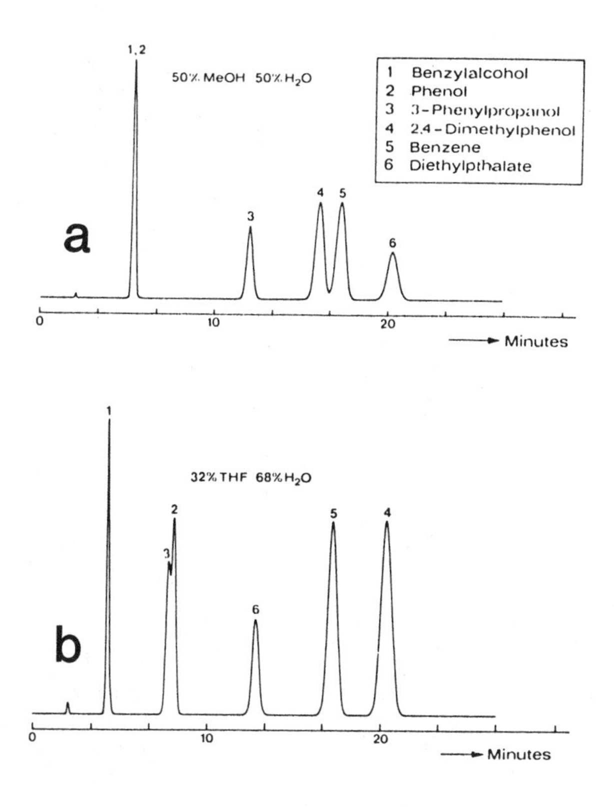


**Fig. 3.5.** The effect of changing the strong solvent in reversed-phase separation of a six-component mixture. (a) methanol/water mobile phase; (b) THF/water mobile phase. Reprinted from ref. (4) with permission.

the chromatogram. Other kinds of changes in experimental conditions can rearrange the bands in the chromatogram. This is illustrated in Fig. 3.5. In the chromatogram of Fig. 3.5A, with methanol/water as the mobile phase, the sequence of retention times for the six bands is

$$1 = 2 < 3 < 4 < 5 < 6$$

For THF/water as mobile phase, this retention sequence changes to

$$1 < 2 \cong 3 < 6 < 5 < 4$$

that is, retention is quite different. Several kinds of changes in experimental LC conditions can cause changes in the relative retention or position of bands within the chromatogram:

1. Change in the strong solvent (e.g., Fig. 3.5)
2. Change in pH
3. Change in the column packing
4. Change in temperature
5. Change in mobile phase additives (e.g., ion-pairing reagents, triethylamine, etc.)

### *Change in Retention*

Troubleshooting in LC often involves determining sample retention. The problem may be that retention has changed, so that the LC method no longer works (run times too long, bands not separated, etc.). Or small changes in retention may accompany some other problem. In this case, the change in retention may tell us something about the cause of the main problem. For these and other reasons, it is important for us to know how any change in separation conditions will affect values of $k'$ or $t_R$. Table 3.4 summarizes this information. A change in flow rate $F$ is seen to decrease $t_0$ and run-time proportionately, but has no effect on relative retention (band spacing). A similar pattern is observed for a change in column length or diameter; a resulting increase in column volume $V_m$ increases $t_0$ and run-time proportionately, but has no effect on relative retention. Increasing the %v of the strong solvent (e.g., from 30 to 50 %v) has no effect on $t_0$, reduces run-time, and can have a moderate effect on relative retention. The remaining variables of Table 3.4 have little effect on $t_0$, but usually change run-time and relative retention significantly.

## 3.3. Bandwidth and Separation

So far we have not said much about molecular spreading (e.g., triangles of Fig. 3.2C) and bandwidth. Bandwidth is one of the most important separation parameters we deal with in LC. When all the bands in the chromatogram are sufficiently narrow, the overall separation generally will be good. As the bands become

Table 3.4.
Effect of Different Conditions on Sample Retention

| Change in separation | Effect on retention time | | |
|---|---|---|---|
| | $t_0$ | Run time | Band spacing |
| Flow rate, $F$ | $1/F$ | $1/F$ | None |
| Column volume, $V_m$ | $V_m$ | $V_m$ | None |
| Increase in vol-% strong solvent | None | Decrease | Small change |
| New strong solvent | None | Changes | Changes |
| pH value | None | Changes | Changes |
| Column packing (e.g., cyano vs C18) | Little | Changes | Changes |
| Increase temperature | None | Decrease | Small change |
| New mobile phase additives | None | Changes | Changes |

broader, the separation steadily deteriorates. A general objective of LC method development is to obtain adequately narrow bands for compounds of interest (analytes). Likewise, troubleshooting often involves restoring bands to their original (narrow) widths.

### *Column Plate Number N*

Bandwidth is commonly assessed in LC by measuring the *Plate Number N* for one or more bands in the chromatogram. This is often done as illustrated in Fig. 3.6. For the tangent method, a ruler is placed along each side of the band and a tangent is drawn to the baseline. The resulting width measured on the baseline (baseline bandwidth) is defined here as $W_{TAN}$. Now the column plate number (for this first band) is given as

$$N = 16\,(V/W_{TAN})^2 \qquad (3.6)$$

where $V$ is equal to $t_R$.

The resulting plate number $N$ is a good measure of column performance—the ability of the column to provide narrow bands

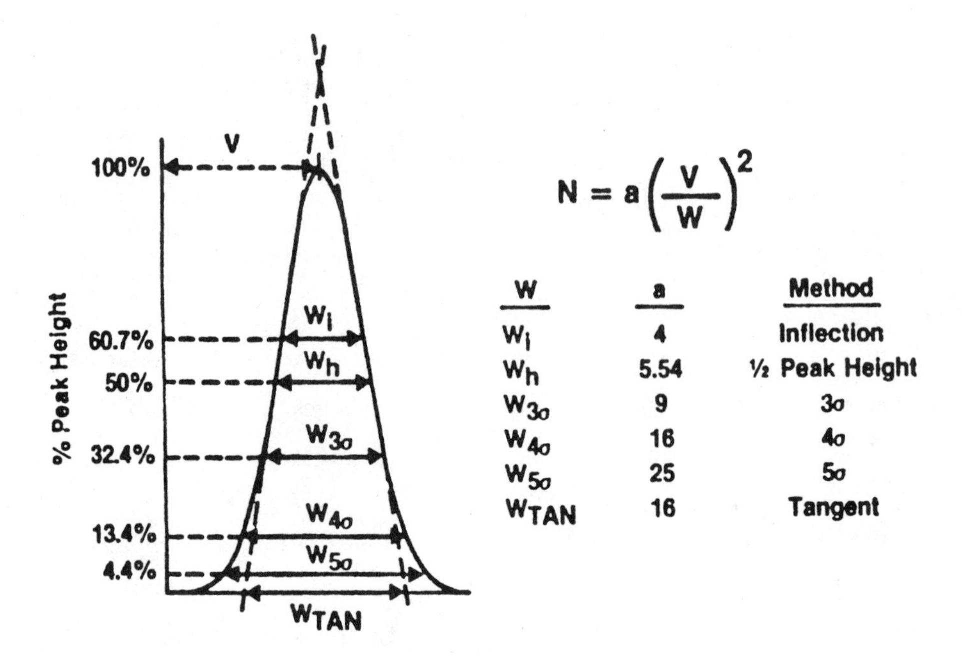


**Fig. 3.6.** Measurement of column plate number $N$ for bands within the chromatogram. $V$ is the retention time of the peak $(= t_R)$, $W$ is the peak width, and $a$ is a constant. Courtesy of Waters, Division of Millipore.

for different compounds. Often values of $N$ for different bands within a particular chromatogram are about the same. So if we measure $N$ for one band we can assume $N$ will be about the same for other bands as well. That is, one parameter ($N$) tells us whether all the bands are relatively narrow or wide. Because $N$ is roughly constant, this means (see Eq. 3.6) that $(t_R/W)$ is about the same for all the bands in the chromatogram, which means that bandwidth $W$ increases with retention time $t_R$. This can be seen in Figs. 3.2, 3.4, and 3.5, and is the normal pattern.

Although many people use Eq. 3.6 to measure values of $N$, it is not very precise. Different workers will draw the lines on each side of the band (as in Fig. 3.6) in different ways, giving different values of $W$ and of $N$. A better approach is to measure the bandwidth at the band half-height ($W_h$ in Fig. 3.6), by (a) dropping a line from the peak maximum perpendicular to the baseline, (b) dividing this distance in half, and (c) drawing a line parallel to the baseline halfway between the top and bottom of the band. The

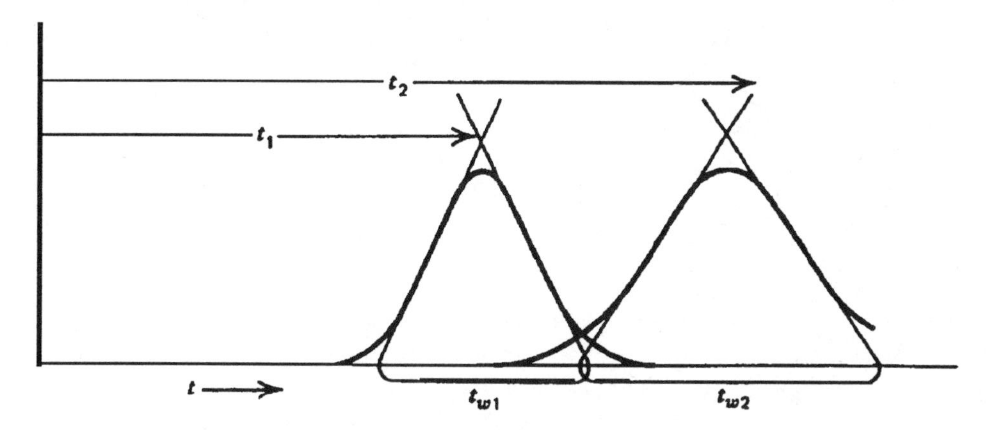


**Fig. 3.7.** Measurement of resolution. $t_1$, $t_2$ are retention times and $t_{w1}$, $t_{w2}$ are bandwidths of bands 1 and 2. Reprinted from ref. (1) with permission.

width of the band at this point can be defined as $W_{0.5}$, and now plate number is given as

$$N = 5.54\ (t_R/W_{0.5})^2 \tag{3.7}$$

Values of $N$ measured in this way are much more accurate than values from Eq. 3.6. Calculations of $N$ based on $W$ at several other points on the peak are shown also in Fig. 3.6.

The value of $N$ depends on experimental conditions. $N$ increases in proportion to column length, generally increases for lower flow rates, and is larger (other factors equal) for columns packed with smaller particles (e.g., 3- vs 5-µm particles). The value of $N$ can be predicted quite accurately for a given separation, using the DryLab I computer program from LC Resources[5]: this in turn can be useful in troubleshooting LC problems of various kinds. Typical $N$-values in LC are 2000 to 20,000 "plates."

### *Resolution $R_s$*

The main objective of an LC separation is to separate sample components of interest; that is, to resolve individual compounds. We can define the degree of separation of two adjacent bands in the chromatogram by the *Resolution* $R_s$, illustrated in Fig. 3.7. Resolution is defined as

$$R_s = 2\ (t_2 - t_1)/(t_{w1} + t_{w2}) \tag{3.8}$$

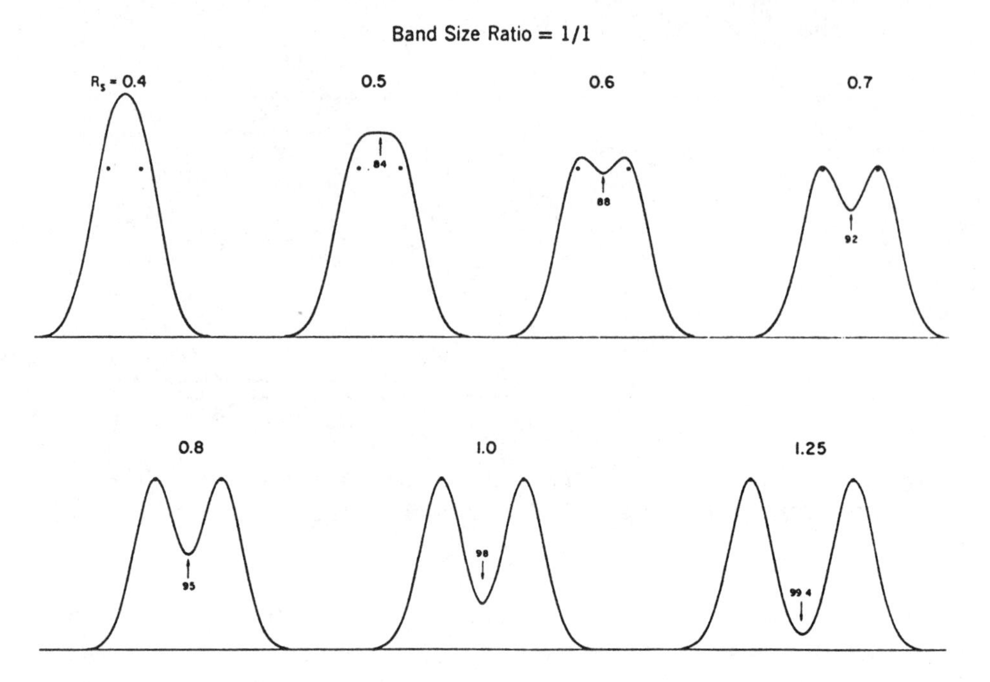


**Fig. 3.8.** Standard resolution curves. Relative band size is 1/1. Reprinted from ref. (1) with permission.

Resolution increases with the distance between the two bands ($t_2 - t_1$) and decreases with increasing average bandwidth. Examples of separation for different $R_s$-values and different relative sizes of the two bands are shown in Figs. 3.8–3.12. Separation is seen to improve for larger values of $R_s$, and is generally better (for the same value of $R_s$) when the bands are of more equal size. Figs. 3.8–3.12 allow values of $R_s$ to be estimated by comparing actual band-pairs with the "standard resolution curves" of these figures.

Resolution $R_s$ depends on those experimental conditions that affect retention time (Table 3.4) or column plate number $N$. We can relate $R_s$ to experimental conditions as (1)

$$R_s = (1/4)\,(\alpha - 1)\;\; N^{0.5}\;\; [k'/(1 + k')] \qquad (3.9)$$

(i) (ii) (iii)

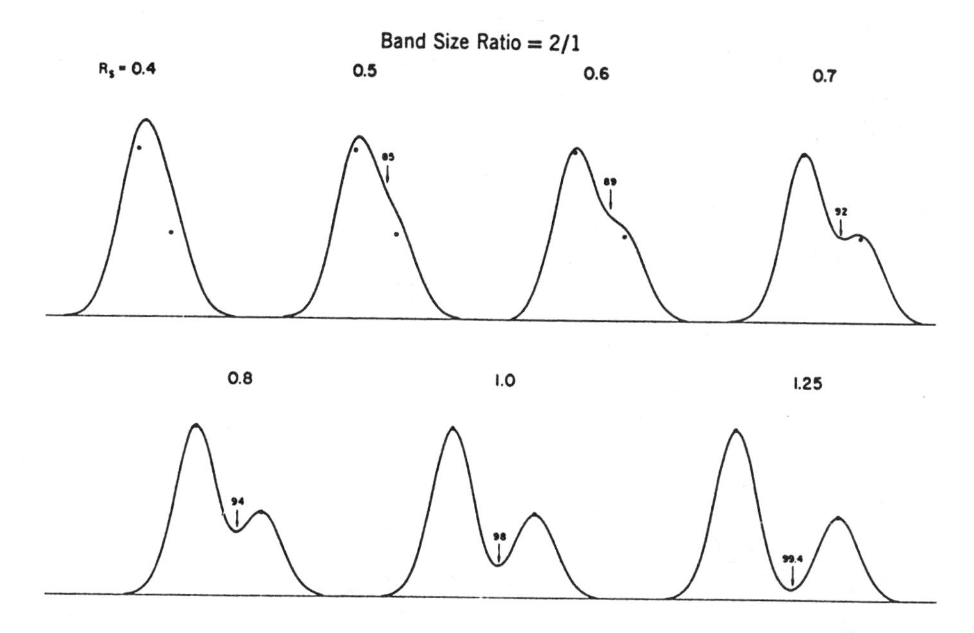


**Fig. 3.9.** Standard resolution curves. Relative band size is 2/1. Reprinted from ref. (1) with permission.

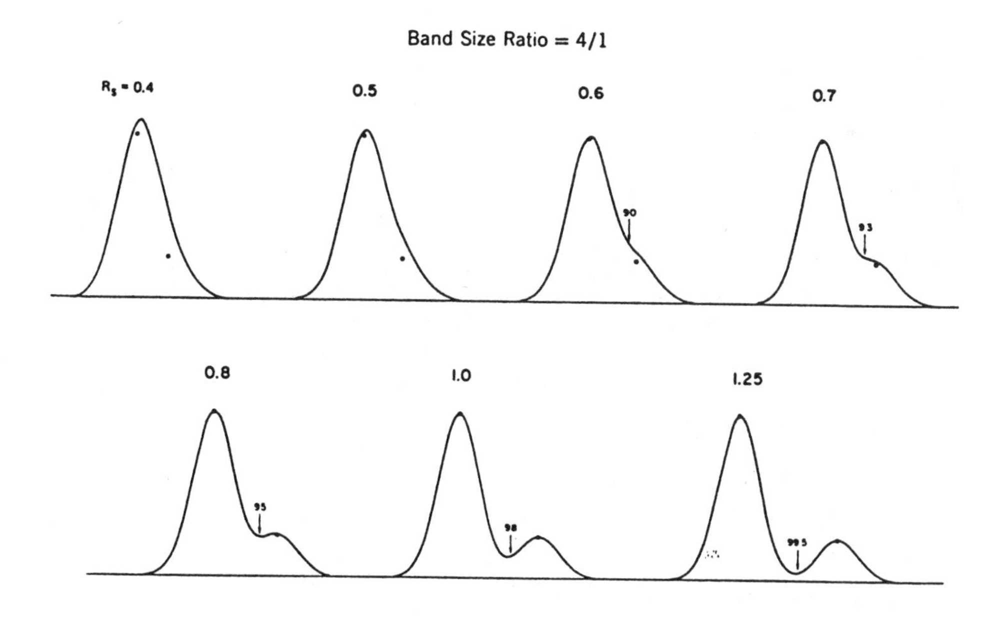


**Fig. 3.10.** Standard resolution curves. Relative band size is 4/1. Reprinted from ref. (1) with permission.

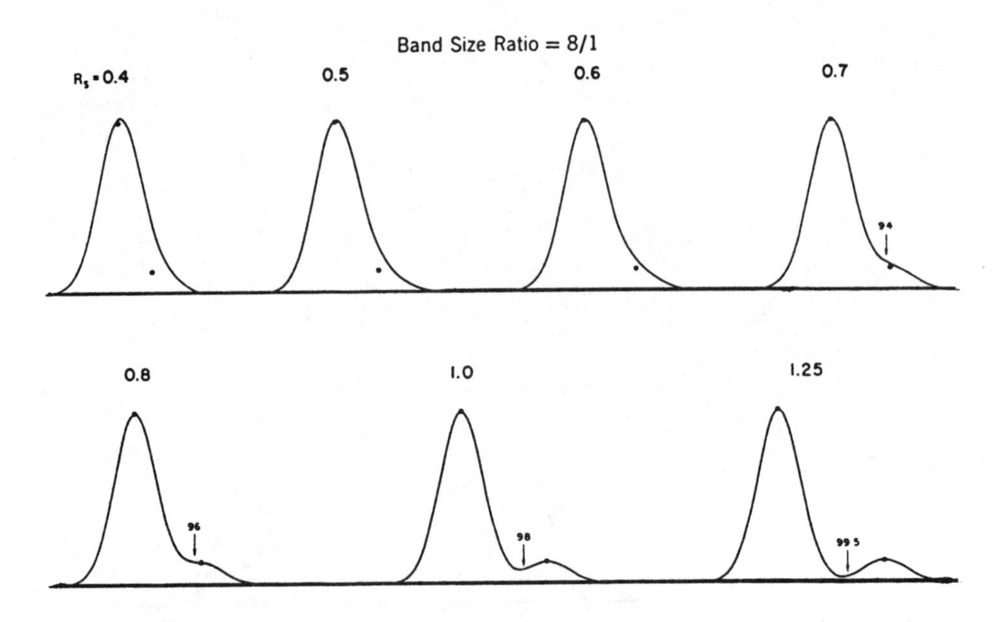


**Fig. 3.11.** Standard resolution curves. Relative band size is 8/1. Reprinted from ref. (1) with permission.

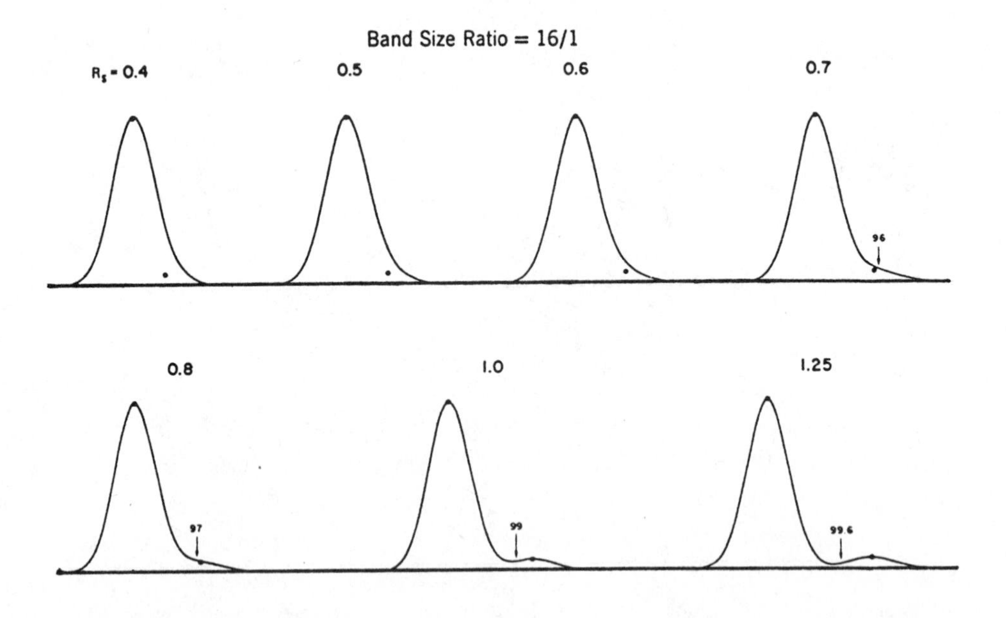


**Fig. 3.12.** Standard resolution curves. Relative band size is 16/1. Reprinted from ref. (1) with permission

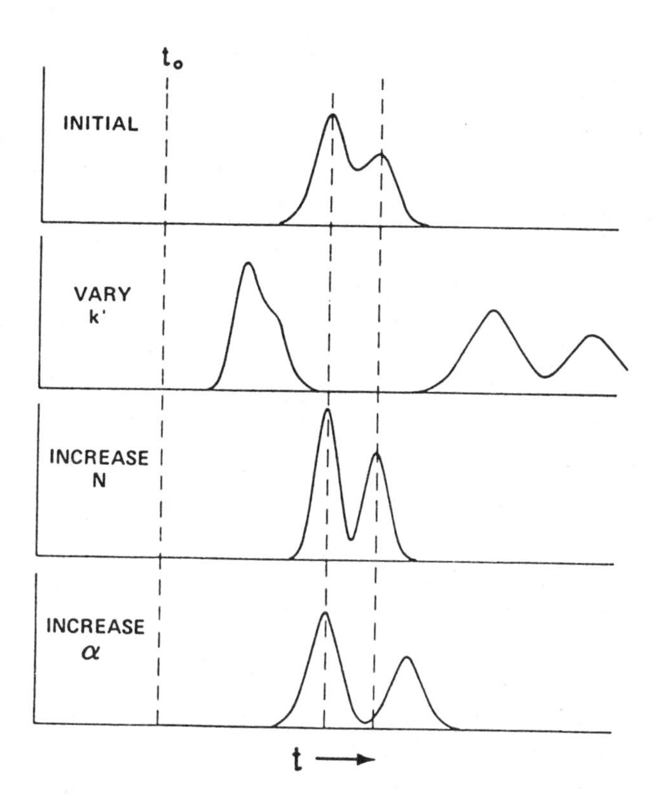


**Fig. 3.13.** Effect of different separation conditions on LC resolution (Eq. 3.9). Reprinted from ref. (1) with permission.

In Eq. 3.9, $\alpha$ is the *Separation Factor,* equal to the ratio of $k'$-values for the two adjacent bands. Resolution is seen to be a function of three different factors: (i) the separation factor $\alpha$ or band spacing, discussed in a preceding section; (ii) the column plate number $N$; and (iii) the capacity factor $k'$. The result of change in each of these three parameters is illustrated in Fig. 3.13. An increase in $k'$ increases $R_s$, but only when the initial value of $k'$ is small ($k' < 10$). An increase in $N$ narrows the bands and also increases resolution. An increase in $\alpha$ moves the band centers apart and further increases $R_s$. Change in $k'$ (run-time) and $\alpha$ (band spacing) as a function of experimental conditions is summarized in Table 3.4. Change in $N$ (and $R_s$) with conditions can be accurately predicted using the DryLab I computer program.[5]

## 3.4. Extra-Column Band Broadening

Separation in LC usually is determined by the experimental conditions reviewed in the preceding section. However, the LC equipment can also contribute to resolution or separation (in a negative way). So far we have assumed that molecular spreading and band broadening occur only within the column (Fig. 3.2). In most cases in which well-designed equipment is used this is approximately the case. However, it is possible for a large part of the band broadening that occurs to take place outside the column. This will be true whenever there are significant "mixing volumes" in certain parts of the LC system. We will next examine this problem. For a detailed discussion, see references 6–10.

### *Equipment Contributions to Band Broadening*

The importance of equipment design in minimizing band broadening is illustrated in Fig. 3.14. Here a small-volume column was used, which we will see increases the importance of extra-column band broadening. In the chromatogram of Fig. 3.14A, a conventional LC system was employed, which contributes significant extra-column band broadening. In Fig. 3.14B, a special low-dead-volume system was used. The difference in bandwidths for these two chromatograms is apparent; the chromatogram in Fig. 3.14B shows generally narrower bands, especially at the beginning of the chromatogram.

In assessing extra-column effects, it is important to measure bandwidths in volume units ($W_v$):

$$W_v = W F \qquad (3.10)$$

That is, the bandwidth in mL ($W_v$) equals the width $W$ (measured in seconds) times the flow rate F (mL/sec). Likewise we can express band retention in volume units (retention volume $V_R$):

$$\begin{aligned} V_R &= t_R F \\ &= V_m (1 + k') \end{aligned} \qquad (3.11)$$

Bandwidth $W_v$ also can be written as

$$W_v = 4V_m (1 + k')/N^{0.5} \qquad (3.12)$$

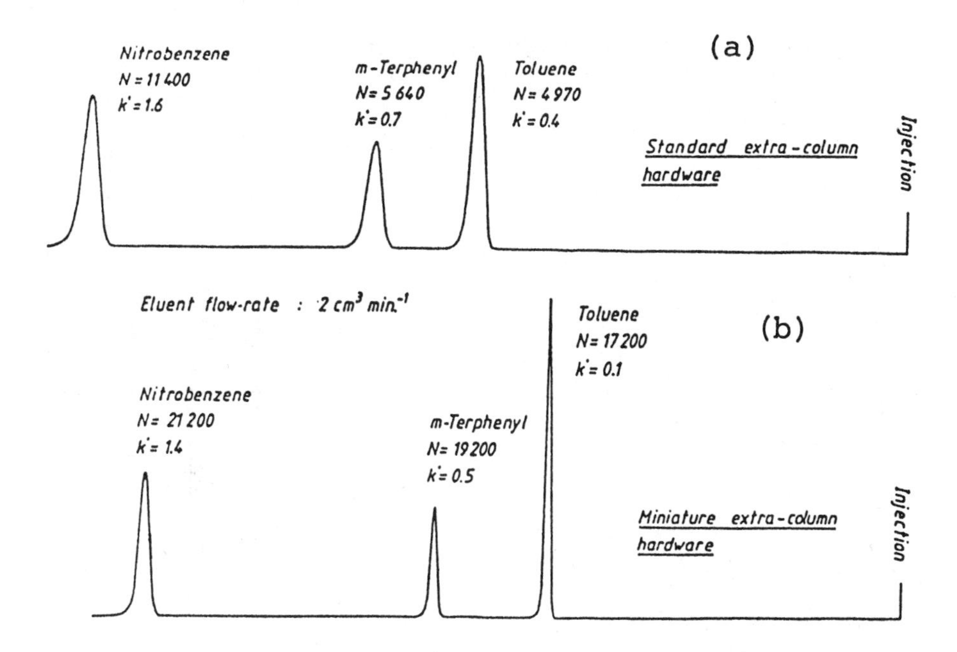


**Fig. 3.14.** Extra-column band broadening differences between two different LC systems. A small-volume column, which is more sensitive to extra-column band broadening, is used. (a) conventional LC system; (b) low-dead-volume LC system. Reprinted from ref. (10) with permission.

Extra-column contributions to bandwidth can be defined:

$W_s$ contribution from volume of injected sample

$W_t$ contribution from connecting tubing

$W_f$ contributions from any fittings

$W_d$ contribution from detector volumes

Usually we can ignore the effect of fittings (or include it in with other parts of the system). The total bandwidth $W_T$ which results from both the column and the LC system is then

$$W_T^2 = W_v^2 + W_s^2 + W_t^2 + W_d^2$$

$$= W_v^2 + W_{ec}^2 \qquad (3.13)$$

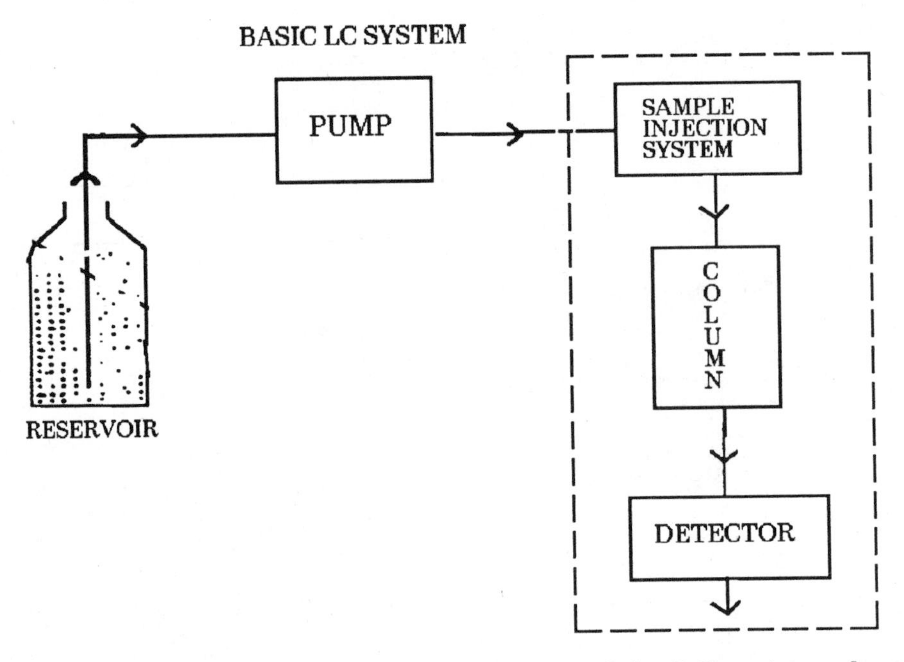


**Fig. 3.15.** The dashed box shows the part of the LC system where extra-column band broadening can occur.

Here, $W_{ec}$ is the total extra-column band broadening contribution to $W_T$. Whenever $W_s$, $W_t$, or $W_d$ is less than 1/3 the value of $W_T$, the extra-column effect is neglible; that is, removal of this extra-column effect entirely would reduce bandwidth by no more than 5%. Therefore, the amount of extra-column band broadening $W_{ec}$ allowed depends on the band volume $W_T$ or $W_v$. Bands that are wider (larger volume $W_v$) are less sensitive to system band broadening. Equation 3.12 tells us that columns of smaller volume ($V_m$), peaks of smaller $k'$, and columns with a larger plate number $N$ will all be more susceptible to extra-column effects.

We will further discuss individual contributions to extra-column band broadening in a moment. First, however, it must be pointed out that extra-column band broadening occurs only in that part of the LC system through which the sample passes. This is represented schematically in Fig. 3.15—inside the dashed region that comprises the sample injector, column, detector, and interconnecting lines. The remainder of the system, including lines to waste, lines between the reservoir, pump, and sample injector, and the pump itself, does not affect bandwidth.

### *Band Broadening in the Detector*

For normal flow rates (1–3 mL/m) and well-designed detectors, the band broadening in the detector ($W_d$, 4σ) appears to be roughly equal to eight times the volume of the flowcell. Many flowcells have a volume of 8 μL (1 mm id x 10 mm long), which means that $W_d$ is then equal to 64 μL. The value of $W_d$ increases at lower flow rates, and decreases for higher flows.[6] For well-designed LC systems having 8-μL flowcells, the flowcell itself is often the main contributor to extra-column band broadening. It is possible to decrease $W_d$ by using a smaller flowcell.

The value of $W_d$ also depends on the design of the LC detector: how much tubing is used to connect the inlet and outlet ports with the flowcell, the geometry of the flowcell, etc. Some detectors have a poor design, so that their $W_d$ values are much larger than eight times the volume of the flowcell.

### *Band Broadening in Connecting Tubing*

Band broadening due to the connecting tubing ($W_t$) increases rapidly as tubing diameter is increased. Therefore LC systems are usually constructed with 0.01-in. id tubing between the sample injector and the detector. Narrower-bore tubing is even better from an extra-column band broadening standpoint, but unfortunately narrower-diameter tubing also tends to become blocked. The amount of tubing that is allowable for a 5% loss in N in the LC system (as a function of column volume $V_m = V_o$, tubing id, and plate number $N$) is shown in Fig. 3.16. For a typical column (e.g., 15 x 0.46 cm, 5-μm particles) $V_m$ will equal about 1.5 mL and $N$ will be about 10,000. For this situation, about 7 cm of 0.01-in. id tubing is acceptable (Fig. 3.16a).

Additional information on band broadening in connecting tubing can be found in Table 8.3 and the accompanying text.

### *Other Contributions to Extra-Column Band Broadening*

*Band Broadening Caused by Sample Volume.* If the sample is dissolved in the mobile phase or a solvent of equal strength (see Table 3.3), the quantity $W_s$ is equal to about twice the sample

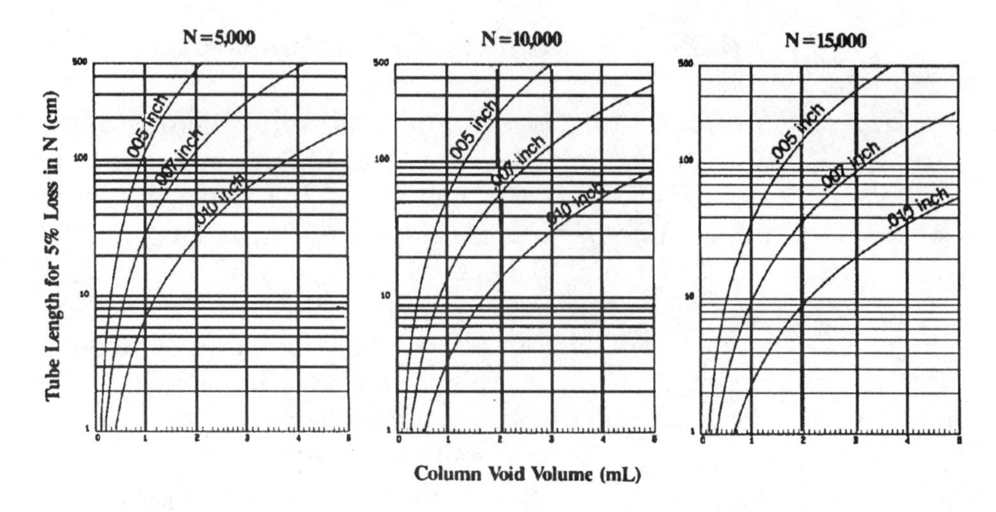


**Fig. 3.16.** The approximate length of tubing that causes a 5% loss of observed plate number when $N = 5000$, $N = 10{,}000$, and $N = 15{,}000$. The curves are valid only when the peak is $k' = 1$, flow rate = 1 mL/min, diffusion coefficient = $10^{-5}$ cm$^2$s$^{-1}$ (typical for small solutes and common mobile phases), and there is no contribution to dispersion caused by the injector, detector, or other tubing. Reprinted from ref. (6a) with permission

volume.* This means that the maximum volume of injected sample should not exceed 1/6 of the value of $W_T$ ($W_v$) for the first peak of interest (lowest value of $k'$ and $W_v$). However, if the sample is dissolved in a weaker solvent, correspondingly larger injection volumes can be used—without any effect on bandwidth.

*Detector Time-Constant.* The actual volume of the sample band can be increased further by a slow response of the detector. Usually the time-constant $t_c$ of the detector is adjustable, so that either a fast or slow response can be chosen. A fast response (smaller value of $t_c$) gives less band broadening, but a noisier baseline. The contribution of the detector time-constant to band broadening $W_{tc}$ equals $4\, t_c\, F$, where $W_{tc}$ is in mL, and $t_c$ and $F$ are measured in the same time units (sec or min). The time constant should be less than 1/12 the baseline width of the first band of interest (smaller $k'$); a time constant of 0.25 sec usually presents no problem.

*In principle the value of *Ws* should equal $(4/3)^{0.5}$ times sample volume, for the plug-injection of sample onto the column. However, laminar flow within the injector loop normally broadens the sample band by a factor of two.

### *Measuring the Extra-Column Band Broadening of the LC System*

Poor resolution sometimes is observed in an LC separation, and it may be suspected that extra-column band broadening is contributing to the problem. In this case we can determine the system extra-column effect, $W_{ec}$, in various ways.[10] One way is simply to disconnect the column from the system, connect the column inlet and outlet lines, and inject a sample that will give a measurable peak. Because the peak will pass through the detector quite rapidly, the fastest-possible chart speed must be used for the recorder. The value of $W_{ec}$ (apart from detector time-constant) usually is not very flow-dependent, so that lower flow rates can be used to give a more easily measured band. The baseline bandwidth of the peak (in seconds) can then be converted to a $W_{ec}$ value by multiplying the bandwidth by the flow rate (mL/sec).

An operational approach for measuring $W_{ec}$ is given in (5), which uses the DryLab I[5] program to analyze experimental bandwidth data as a function of $k'$.

If a low-dead-volume LC system is available (so-called "microbore" system), then chromatograms for the same separation can be compared to see if extra-column effects are important in the standard LC system. If it is assumed that $W_T = W_v$ for the low-volume system, then Eq. 3.13 can be used to estimate $W_{ec}$ for the standard system.

Some autosamplers contribute large extra-column band-broadening effects. If this is suspected to be the case, the autosampler can be disconnected and replaced with a manual injector, in order to measure $W_{ec}$ for the autosampler (using Eq. 3.13). Replacement or redesign of the autosampler may be required for some applications.

## 3.5. LC Method Development

This book is not intended to teach you how to develop an LC method (see ref. 1 for general principles, ref. 11 for a detailed discussion). However, a brief review of what is involved in method development will prove useful in various ways for troubleshooting for LC problems. The steps in developing an LC method are as follows:

1. Select an appropriate LC method for the sample
2. Select a suitable column
3. Select conditions that give good $k'$ values
4. Select conditions that give good band spacing ($\alpha$-values)
5. Select the right "column conditions" (optimum $N$-value)
6. Solve any special problems with "real samples"
7. Validate the method

### *Selecting the LC Method*

The LC method that you choose depends on the sample, as summarized in Table 3.5. For mixtures of typical small molecules, there are three methods that should be tried first: reversed-phase, ion-pair, and normal-phase (polar-bonded-phase) chromatography. Each of these three methods has broad applicability, and each can be "optimized" in a straightforward way. An initial preference for one of these three methods often is not clear-

Table 3.5
The Choice of an LC Method

| Sample characteristics | Possible methods |
|---|---|
| Neutral compounds | Reversed-phase<br>Polar bonded phase |
| Acidic or basic compounds<br>Compounds | Reversed-phase<br>Ion-pair<br>Polar-bonded-phase |
| Synthetic polymers | Size-exclusion |
| Proteins | Reversed-phase<br>Hydrophobic-interaction<br>Ion-exchange<br>Size-exclusion |
| Chiral isomers | Chiral columns (reversed- or normal-phase) |

cut, however the choices shown in Table 3.5 usually are in order of decreasing promise. Thus, for acidic compounds, reversed-phase LC usually is the best starting point, followed by ion-pair, then polar-bonded-phase LC. If the first choice does not work (for any reason), then the other two methods can be tried.

Some samples fall into a "special" category. The examples shown in Table 3.5 include high-molecular-weight synthetic polymers, as well as biological macromolecules such as proteins. Special columns are available for each of the LC methods listed in Table 3.5 for these samples. Chiral isomers are another example of samples that require special columns.

### *Selecting the Column*

When you are beginning method development based on a particular LC method, there is usually a wide range of column types to choose from. Columns vary in length, particle size, and the nature of the column packing. Generally any available column can be used initially; e.g., for reversed-phase, a 15-cm column of 5-μm C8 packing. Each laboratory will have columns it prefers to start with, based on its experience. Now that Cartridge columns are widely available (see Chapter 11), there is an advantage to begin method development with columns of this type. They are easily connected together to vary column length, which is an important consideration (see below).

### *Selecting Solvent Strength*

The next step is to select a mobile phase that provides a good range of $k'$ values. Usually it is advantageous to maintain $k'$ values for bands of interest in the range $1 < k' < 10$ (although $0.5 < k' < 20$ can be tolerated); this provides a good compromise between good resolution, short run-times, and narrow, easily-detectable bands. Trial-and-error selection from a list of compositions such as those of Table 3.3 can be used. Figure 3.4 provides a further example, where 40% methanol/water is the best choice for this sample.

A much more efficient way to find the proper solvent strength is to use one of the DryLab solvent strength optimization programs (e.g., DryLab I, or G; LC Resources; refs. 5, 12). In this case, two initial gradient or isocratic runs are made, and the computer is used to determine the optimum solvent strength rather than doing the actual experiments at the bench. Alternatively, solvent-strength conversions (e.g., between methanol and THF) can be estimated using a solvent-strength nomograph such as is found in Chapter 9 of ref. (11).

### *Varying Band Spacing*

Here a definite strategy can be used for each LC method. The usual approach is to vary mobile-phase conditions, using three different "active" solvents, plus a fourth solvent to vary solvent strength. For reversed-phase LC, the three preferred active solvents are methanol, acetonitrile, and tetrahydrofuran (THF).[13] Separation then is mapped as a function of varying concentrations of the active solvents, while holding solvent strength constant using the fourth solvent (water for reversed-phase LC). This is illustrated for the two runs of Fig. 3.5, where band spacing is different in each case, but neither run is satisfactory. Figure 3.17 shows the results of mixing the two mobile phases of Fig. 3.5 in different proportions, with an excellent separation resulting in Fig. 3.17B. Similar separation strategies have been reported for ion-pair chromatography[14] and normal-phase chromatography.[15, 16]

DryLab software programs[5,12] also aid in the selection of the solvent composition for optimum band spacing. With just two initial runs, the programs can predict band spacing under any mixture of the initial solvents. Although these programs are not as powerful as the solvent mapping approach discussed above, they can be used in most cases. In addition, DryLab programs require only two initial runs, as opposed with a minimum of seven runs for solvent mapping.

### *Optimizing Column Conditions*

After the mobile phase composition has been selected for obtaining a good band spacing, as in Fig. 3.17B, the separation can be improved further by varying "column conditions": column

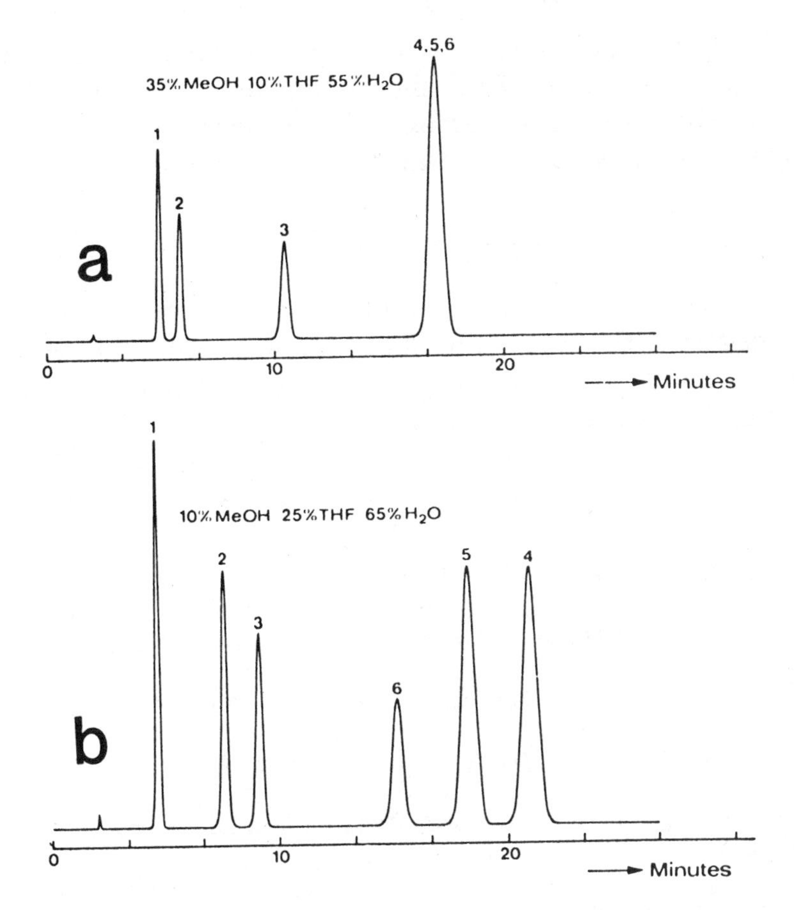


**Fig. 3.17.** Mapping the effect of solvent composition on band spacing. LC system of Fig. 3.5. (a) 35 %v methanol/10 %v THF/water; (b) 10 %v methanol/25 %v THF/water. Reprinted from ref. (4) with permission.

length, particle size, and flow rate. In this way, an adequate separation can be achieved with a minimum run-time (for faster analyses), acceptable pressure and other advantages. The Dry-Lab I computer program[5] is especially designed to facilitate this process.

## *Special Problems*

Some samples require a pretreatment before they can be injected for LC analysis (Chapter 17). Other samples may require the use of gradient elution (Chapter 17), an LC procedure in which the composition of the mobile phase changes during each run, requiring special equipment. Still other problems may arise during method development that require a special approach.

### *Method Validation*

When the method has finally been developed, it is necessary to demonstrate that it is reliable, gives acceptable accuracy and precision, and is suitably rugged for routine operation. Method validation is described in Chapter 16.

## 3.6. References

[1]Snyder, L. R. and Kirkland, J. J. (1979) *Introduction to Modern Liquid Chromatography,* 2nd ed., Wiley-Interscience, New York.

[2]High Purity Solvent Guide (1984) Burdick and Jackson Laboratories, Muskegon, MI.

[3]HPLC Solvent Reference Manual (1985) J.T. Baker Chemical Co., Phillipsburg, PA,

[4]Schoenmakers, P. J., Billiet, H. A. H., and de Galan, L. (1981) *J. Chromatogr.* **218,** 261.

[5]Snyder, L. R. and Dolan, J. W. (1987) *DryLab I User's Manual,* LC Resources, Lafayette, CA.

[6]Martin, M., Eon, C., and Guiochon, G. (1975) *J. Chromatogr.* **108,** 229.

[6a]Bakalyar, S. R., Olsen, K., Spruce, B., and Bragg, B. G. (1988) *Technical Notes 9,* Rheodyne, Cotati, CA.

[7]Guiochon, G. *High-Performance Liquid Chromatography. Advances and Perspectives* **2,** (Cs. Horvath, ed.) Academic, New York, p. 1.

[8]Hupe, K. -P., Jonker, R. J., and Rozing, G. (1984) *J. Chromatogr.* **285,** 253.

[9]Scott, R. P. W. (1983) *Adv. Chromatogr.* **22,** 247.

[10]Freebairn, K. W. and Knox, J. H. (1984) *Chromatographia* **19,** 37.

[11]Snyder, L. R., Glajch, J. L., and Kirkland, J. J. (1988) *Practical HPLC Method Development,* Wiley, New York.

[12]Snyder, L. R. and Dolan, J. W. (1987) *DryLab G User's Manual,* LC Resources, Lafayette, CA.

[13]Glajch, J. L., Kirkland, J. J., Squire, K. K., and Minor, J. M. (1980) *J. Chromatogr.* **199,** 233.

[14]Goldberg, A. P., Nowakowska, E., Antle, P. E., and Snyder, L. R. (1984) *J. Chromatogr.* **316,** 241.

[15]Glajch, J. L., Kirkland, J. J., and Snyder, L. R. (1982) *J. Chromatogr.* **238,** 299.

[16]de Smet, M., Hoogewijs, G. Puttmans, M., and Massart, D. L. (1984) *Anal. Chem.* **56,** 2662.

## Chapter 4

# PRINCIPLES OF TROUBLESHOOTING



## Introduction

The goal of troubleshooting LC systems is to identify a problem early, isolate it quickly, and correct the problem so that downtime is minimized. First, take a few minutes each day to look over the system. You will get a good impression of the system as it operates normally, and this will help you recognize when it is not running properly. When a problem arises, isolate the source of trouble in a logical manner as discussed in this chapter. Finally, correct the problem with the help of (a) this book, (b) the operator and service manuals for the system, and, if necessary, (c) the manufacturer's service department.

In this chapter we discuss general techniques that can speed up problem isolation. Chapter 2, Logical Approaches to Troubleshooting, gives step-by-step procedures to isolate LC problems. You can skip the present chapter and follow the steps outlined in Chapter 2 to locate LC problems, but you will understand the process better if you first review the material presented here.

## 4.1. Basic Rules

There are several common practices that permeate good troubleshooting habits. These are summarized in the "rules of thumb" listed below. If you make these a part of your maintenance and troubleshooting protocol, you will find that the troubleshooting process is greatly simplified.

### *Rule of One*

Make only one change at a time. The "shot-gun" approach of changing many things at once may correct the problem, but it is inefficient. It does not give you much information to use for problem prevention, or help correct the same problem next time. For example, to eliminate a peak-tailing problem, it is possible to modify the mobile phase, and to replace the guard and analytical columns as a single step. If this solves the problem—fine, but which change made the difference?

### *Rule of Two*

The rule-of-two states that the observation of a problem or a solution should be confirmed before any further action is taken. In other words, make sure that a problem exists before you try to correct it. For example, if you see that the peak height for an internal standard is lower than expected, reinject the standard to confirm that the problem is reproducible and not caused by a one-time error (e.g., a bubble in the sample loop). The rule-of-two should also be used whenever system changes are made, to assure you that equilibrium has been reached. For example, when the mobile phase is changed, inject a standard twice to confirm that the retention times are stable before proceeding with an assay. When extra peaks are found in gradient runs, run a gradient blank to check for artifacts (does the problem really exist?). Adherence to this rule will help you keep from making unnecessary changes and will help you confirm that a corrective measure really solves the problem.

### *Substitution Rule*

Substituting a known good part for a questionable one is one of the fastest ways to confirm the source of a problem. For example, if you suspect that the detector is the cause of a noisy baseline, swap the detector with one that you know works properly; if the symptoms go away, there is strong evidence that the old detector is not working properly. The substitution rule can be applied at many different levels in LC troubleshooting—from substituting an entire module to exchanging an integrated circuit chip on a printed circuit board.

### *Put It Back*

The put-it-back rule should be used in conjunction with the substitution rule. When a good part is substituted for a suspect one and no change in symptoms is seen, the original part is probably good. In this case, reinstall the original part. This will minimize the expense of troubleshooting and will prevent the accumulation of used parts. This rule assumes that an LC problem is caused by a single change in the system, which is true in nearly every case. This rule *should not* be used in cases where: (a) parts are damaged when they are removed (e.g., pump seals), (b) parts are inexpensive (e.g., in-line filters), (c) reinstallation risks damaging a module (e.g., column frit replacement), or (d) parts are scheduled for replacement anyway (e.g., preventive maintenance of pump seals, detector lamps).

## *Reference Conditions*

It is important to have a point of reference so that you can tell whether or not any particular LC is operating properly. In this book, we use two sets of reference conditions: (1) standard reference conditions, and (2) assay reference conditions.

*Standard Reference Conditions (Column Test Conditions).* These are conditions that can be reproduced easily from system to system and lab to lab. Data taken under these conditions are useful to help distinguish between problems with an assay and problems with the LC system. For example, if the system pressure is high under the conditions for an assay, but normal under standard reference conditions, the excessive pressure is caused by one of the assay variables (such as flow rate or mobile phase composition). Data taken under standard reference conditions will also help you communicate at a common level with a manufacturer's service or support personnel when outside help is needed. These conditions are most often the ones under which a new column is tested. For example, if you use a C18 column, the recommended conditions (see Table 4.1) are: 70 %v methanol/ water mobile phase at a flow rate of 1.0 mL/m, with a UV detector set at 254 nm. The test solutes are those used for column testing. (Standard reference conditions for other LC methods are shown in Table 4.1.) You should run standard reference chromatograms at regular intervals, for example, each time a new column is installed.

*Assay reference conditions* are used to determine proper system operation on a daily basis. These conditions should be made a convenient part of each day's run. Most users find that copies of the chromatograms for the first two calibrators (rule-of-two) run each day are a sufficient record for the assay reference. It is convenient to record the system pressure and other operating conditions on the chromatogram itself. Daily examination of the retention times, bandwidths, and system pressure will assure you that the system is operating properly before samples are run. If problems occur in the future, you can trace changes in peak shape, column plate number, or other parameters by referring to the chrom-

Table 4.1
Standard Reference (Column Test) Conditions.[a]

| | |
|---|---|
| *Reversed-phase* | |
| Mobile phase | 70%v methanol/water |
| Column | C18 |
| Flow rate | 1 mL/min |
| Detection | UV, 254 nm |
| Sample | Uracil (for $t_0$)<br>Phenol<br>Acetophenone<br>Nitrobenzene<br>Methylbenzoate<br>Anisole<br>Toluene |
| *Polar-bonded-phase* | |
| Mobile phase | 75%v hexane/isopropanol |
| Column | Cyano |
| Flow rate | 1 mL/min |
| Detection | UV, 254 nm |
| Sample | Nitrobenzene<br>Benzyl alcohol<br>2,4-Dinitrotoluene<br>*p*-Nitrobenzyl alcohol |
| *Normal-phase (silica)* | |
| Mobile phase | 95%v Hexane, 4%v $CH_2Cl_2$,<br>1%v IPA |
| Column | Silica |
| Flow rate | 1 mL/min |
| Detection | UV, 254 nm |
| Sample | 2-Phenyl-2-propanol<br>Methyl benzyl alcohol<br>Cinnamyl alcohol |

[a]These are typical test conditions used for new columns

atograms that you have on file. Whenever there is a system problem whose solution is not obvious, rerun a standard chromatogram under the assay reference conditions. Now you will have a uniform reference point from which to isolate the problem.

### *Write It Down*

This rule reminds you to keep a written record as you troubleshoot an LC problem so that you are able to determine cause-and-effect relationships for any changes you make during system troubleshooting. Jot down notes each time you make an observation or change in the system. The best place for these notes is in your lab notebook or the system logbook, but you may prefer to use a separate pad of paper. The objective of these notes is to help you keep track of troubleshooting changes and their affect on the problem. Make the notes brief—they are only meant for temporary use. For example, if a bad guard column is suspected to be causing tailing peaks, you might write "New guard column, pressure 1875—> 1400 psi, tailing same." It is a good idea to mark parts with a felt-tipped pen or an adhesive label to prevent confusing parts which are exchanged. When you have solved the problem, record the symptom, the cause, and the solution in the system recordbook (which also contains all system use and maintenance records). This will: (a) help other users know what changes were made in the system, (b) help you establish failure patterns, and (c) provide information to help you solve the same problem faster the next time it occurs.

### *Crystal Ball*

If you had a crystal ball, you could tell in advance when an LC failure was going to occur. With troubleshooting practice and good preventive maintenance habits, you should be able to predict many routine failures. Time and money invested in regular preventive maintenance will pay you back in improved system reliability and reduced operating costs. Reliability is improved because failure is controlled (or prevented) and secondary failure is eliminated. By secondary failure, we mean the failure of a second part because another part failed. For example, if a pump seal is allowed to fail, mobile-phase leakage can cause secondary failure by corrosion of other pump parts. Operating costs are reduced, because the maintenance schedule is controlled by the operator, not the system. For example, you can choose to replace the detector lamp at the beginning or end of a shift with minimal impact on the day's sample load. If you wait for the lamp to fail, it may disrupt or invalidate an entire day's run—costing much more than the price of a lamp.

### *Buffer Rule*

The buffer rule is to remind you to remove mobile-phase buffers from the LC when it is not in use. Buffer residues in the LC system can cause corrosion, abrasion, and blockages. Physiological buffers are highly susceptible to algae and bacterial growth. It is imperative that buffered (or salt-containing) mobile phases are flushed out of the LC when it is not in use. The ideal flushing solution is mobile phase of a similar composition, but without buffer; water is also OK. (But do not store the system in pure water because of potential bacterial growth; use at least 10% organic or 0.02–0.05% azide.) *Do not* flush buffers from the LC with pure organic solvents; they can cause buffer precipitation in the system.

## 4.2. Logical Problem Isolation

The logical isolation of LC problems is key to rapidly correcting system failure. Some malfunctions, such as dripping fittings, are obvious and straightforward to correct. Other observations, such as tailing peaks in the chromatogram, may be more elusive. The problems that are not immediately obvious often require the most effort in troubleshooting. It is important to follow a regular pattern to isolate these problems, rather than to randomly check one component, then another. We will look next at how to locate the problem.

### *Once-Over*

*First,* when a problem becomes apparent, give the system a quick examination. That is, visually track the flow path from the reservoir through the system to waste, to see if there are any obvious causes for the problem. You may find bubbles in the inlet line to the pump, a leaky fitting, an excessive pressure reading, or another obvious abnormality that will help to isolate the cause of the problem. *Second,* verify that the system is set properly for the method being used. Check the mobile phase, flow rate, pressure, column type, and detector and recorder settings against those for the method procedure. These two steps should only take a minute or two, but will often reveal the cause of system (or operator) failure.

### *System Changes*

*Next,* determine whether any recent changes have been made in the system that might account for the failure. Has maintenance just been performed, a module replaced, new mobile phase added, an unusual sample assayed, a method changed, or has a power interruption occurred? If there are other operators in the lab, check with them to see whether they changed any part of the system. Each change will be apparent if a system record is kept for each LC system (*see* Chapter 5 for suggestions on recordkeeping). If any system changes have been made, you may be able to quickly recognize and correct the problem. If the problem is still unresolved, then the knowledge that a system change has occurred can be helpful as you continue to track down the source of the problem.

### *Reference Conditions*

For problems that appear as changes in the chromatogram, the next step is to run an assay reference chromatogram (see the rules above). In this way, you can determine whether or not the problem is sample-related. If the reference chromatogram is OK, then the problem is somehow related to the sample. If both the reference and sample chromatograms are bad, then a system change has caused the problem.

When problems do not show up as changes in the chromatogram (e.g., pressure changes), it is not necessary to repeat the reference chromatogram. Instead, verify that the proper mobile phase and flow rate are being used and proceed with step-wise isolation of the problem.

### *Summary*

The Troubleshooting Tree in Chapter 2 leads you through a series of steps that can help isolate the problem; these steps are not repeated here. The practices discussed earlier in the "Basic Rules" section should be used in conjunction with the Troubleshooting Tree.

Recordkeeping during problem isolation is essential (recordkeeping rule). It is a natural tendency to skip notetaking and keep all the information in your head; but even experts can get things mixed up. Make only one change at a time (rule of one). Once a change is found not to correct a problem, reverse the change if appropriate (put-it-back rule), thus restoring the system to its original condition. This saves unnecessarily replacing expensive system parts. Label or discard all parts that you replace, so that you can identify the new ones as well as the old ones.

Once a problem has been isolated and corrected, write a short description of the problem and its solution in the system recordbook (recordkeeping rule). Be sure to include part numbers, serial numbers, and other descriptive information, as well as a list of all the parts that were replaced. Don't forget to order replacement parts for those that you used from stock.

Logical problem-isolation requires common sense and all the tools you have available to rapidly solve LC problems. When a problem is observed, give the system a quick visual check to see whether the cause of the failure is obvious, but do not get sidetracked from a systematic approach to troubleshooting. Next, check the system logbook to see whether recent changes might have caused the problem. Run a reference chromatogram to compare with earlier ones so that the magnitude of the problem can be determined under standardized conditions. As you isolate the problem, (1) use the Troubleshooting Logic Tree (Chapter 2) as a guide, (2) keep notes on the changes you make, and (3) use module substitution to help isolate the problem. After the problem is isolated and corrected, do not forget to update the system records with a summary of the problem and its solution. The important steps of the above discussion are summarized in Table 4.2.

Table 4.2
Steps in Logical HPLC Problem Isolation.

*Once-Over*
- Visual check
- Confirm method settings
  - Mobile phase
  - Flow rate
  - Column type
  - Detector settings
  - Recorder settings

*Recent System Changes*
- Maintenance
- Module substitution
- New mobile phase
- Unusual samples
- New method
- Power interruption
- New operator

*Assay Reference Conditions*
- Rerun calibrator
- Compare with earlier records

*Step-wise Isolation*
- Take notes
- One change at a time
- Reverse ineffective changes
- Label all parts
- Record problem solution

*Module Substitution*
- Entire module or subassembly
- Impacts purchase decisions

## Chapter 5

# PREVENTION OF PROBLEMS

## Introduction

The easiest way to troubleshoot an LC system is to not have to do it at all. Of course, it is not possible to eliminate all problems with the LC, but conscientious preventive maintenance will minimize the number of problems that do occur.

There are three approaches to troubleshooting:

1. Wait until breakdown
2. Preventive maintenance
3. Anticipate problems

*First,* you can wait until a breakdown occurs. *Second,* you can practice regular preventive maintenance in order to prevent as many problems as possible. *Third,* you can pay careful attention to the normal operation of the system, so that you anticipate problems before they occur.

In practice, all three methods should be used for best results. For example, problems such as circuit board failures are difficult to predict, so these parts are replaced only when they fail (wait until breakdown). On the other hand, problems with pump seal deterioration can be eliminated if the seals are replaced on a regular basis before they fail (preventive maintenance). If the system backpressure starts to increase because of a blocked guard column, there is enough time to install a new guard column at the beginning of the next shift (anticipate problems). In this way, downtime is virtually eliminated. The rest of this chapter covers various ways to apply these three troubleshooting techniques in order to minimize downtime.

## 5.1. Recordkeeping

The goal of preventive maintenance is to minimize problems with the LC system, and to avoid total-system-failure during a run or shift. Troubleshooting is especially difficult in a "crisis atmosphere." Though it is impossible to prevent problems completely, we can minimize their impact. To minimize downtime, proper recordkeeping must go hand-in-hand with preventive maintenance. This means that all maintenance for a given LC system should be recorded in the system notebook along with reference chromatograms. Each laboratory will have its own requirements and policies for recordkeeping, so the following discussion may not apply completely for your lab. We recommend at least three types of records: system records, column records, and personal or assay records. These are discussed here as three separate notebooks, although you can combine or modify these to suit your own needs.

### *System Logbook*

The most convenient way to keep system-level records is to use a dedicated notebook for each LC system in the lab, or a single notebook with a section for each LC. The information to be included in this notebook is listed in Table 5.1: (a) what specific components make up a system, (b) when changes are made or maintenance is performed, and (c) other relevant information. This is mainly a record of the system hardware, maintenance, and operation under standard reference conditions (column test, see Chapter 4). The list of data should be kept short, so that it is convenient to record and update, but thorough enough to provide the necessary information when troubleshooting system problems. For example, you might be able to trace erratic system pressure (following pump maintenance) to a pump seal improperly installed by someone on a previous shift.

Table 5.1
Data to Include in the System Logbook

Module identification: brand, model, serial numbers, purchase (or installation) date, and warranty information for each module in the system (e.g., autosampler, pump, injector, column, detector, data system).

Reference chromatograms[a] and operating parameters: e.g., chromatogram under test conditions of Table 4.1; note flow, pressure, temperature, mobile phase, detector settings, sample size.

Maintenance performed: what, when, by whom; include routine preventive maintenance as well as breakdown repairs.

Column replacement: note brand, type, serial number; run a new reference chromatogram[a]; cross-reference column log.

Module replacement: when, why, model, serial number; run a new reference chromatogram.[a]

[a]In most cases this system reference chromatogram is simply a column test chromatogram, run under the column manufacturer's suggested conditions.

## *Column Logbook*

The column logbook is a central organizer for all columns in the lab. It should include (a) a section for each column type (e.g., C18, C8), and (b) the information found in Table 5.2. This information helps to trace column history in order to find ways to extend the lifetime of this expensive and vital part of the LC system.

A column might follow this scenario: (a) The column is received from the manufacturer, (b) the vital data, including the manufacturer's test results and the lab's repeat of this test, are entered into the column log by the technician responsible for column inventory. Next, (c) the column is installed on an LC used for routine assay of product uniformity, and (d) it is returned following three weeks of use, having successfully run 623 samples and standards. At this time (e) the column is re-evaluated and it is found to be within 15% of its original plate number, with satisfactory asymmetry and selectivity. Then, (f) after the column is installed on another LC for a second assay procedure, it fails before 100 samples are run. With simple records to keep track of column history, the column technician can begin to piece together failure patterns—and correlate column failure to a particular system, assay procedure, operator, or mobile phase. Once this pattern is determined, changes can be made to reduce the failure rate by adjusting equipment, modifying an assay procedure, or better-training an operator.

For convenience, the column type and serial number plus the type and number of samples run should be recorded in both the system- and the assay-logbook. Of course, unnecessary duplication should be avoided.

Table 5.2
Items to Include in the Column Logbook

| |
|---|
| Date received / first used. |
| Manufacturer's specifications: test mix (sample), mobile phase, flow rate, *N*, *As*, $R_s$, $t_R$, $t_0$. |
| Performance when new: manufacturer's actual results (if given) plus repeat in your lab. Results of other standard lab tests, if any. |
| Record of use: instrument, operator, number and type of samples. |
| Storage information: solvent (no buffers!), capped or not, extended time out of service (e.g. > 2 wk). |
| Maintenance performed: when, why, by whom; e.g., backflush, frit replacement. |
| Re-evaluation (if performed): repeat manufacturer's or lab's standard test. |
| Cause or mode of failure. |
| Summary of use: column life in months, and number of samples, cause of failure, suggestions for extending life. |

## *Assay Logbook*

This record, which is most often kept in a personal lab notebook, includes all the details and results for a given analysis. It should cross-reference any system- or column logbooks in order to minimize repetition. Copies of the assay reference chromatograms (see Chap. 4) should be kept in the assay record. (To reduce the bulk of the assay record notebook, you may want to keep the reference chromatograms chronologically in a file folder.) A person wishing to duplicate the results will be able to find all the necessary information, including the sample ID number and solvent batch numbers, so that the assay can be repeated if necessary. This information can be especially useful for tracing the source of a problem that shows up as a change in the chromatogram. For example, the assay record should help you to trace chromatographic problems to a bad batch of solvent, failure of a water purification system, column failure, or some other problem. The assay record, along with the column log, will help you distinguish between normal column aging and early failure from another cause.

Table 5.3
Items to Include in the Assay Logbook

*Information necessary to repeat the assay:*

1. Mobile phase recipe: volumes, weights, pH, filtering, storage, etc.
2. Sample pretreatment method
3. Assay procedure: sample size, standardization, run sequence
4. Data analysis procedure

*Equipment*

1. Configuration: modules and serial numbers
2. Operating conditions: flow rate, temperature, detector, and data processor settings, run time

Reference chromatograms: calibrator or controls run daily

Reagent formulation notes: actual weights and volumes used, lot numbers of reagents

All sample data: ID, areas, heights, $t_R$, results, reports

General observations: problems, maintenance, cross references to other records, etc.

## 5.2. Selection of Initial Components

The choice of modules for an LC system affects its future reliability. There are many tradeoffs to evaluate when purchasing an LC system; because the requirements of each lab are different, there are no hard-and-fast rules for selecting equipment. A discussion of some of these tradeoffs is presented here, so that you can include them in a purchasing decision.

### *Modular vs Integrated Systems*

LC systems come in two forms: (a) *Modular systems* are composed of a set of modules purchased from one or more manufacturers; (b) *Integrated systems* have most (or all) of the components of the LC system organized within one cabinet supplied by a single vendor.

Modular systems are popular because the system uses selected ("better") components from different vendors. This allows you to pick the best individual pump, injector, column, and detector, so that the system is optimized for your needs. Modular systems often are easier to rearrange for different applications. Thus, a two-pump gradient system can be converted to a single-pump isocratic system, with the extra pump available for a second system; detectors can be moved from one system to another when application requirements change. Troubleshooting system problems via module substitution (substitution rule) often is more convenient with a modular system.

Modular systems have drawbacks, however. They typically require more bench space than integrated units. Because they are more spread out, modular systems also can be more difficult to move from one location to another. When all the modules are not from the same vendor, service support can be harder to obtain, compared to a modular- or integrated system from a single vendor. Modular systems are less likely to have system-level diagnostics because each module is designed to operate independently. However, modular systems from a single vendor can include a controller that provides diagnostic capabilities. When assembling a modular system, care should be taken to select modules that will function together properly. Mismatched components can degrade system performance.

Most integrated systems have a smaller "footprint," so they occupy less bench space than modular systems (often a critical factor in crowded labs). Compact units, however, may reduce the accessibility of individual components for maintenance and repair. Central diagnostics and system control make integrated LC systems easy to use and troubleshoot. Failure of one module in the system, however, may make the entire system unusable, whereas a modular system can be placed back in service by substituting a good module for a faulty one. The components of an integrated system are designed to work together, so performance problems arising from component mismatch are rarely encountered. Integrated systems often are more flexible for methods development, allowing sequential runs under different conditions (e.g., gradients, temperatures, detector settings, column selection) because all components of the system can be controlled from one point.

### *One vs Several Manufacturers*

Choosing between a single-vendor and multi-vendor modular system usually is based on service and performance requirements. Components generally should be purchased from the same vendor if you need to depend on vendor service to keep the system operating properly. Single-vendor LC systems eliminate problems created when service calls are required on multi-vendor units, because service engineers are often reluctant to work on another vendor's equipment. In some locations, however, independent service technicians will repair any brand of module in an LC system. The expense and inconvenience of service calls can be reduced greatly by using the troubleshooting techniques presented in this book. By isolating the problem to the module level, a substitute module can be used temporarily. Meanwhile, the faulty module can be returned to the vendor for repair, and the system can continue to be used.

Service and interchangeability factors should also be considered when more than one LC unit is purchased for the same facility. Some labs purchase all their LCs from a single vendor, because it (a) gives them a ready supply of modules for substitution during troubleshooting, (b) results in faster service response from the vendor, (c) reduces spare parts inventory, and (d) reduces the breadth of skills required to troubleshoot systems from several vendors. Service contracts may also be more attractive when several systems from a single manufacturer are located at the same facility. Ease of access for service varies greatly among vendors for both integrated and modular systems.

### *System Performance*

System performance should be the most important factor when choosing an LC, but do not be blinded by "specification-itis." Most LC assays do not require instrumentation that performs at the cutting edge of technology. This means that, if a simple isocratic system is all that is required for the assays you anticipate, you may be wasting money and often adding unnecessary complexity by purchasing a ternary gradient system. Do keep in mind, however, that a gradient system offers convenience and flexibility not available with an isocratic system; for example mobile-phase blending, and method-development assistance. (If a gradient system is selected, be sure that you understand the tradeoffs between low- and high-pressure mixing that we discuss in Section 7.3.) In this context, performance includes not only the physical specifications (e.g., flow rate, detector sensitivity), but also ease of use, automation, and other factors that relate to the effectiveness of the system in performing the required assays. So, when selecting an LC system, project the lab's future needs and take them into account, but don't go for overkill "just in case"—all purchases should be cost-justified.

### *Which Manufacturer?*

This is probably the most common question when an LC purchase is being made. Actually, it is difficult to make a bad choice when selecting LC equipment today; that is, most vendors offer LC equipment that will accommodate the routine assay requirements of most labs. Even if you have demanding performance criteria, you generally can find several vendors who can fill your needs. In addition to performance specifications, remember to keep other factors in mind when selecting a manufacturer. A strong service reputation is the number-one requirement for post-purchase support. Do you require on-site service? If you are willing to return malfunctioning modules to the factory or a regional service center for repair, a vendor with local service capabilities may not be needed. Is the sales person pleasant and

Table 5.4
Considerations for LC System Purchases

| Considerations |
| --- |
| 1. *Type of system required* |
| Modular vs integrated |
| 2. *Specific assay requirements* |
| Microbe, prep, routine, research, etc. |
| Biological inertness |
| Automated or unattended operation |
| 3. *System components* |
| One vendor or several |
| 4. *Desired vendor* |
| Service reputation |
| Other systems already owned |
| Local or factory service |
| Good relationships with sales and service personnel |
| Good technical support available |
| 5. *Other factors* |
| Familiarity with a specific brand |
| Ease of maintenance, troubleshooting |
| Compatibility with existing data systems |
| Amount of space available |
| Cost |
| Growth for future needs |

responsive? This person often is your first-line contact with the company and will facilitate a rapid response from the vendor. Do you have other instrumentation from the same vendor? Multiple instrument installations from the same vendor, whether the units are LCs or not, often improve vendor responsiveness. In the final analysis, personal experience or qualified references are the best source of information of the type of support that a manufacutrer will give you.

## *Summary*

When choosing an LC system, carefully match system performance, vendor support, and operating convenience with your assay requirements, in-house service skills, and operator experience. A summary of things to look for in an LC purchase is given is Table 5.4. Additional information on specific modules is given in later chapters on individual modules.

## 5.3. Routine Preventive Maintenance

Regular maintenance of the LC is important in order to prevent problems at inopportune times. The specific procedures to perform and the interval between maintenance periods should be based on records of LC usage in the lab. Some manufacturers also list recommended service intervals in the system manual. In the chapters on the various LC modules, specific items are suggested for inclusion in the preventive maintenance schedule. These are only guidelines, and should be modified when necessary to meet your needs. The first rule for a preventive maintenance schedule is to make the list short enough to be followed with ease. It is better to perform a few critical tasks regularly, than to have complex procedures that are ignored because they are too much trouble.

The following discussion highlights specific areas of routine maintenance. See the module-specific chapters for more details.

### *Reservoir*

Cleanliness is key to mobile-phase-reservoir maintenance. HPLC-grade solvents and reagents should be used whenever possible. Mobile phases containing buffer salts or non-HPLC-grade solvents should be filtered through an 0.5-μm filter to remove particulate matter. Additionally, the reservoir should be loosely capped and an inlet-line frit ("sinker") should be used to prevent dust or other particulates from reaching the pump. Be careful to avoid cross-contamination when changing solvents. Aged mobile phases should be discarded regularly (e.g., daily) to prevent problems caused by microbial growth or compositional changes. The reservoir container should be washed or replaced periodically (e.g., every month) to prevent the buildup of contaminants.

### *Pump*

The moving parts of the pump make it susceptible to wear. Preventive maintenance will reduce or correct this problem, making regular pump maintenance especially important. The pump seal is the most rapidly wearing part of the pump. Because seal failure can cause other LC system problems, it is recommended to replace the pump seal(s) every three months. Buffered mobile phases should be flushed from the pump daily to prevent buildup of buffer salts (buffer rule). Store the pump in nonbuffered mobile phase or pure-organic solvent. Check valves can leak or stick, causing flow and/or pressure problems. These problems can be minimized by (a) using HPLC-grade solvents, (b) using an inlet-line frit, (c) degassing the mobile phase, (d) replacing the pump seal(s) regularly, and (e) flushing the pump daily with nonbuffered mobile phase. Use of helium sparging with a positive reservoir pressure (e.g., Kontes, Perkin-Elmer) can help the check valves seat properly for more reliable operation at low flow rates.

### *Injector*

Little maintenance is required to keep the injection valve operating properly. Rinse the sample loop at least daily to remove residual sample. Protect the rotor seal from buildup of abrasive salt crystals by storing the system in nonbuffered mobile phase when not in use. Maintain adequate air pressure for air-operated valves. Do not use syringes with sharp needles (e.g., gas chromatography syringes).

### *Column*

Protect the analytical column from premature aging caused by (a) chemical attack, (b) particulate buildup, and (c) pressure shock by using the appropriate protective measures. A precolumn (saturator column) should be used whenever the mobile-phase pH is greater than 7. An 0.2- or 0.5-μm in-line filter will keep particulates from blocking the frit on the analytical column. A guard column should be used for most assays to protect the analytical column from buildup of chemical contaminants. Proper sample preparation and occasional column cleaning also will extend column life.

Pressure shocks can be minimized by using a pump with built-in pulse damping, or by adding an auxiliary pulse damper to a nondamped pump.

Store columns properly when not in use. Buffers should be flushed from the columns to minimize microbial growth. Store columns with >10% organic as a solvent (or with azide when organic must be avoided); cap endfittings to prevent solvent evaporation.

### *Detector*

Preventive maintenance for LC detectors primarily involves keeping the detector clean. Well-intentioned maintenance of detector cells often makes operation worse; so do not routinely clean the cell other than a daily washout with a strong mobile phase. Acid washing and electrode polishing should be deferred until problems are encountered, unless experience in your lab shows that these procedures are definitely beneficial. For all detectors, flush any buffer salts from the detector cell during the system-flush at the end of each day. Be sure all buffer salts are removed from the detector cell before nonpolar organic solvents are introduced, or buffer precipitation can ruin the cell by blocking the heat exchanger tubing or the cell itself. To prevent bubble formation in the detector cell, thoroughly degas mobile phases. Detector lamps have finite lifetimes, so generally it is not wise to leave the lamp on when the detector is not in use.

### *Recorders and Data Systems*

Very little maintenance is required for data recording devices. Replace or replenish the ink supply when the recorder trace begins to be faint. Be sure to lift the pen from the paper and cap it when the recorder is not in use to prevent premature pen failure. Dot-matrix, thermal, and impact printers seldom require printer-head maintenance.

Check the chart-paper supply for all recorders and data systems before the instrument is started for the day or left for a period of unattended operation. It is better to discard a little paper, than to encounter the frustration and downtime when sample data are lost due to running out of chart paper.

If data are being stored on a disk, be sure that adequate storage space is available for unattended runs.

### *Operating Limits*

Most LC pumps have built-in upper-pressure-limit switches that automatically turn off the pump when the system backpressure reaches a selected value. Upper-limit switches can be used to protect resin-based columns from overpressure, and also protect other hardware in the system. For example, when the upper limit is set to 2000 psi rather than 6000 psi, pump seals and injection valves last longer, and there are fewer problems created by buffer salts leaking and crystallizing in fittings, check valves, and other parts of the system. When choosing an upper pressure-limit, balance the advantages and disadvantages of a reduced pressure-limit. A lower pressure-limit reduces wear on the autosampler, injector, and pump; however, it increases the frequency of changing frits, filters, and guard columns, because blockage of these components causes the system to reach the maximum pressure more quickly than with a 5000 psi limit. A lower upper-pressure-limit can require lower flow rates, which translate into longer run times. A good balance is reached if the upper pressure-limit is set at (a) about 2000–2500 psi, or (b) 1000 psi above the operating pressure with a new column plus associated protection devices (filters, precolumns, etc.). For example, if an assay procedure runs at 1900 psi with a new column, an upper limit of 3000 psi offers protection against excessive pressure, yet allows for a significant buildup of pressure as contaminants collect in the system. When selecting the upper pressure-limit for a gradient assay, remember that the midgradient or final backpressure can be double the starting pressure for some mobile phases (you don't want the system to shut off in the middle of the run!).

Lower-pressure-limit switches can be used to shut off the pump automatically in case the reservoir runs dry or a massive leak occurs. If the lower-limit switch is used, set it at about 50–100 psi; otherwise disengage it by setting the lower limit to zero.

## 5.4. Early Replacement of Critical Parts

Replacing a part shortly before it fails will save time in future maintenance and troubleshooting (crystal ball rule), because replacement can be scheduled to coincide with other system maintenance. The cost of parts and labor, and the impact of unexpected downtime should be evaluated when determining which parts to replace routinely prior to failure. These parts can be classified into two categories: (a) parts to replace on a calendar schedule, and (b) parts to replace when the first indication of failure is observed. The parts in each category will vary from lab-to-lab, and system-to-system.

### *Regular Replacement*

Some parts have a finite and predictable lifetime, for example, pump seals typically last 3–6 months. Such parts should be replaced on a schedule based on their lifetime; e.g., replace pump seals every 3 months. The lifetime of other parts such as the precolumn, in-line filter, and guard column need to be determined empirically for each lab. System use-and-maintenance records will help you select which parts to replace regularly. For example, in-line filters may last only a few days for one application, but may last two months for other applications. Specific recommendations for each module are included in the preventive maintenance section of the appropriate module chapter.

### *Pre-Failure Replacement*

Often it is easy to recognize a part of the LC system that is about to fail. If these symptoms can be identified, a replacement part can be ordered, and the part can be replaced before a catastrophic failure occurs in the middle of a sample run. If the problem is discovered early, a part can sometimes be repaired temporarily, so that it will operate for another day or two until a replacement arrives. A key to identifying these pre-failure conditions is to know how the particular LC system operates normally, so that you can quickly spot abnormal system behavior. See the individule module chapters for specific parts to watch for pre-failure replacement.

## 5.5. Spare Parts Supply

The contents of your spare parts supply will depend on the importance of downtime and the number of similar LCs in your lab. We have classified spare parts into three categories: (a) regular spares, (b) backup spares, and (c) as-needed parts. Your specific needs may result in shifting some items from one category to another. Table 5.5 is a suggested minimum list. Consult the operator's manual for additional spares suggested by the manufacturer.

Table 5.5
Recommended Spare Parts

| Regular | Backup | As-needed |
|---|---|---|
| Sampler parts | Check valves | Lamps |
| Precolumn | Fittings | Injectors |
| Guard column | Tubing | Circuit boards |
| Pump seals | Column | |
| Column frits | Pump head assay | |
| Inlet filters | | |
| In-line filter frits | | |
| Fuses | | |

### *Regular Spares*

This category contains consumable items needed for day-to-day LC operation. These spares are handy for unexpected failures that can interrupt analysis of a critical group of samples. Some parts, such as fuses and autosampler needles, can fail at any time, so spares should be kept in order to minimize downtime. Other parts, such as pump seals or detector lamps, should be replaced regularly; however, these parts can also fail earlier than expected.

### *Backup Spares*

These parts are needed less freqently than the previous category. For example, check valves fail only occasionally, but render the LC useless when they do; keep at least one inlet and one outlet check-valve on hand for emergencies. A partitioned plastic box can be used to conveniently organize the fittings. A spare analytical column can be used to substitute for a questionable one during troubleshooting, as well as serving as a backup for the column in regular use. Because these parts are in less demand, a small number of spares can be kept as a central supply for all the systems in the lab.

### *As-Needed Parts*

Many parts of the LC system have very low failure rates (e.g., circuit boards) and some have limited shelf lives (e.g., some detector lamps); it is usually better to order these parts by overnight package express than to maintain a large inventory of parts that may never be needed. Most LC manufacturers have an expedited parts-delivery system designed to handle emergency needs. Other spares are listed in the individual module chapters.

### *Strategy*

When determining a spare-parts strategy, consider the following:

1. When many LCs are in use, it is easier to justify purchasing fewer failure-prone parts as backups than if a single LC is used.

2. The impact of downtime should be considered; if it is imperative that the LC be operational, even overnight parts delivery may be too slow.

3. It is advisable to check ahead of time with your parts supplier to find out the policy and ordering procedures for emergency parts. It may also be necessary to arrange a procedure with your purchasing department that allows you to order certain emergency parts before a purchase order is completely processed.

4. Some modules, such as autosamplers or specialty detectors, may require special parts; be sure to include these in your spare-parts inventory.

5. Always order replacements for parts that are used at any maintence sessions, so that they will be available the next time they are needed.

## 5.6. Tool Kit

Proper tools are essential for preventive maintenance and repair. Table 5.6 is a suggested minimum list. Unfortunately, tools "disappear" just when they are needed most. Some labs address this problem by assigning personal tool kits that contain most of the common tools. A less expensive, and more convenient, alternative is to give each user a pair of 1/4- x 5/16-in. open-end

Table 5.6
Recommended Tools

| |
|---|
| *Screwdrivers* (general purpose) |
| 1/8-in. slotted |
| 1/4-in. slotted |
| No. 2 Phillips |
| *Wrenches* |
| 10- or 12-in. adjustable (compressed gas tanks) |
| Open-end (fittings, columns, check valves, etc.) |
| Two 1/4- x 5/16-in. (tube fittings) |
| 3/8, 7/16, 1/2, 5/8, 9/16, and 11/16-in. |
| Hex key (Allen) set: 1/16-in. through 1/4-in. |
| *Pliers* |
| 6-in. Slip-joint |
| 5-in. Long-nose |
| Wire cutter/stripper |
| *Misc.* |
| Knife-edge file (tube cutting & dressing) |
| Tube cutter (e.g., Terry Tool) |
| Magnifying glass (general purpose) |
| Knife |
| Ruler |
| Pump seal-insertion tool (if required) |
| Plastic tube flaring tool (for Cheminert fittings only) |
| Single-edged razor blade or scalpel |
| *Nice to have, but not essential* |
| Soldering iron |
| VOM (volt-ohm meter) |
| Power tubing cutter (e.g., SSI cutoff machine) |
| Integrated circuit removal tool ("chip puller") |
| Pointed forceps |
| Diagonal cutters |
| Flashlight |
| Magnetic pick-up tool |

wrenches (for tube fittings) and a screwdriver with a reversible tip (for slotted and Phillips screws). Buy brand-name tools (e.g., Xcelite, Craftsman, Snap-On, Indestro, Crescent); the higher quality helps prevent bruised knuckles and extends the life of fittings. These tools can be purchased as kits (e.g., Jensen) or (less expensively) at a local hardware store. *DO NOT* cut corners by using an adjustable wrench instead of an open-end wrench on check valves, tube fittings, and column fittings; damage to the part (and your knuckles) may result if the wrench slips.

## 5.7. Getting Help

Knowing when and where to ask for help can prevent problems resulting from unintentional mistakes. Damage to a module can result if you try to repair or maintain it without the proper tools or skills; this adds unnecessary expense and downtime. For example, repairing electronics problems (other than replacing circuit boards, fuses, or lamps), is beyond the capability of most users. Some integrated circuits can be ruined by the static electricity released when they are touched, and power supplies can carry life-threatening voltages. There are four excellent sources to consult for further information:

*First,* this book has specific information to help you isolate and correct many problems. Familiarize yourself with the organization of *Troubleshooting LC Systems,* and use the index to find the desired section or table quickly.

*Second,* the operator's and service manuals for each module contain detailed procedures for maintenance and repair, usually with exploded diagrams or pictures.

*Third,* often there are others in the lab with who have experience with a particular problem. Sometimes these are not LC specialists, but together you may be able to solve the problem. For example, a spectroscopist may be able to help solve a problem with a photometric detector, or an electronics technician can use a circuit diagram to find an electronic problem.

*Finally,* the LC manufacturer is an underutilized source of problem-solving information. Most manufacturers offer free telephone consultation to help solve problems without the necessity of a service call (ask for "technical support"). The telephone number of the service department should be listed in the service manual.

# Chapter 6

# RESERVOIRS AND DEGASSING

## Introduction

Mobile-phase reservoirs hold the premixed mobile phases or mobile-phase components that will be pumped into the LC system. Reservoirs range from the original solvent containers to specialized reservoirs for a particular LC model.

Reservoirs are relatively simple devices, but they are subject to certain problems. Reservoir contamination can cause blocked frits, poor pump performance, extra peaks, or excessive noise in the chromatogram. Excessive washout volumes can be encountered in certain cases when improper reservoir components are used. It often is necessary to degas the mobile phase in order to avoid erratic pump delivery or detector noise.

The basic parts of the mobile phase reservoir are shown in Fig. 6.1: the reservoir container, the solvent inlet line, and inlet frit. Means to facilitate mobile phase degassing are often a part of the reservoir (e.g., Fig. 6.6). Because selection of a degassing technique often influences the choice of a reservior container, degassing is discussed first in this chapter.

## 6.1. Degassing

Dissolved air in the mobile phase can interfere with proper LC operation by (a) causing bubble problems in the pump, or (b) degrading the detector response through baseline offset and/or-signal quenching. These two problems can be controlled with the

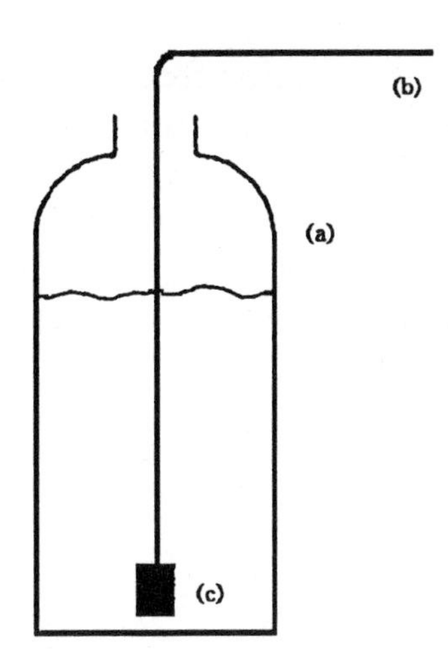


**Fig. 6.1.** Mobile phase reservoir: (a) reservoir container (e.g., 1-L, glass), (b) solvent inlet line (e.g., 1/8-in. od x 1/16-in. id Teflon), (c) inlet frit (e.g., 10-µm porosity). Reprinted from ref. (1a) with permission.

same techniques, yet their causes are somewhat different. Bubbles form when the mobile-phase mixture becomes supersaturated with air. This results when the solubility of air in the mobile-phase mixture is significantly less than in the pure solvents. The major components of air ($N_2$ = 78%, $O_2$ = 21%) each contribute to bubble problems in solvent mixtures. Detector problems (apart form bubbles), on the other hand, are primarily caused by excessive oxygen levels in the mobile phase.

When mixtures (i.e., mobile phases) of air-saturated solvents are made, the air level in the the mixture is often greater than the saturation level, as is shown in Fig. 6.2.[4] This problem is apparent especially when one of the solvents is water (as in reversed-phase systems). This is shown graphically in Fig. 6.3, where the amount of gas evolved upon mixing various proportions of solvents is plotted. If the evolution of gas from such supersaturated mixtures occurs when mobile phases are mixed manually, gas bubbles evolve and a new air-mobile-phase equilibrium is reached. Generally, manually mixed mobile phases may be pumped without bubble problems.

When the mixing takes place in the LC system, however, problems can result. If solvents are mixed before reaching the pump (i.e., low-pressure mixing, see Section 7.3), the pressure is near to atmospheric and bubbles form. When these bubbles are drawn into the pump, they then interfere with piston and check-valve operation; the results are erratic flow and pressure. If, on the other hand, mixing takes place after the pump (i.e., high-pressure mixing, see Section 7.3), the pressure is high enough to keep the bubbles in solution (see further discussion below).

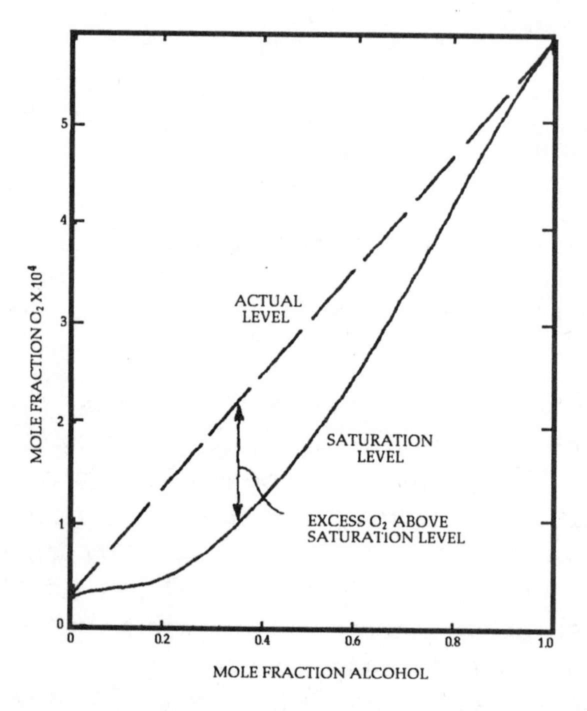


**Fig. 6.2.** Oxygen solubility vs oxygen concentration. The actual level represents the amount of oxygen in the admixtures, starting with air-saturated pure water and air-saturated pure ethanol. Reprinted from ref. (4) with permission.

Although bubble problems can occur in the detector, the primary detector problem created by excess gas in the mobile phase is a result of oxygen interfering with detector response. Bakalyar et al.[4] showed two major problems that can occur in photometric detectors due to oxygen in the mobile phase. These are illustrated in Figs. 6.4 and 6.5. In UV detectors, oxygen can cause a significant baseline rise (see Figs. 6.4a and 6.5a,c). The baseline offset increases with low-wavelength UV detection (e.g., 200 nm). With

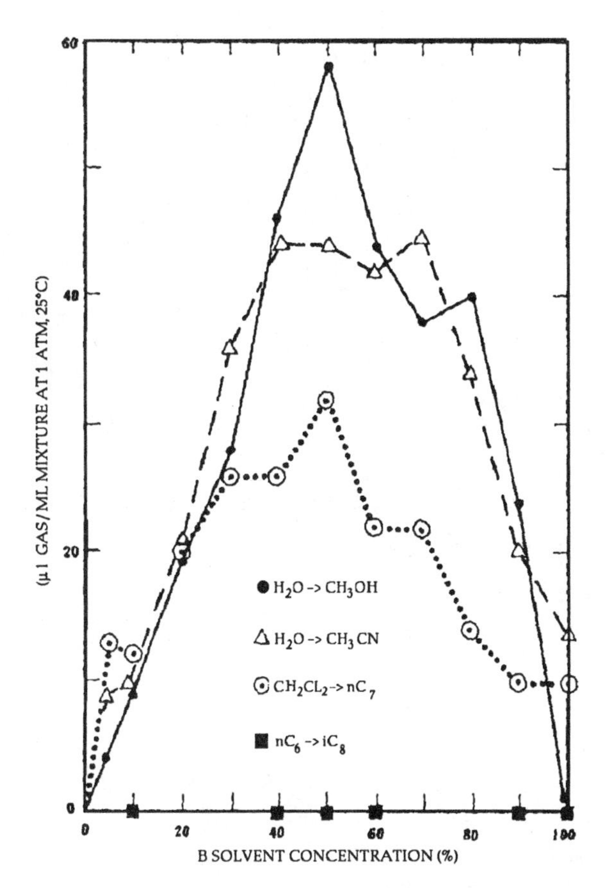


**Fig. 6.3.** Gas evolved vs solvent composition. Each point represents a single measurement, so that the detailed shape of the curves is not to be taken as significant. Reprinted from ref. (4) with permission.

fluorescence detection, oxygen can adversely affect sensitivity by quenching sample fluorescence, as is shown in Figs. 6.4 and 6.5b,d. A small negative baseline shift caused by oxygen quenching of the mobile-phase background-fluorescence is also noted. Oxygen-free mobile phase is required for use of electrochemical detectors in the reductive mode.

It is not always necessary to degas the mobile phase. Degassing usually is not required for most solvents used in normal-phase chromatography or non-aqueous SEC, because the solubility of air in mixtures of these organic solvents (e.g., methylene chloride, hexane, toluene) is similar to that in the pure solvents. Therefore, when the solvents are mixed, no gas bubbles evolve to

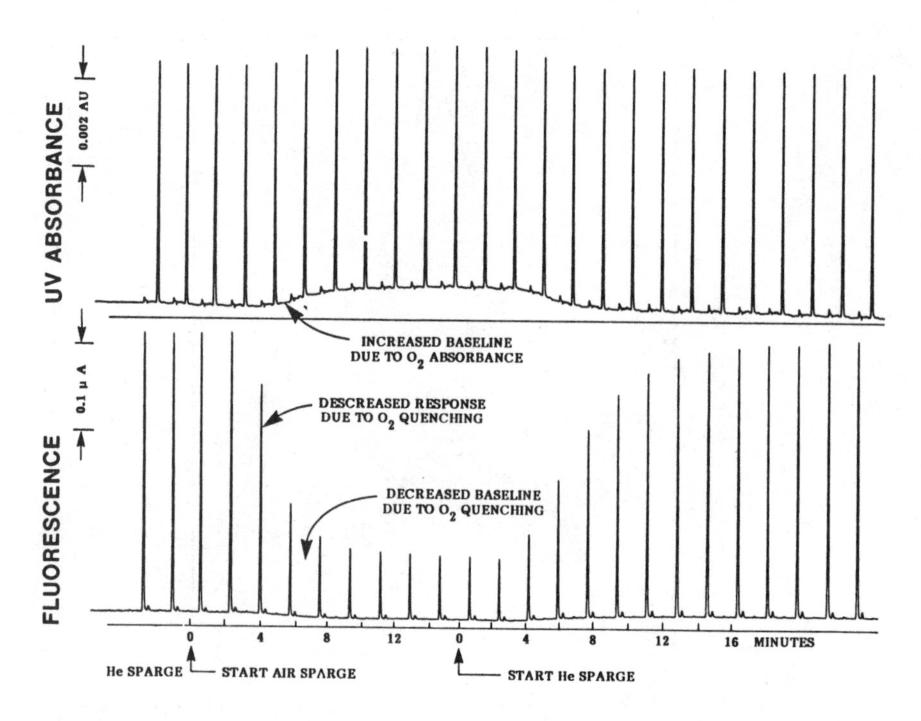


**Fig. 6.4.** UV absorbance detector and fluorescence detector response to mobile-phase oxygen. LiChrosorb RP-8 column, 75/25 acetonitrile, 5.0 mL/min, 25°C. UV detection (upper) at 254 nm; 250 nm excitation, 340 nm emission for fluorescence (lower). Sample = naphthalene. Reprinted from ref. (4) with permission.

cause pump or detector problems. Furthermore, Snyder[5] showed that significant (i.e., >1%) changes in retention can occur when partially water-saturated organic mobile phases (which are used commonly in normal-phase LC) are degassed by helium sparging. This results from selective volatilization of the minor (water) component. Significant retention changes are likely for these mobile phases when the value in the righthand column of Table 6.1 is greater than 1.4.

As was mentioned in the previous section, LC systems that use high-pressure mixing generally do not require degassing for reliable pump operation, because the mobile-phase components are combined after the LC pump(s). The system pressure keeps dissolved gas in solution. Bubbles can form in the detector cell, however, if there is not a sufficient backpressure restriction after

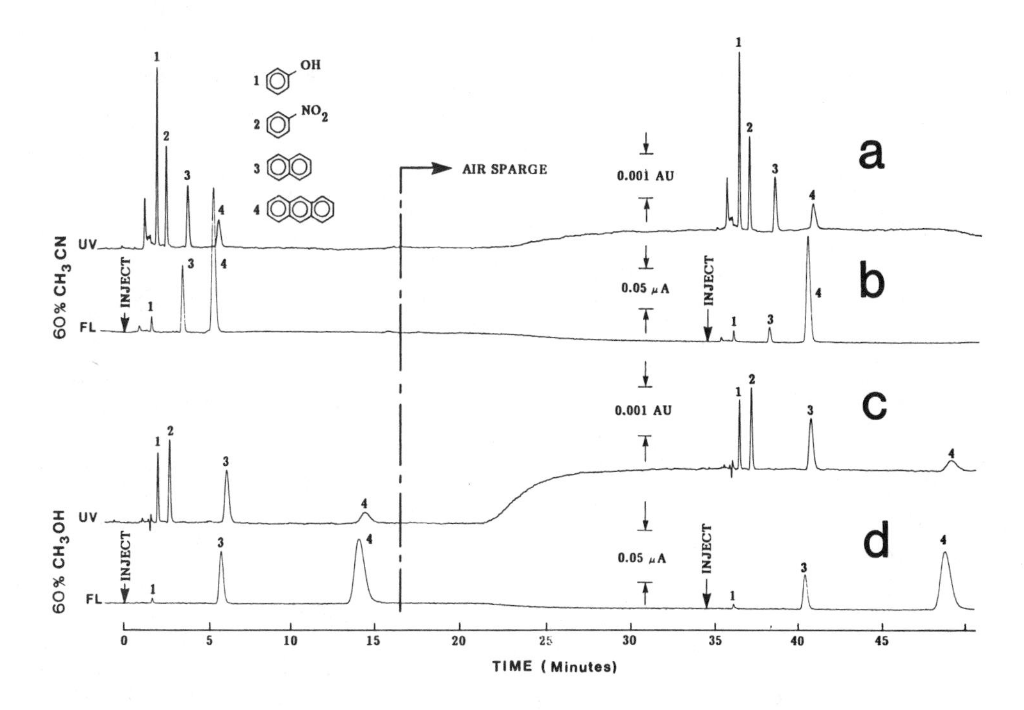


**Fig. 6.5.** Characteristics of UV (a, c) and fluorescence (b, d) detectors. Conditions as in Fig. 6.4, except mobile phase: 60/40 acetonitrile/water or 60/40 methanol/water. Samples as shown. Reprinted from ref. (4) with permission.

the cell. Bubbles lodging in the flowcell usually are indicative of a dirty cell, not a degassing problem. If bubbles are a problem in a clean flowcell, degas the mobile phase and use a pressure restrictor to keep the bubbles in solution until after the detector cell (see Chapter 12).

Manually-premixed mobile phases may not need degassing if the bubbles evolved upon mixing disappear before the mobile phase is used. Although degassing may not be required for manual (off-line) or high-pressure mixing, routine degassing generally improves system reliability. Degassing may also be required for special detectors or for low-noise operation.

Removing dissolved air from the mobile phase can improve chromatographic operation by eliminating bubble problems and improving detector performance. For these reasons, it is advisable to degas the mobile phase for most LC assays. There are four

Table 6.1
Loss in Mobile-Phase Components as a Function of Degassing

| | $K_x$ values | | | | Loss in X (%) for $(V_g/V_m)=(V_g/V_m)_{ox}$*** | |
|---|---|---|---|---|---|---|
| Solvent | He* | $N_2$* | $O_2$* | $(V_g/V_m)_{ox}$** | Solvent | Water |
| water | 105 | 63 | 32 | 0.14 | 0.0003 | — |
| *Organic Solvents* | | | | | | |
| acetonitrile | 28 | 5.6 | 3.6 | 1.3 | 0.033 | — |
| benzene | 47 | 8.1 | 4.4 | 1.0 | 0.036 | 0.84 |
| *n*-butyl chloride | 19 | 3.7 | 2.4 | 1.9 | 0.086 | 1.1 |
| chloroform | 19 | 3.8 | 2.4 | 1.9 | 0.13 | 1.2 |
| cyclohexane | 36 | 5.8 | 3.5 | 1.3 | 0.06 | 6 |
| cyclopentane | | | (3) | 1.5 | 0.20 | 9 |
| ethyl acetate | 20 | 4.0 | 2.6 | 1.8 | 0.07 | 0.03 |
| ethyl ether | 16 | 3.2 | 2.1 | 2.2 | 0.55 | 0.10 |
| *n*-heptane | | | (3) | 1.5 | 0.04 | 6.0 |
| *n*-hexane | 21 | 3.8 | 2.8 | 1.6 | 0.14 | 7 |
| methanol | 27 | 5.9 | 3.9 | 1.1 | 0.03 | — |
| methyl-t-butyl ether | | | (2) | 2.3 | 0.48 | 0.10 |
| methylene chloride | 17 | 3.4 | 2.2 | 2.1 | 0.31 | 0.42 |
| *n*-pentane | | | (3) | 1.5 | 0.39 | 9 |
| tetrahydrofuran | 24 | 4.7 | 3.0 | 1.5 | 0.09 | — |
| 2,2,4-trimethylpentane | | | (3) | 1.5 | 0.07 | 9 |

*$K_x$ = conc. gas in atmosphere/conc. gas in solvent
**Volume of He required per volume of solvent for 99% oxygen removal;
***% Loss of solvent or water during 99% removal of air by degassing. Condensed from ref. 5 (Table I) with permission.

common approaches to the dissolved-air problem: (a) helium degassing, (b) vacuum degassing, (c) sonication, and (d) heating. These techniques and their relative effectiveness are discussed in the following sections.

## *Helium Sparging*

Helium sparging is the most popular degassing method. In this method, helium is bubbled gently through the mobile phase to remove dissolved air, as shown in Fig. 6.6. When properly used, helium removes 80–90% of the dissolved air within 10 min (see Fig. 6.7).[1] Helium de-aerates the mobile phase by drawing unwanted dissolved gases (i.e., oxygen and nitrogen) from solution as they equilibrate with the helium bubbles. Helium itself has a very low solubility in LC solvents, so a helium-sparged solution is nearly gas-free. Redissolution of unwanted gases is prevented by continued sparging with helium or by keeping the mobile phase in

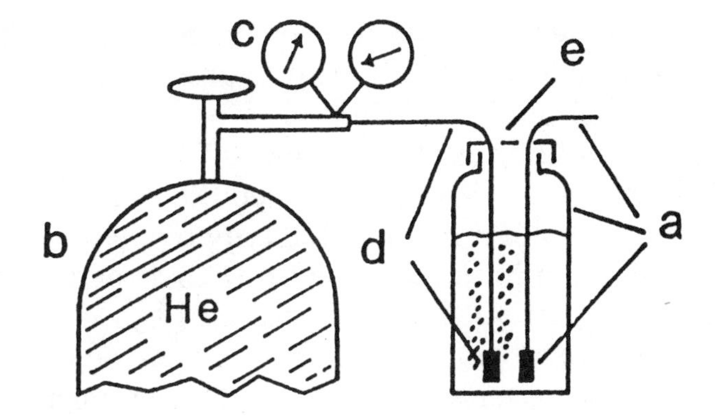


**Fig. 6.6.** Lab-built helium sparging apparatus: (a) reservoir, inlet line, and frit (same as Fig. 6.1), (b) helium supply cylinder, (c) two-stage regulator (second stage at ≅ 5 psig), (d) helium line and sparging frit (e.g., 1/8-in. od x 1/16-in. id Teflon tube with 10-µm porosity frit), (e) vent (1 mm maximum). Reprinted from ref. (1a) with permission.

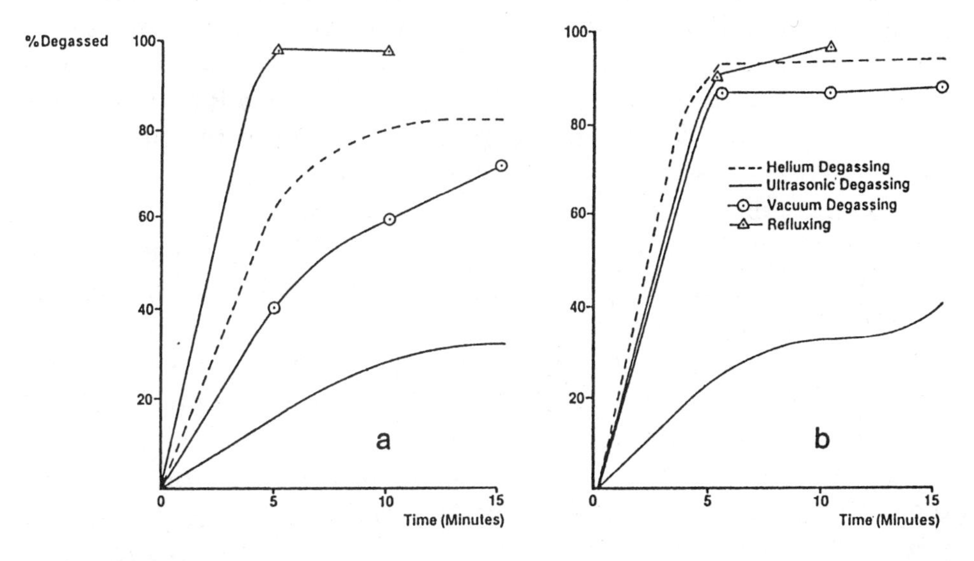


**Fig. 6.7.** Comparison of degassing techniques: (a) oxygen removal from methanol, (b) oxygen removal from hexane. Reprinted from ref. (1) with permission.

a helium atmosphere. (Use adequate ventilation, because the vented helium contains low levels of evaporated mobile phase.) It is important to prevent back-diffusion of air into the helium-sparged solution, or the degassing effectiveness can be reduced

20% or more.[1] Generally, a tightly-fitting reservoir cap with a 1 mm (maximum) vent hole is sufficient to prevent back-diffusion if a slow trickle of helium is maintained. Under these conditions, a helium flow of 4 mL/min is ten times as effective at removing air as a flow of 30 mL/min with an uncapped (20 mm opening) reservoir. Back-diffusion of air is improved only twofold when the flow is varied from 4 mL/min to 120 mL/min with a restricted vent.[1]

In general, retention-time shifts caused by selective volatilization of mobile phase components during sparging are not a significant problem (i.e., <1% change in $t_R$ under normal sparging conditions).[5] See the preceeding discussion and reference (5) for conditions under which helium sparging is not recommended.

A ready-to-use helium sparging system, such as the one shown in Fig. 6.8, can be purchased. Alternatively, a helium sparging system can be constructed from readily-available components, as shown in Fig. 6.6. A Teflon tube (e.g., 1/8- in. od x 1/16- in. id) is connected to the regulator on a helium tank and to an inlet frit (e.g., 10-μm porosity) submerged in the mobile phase reservoir. (Only high-purity helium should be used; oils or other contaminants in the helium can be extracted into the mobile phase.) The second stage of the gas regulator is adjusted (e.g., 5 psig) to provide a freely flowing stream of bubbles (e.g., >30 mL/min) for about 10 min. to degas the solvent. Then the flow is reduced to a trickle (e.g., 4 mL/min) to conserve helium, yet maintain degassed mobile phase. Make sure that the flow of helium bubbles does not contact the inlet filter, or bubbles may enter the inlet line. The gas cylinder should be shut off when not in use in order to minimize helium waste (helium is expensive!).

The helium-degassing systems from some manufacturers (e.g., Kontes, Perkin-Elmer, Omnifit) maintain a positive pressure on the reservoir (similar to the system shown in Fig. 6.8). This increases check-valve reliability, especially at low flow rates. Also, because these systems are closed to the atmosphere, degradation of air-sensitive mobile phases is minimized.

Regardless of the degassing method, or in the absence of degassing, it is normal to see a few small gas bubbles clinging to the sides of the reservoir or to the inlet frit. These bubbles generally are no problem, and can be dislodged by gently tapping the reservoir. Bubbles *should not* be seen moving in the inlet line to

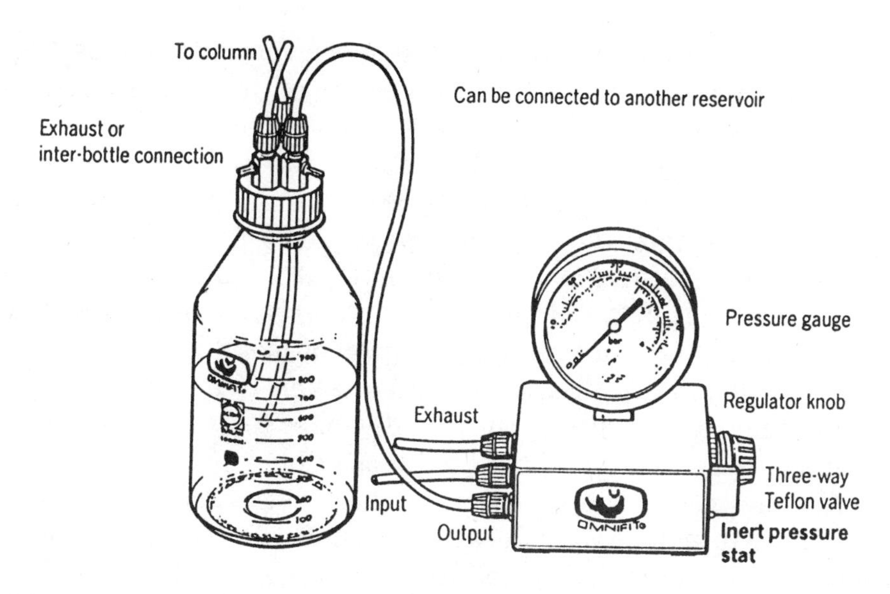


**Fig. 6.8.** Commercial helium sparging apparatus. Courtesy of Omnifit Ltd.

the pump—these are indicative of degassing problems or blockage of the inlet frit. Sometimes, however, a bubble will lodge in the highest point in the inlet line as it loops out of the reservoir into the pump. If this bubble is stationary, most workers ignore it, because it can be difficult to remove.

## *Vacuum Degassing*

Vacuum degassing is the second-most popular degassing method. Dissolved gases are removed by applying a partial vacuum (e.g., 10 mm Hg) to the reservoir, which increases the vapor pressure of the dissolved gases, causing them to bubble out of solution. When the vacuum is released, the mobile phase is left in a relatively gas-free state. In contrast to continuous helium sparging, vacuum degassing is done batchwise, usually once or twice a day. Vacuum degassing is less effective than helium at removing dissolved gas, as seen in Fig. 6.7.[1] In most cases, however, this method is sufficient to eliminate bubble formation at the pump inlet. Following vacuum degassing, air gradually redissolves at the liquid–air interface. If this is a problem, use helium sparging or a system that applies a helium atmosphere after degassing. Some workers[6] feel that more thorough degassing is obtained if vacuum and sonication are used in combination (see discussion below).

If mobile phases are filtered using a vacuum filtration apparatus (e.g., Fig. 6.11), degassing and filtration can be accomplished in a single step. Although vacuum filtration is not as effective as degassing by vacuum alone, it is adequate for many LC systems. Figure 6.9 shows a commercially available filter-degasser assembly that allows filtration of mobile phases directly into most reservoirs. Care should be taken to avoid using stoppers that may contaminate the mobile phase (e.g., use neoprene or Teflon-coated stoppers instead of black rubber).

A vacuum degassing system can be constructed from a heavy-walled vacuum flask as shown in Fig. 6.10. Use a protective shield in case the reservoir collapses. Some suppliers sell plastic-coated bottles (e.g., Omnisolve) or plastic sleeves (e.g., American Burdick and Jackson) for this purpose. Vacuum is supplied from a vacuum pump; a water aspirator is also effective. Be sure to use an explosion-proof vacuum pump, because potentially explosive solvent vapors will be drawn through the pump during operation.

### *Sonication*

Sonication can be used to degas mobile phases, but this technique is only about 30% effective at removing dissolved gas (see Fig. 6.7). Under some conditions, ultrasonic degassing can lead to *increased* air levels in the mobile phase,[1] so this technique should not be used with oxygen-sensitive (e.g., electrochemical) detectors. Williams and Miller[7] found that ultrasonic treatment was ineffective at degassing except for supersaturated solutions.

To degas the mobile phase, the entire reservoir is placed in a sonic (ultrasonic) bath, or a sonic probe is inserted into the reservoir. Sonication in combination with vacuum degassing is more effective than sonication alone.[7] In this case, the reservoir is placed in the sonic bath and vacuum is applied as with vacuum degassing.

### *Heating*

Heating or boiling the mobile phase is the most effective (see Fig. 6.7), but least convenient method of removing dissolved gas.

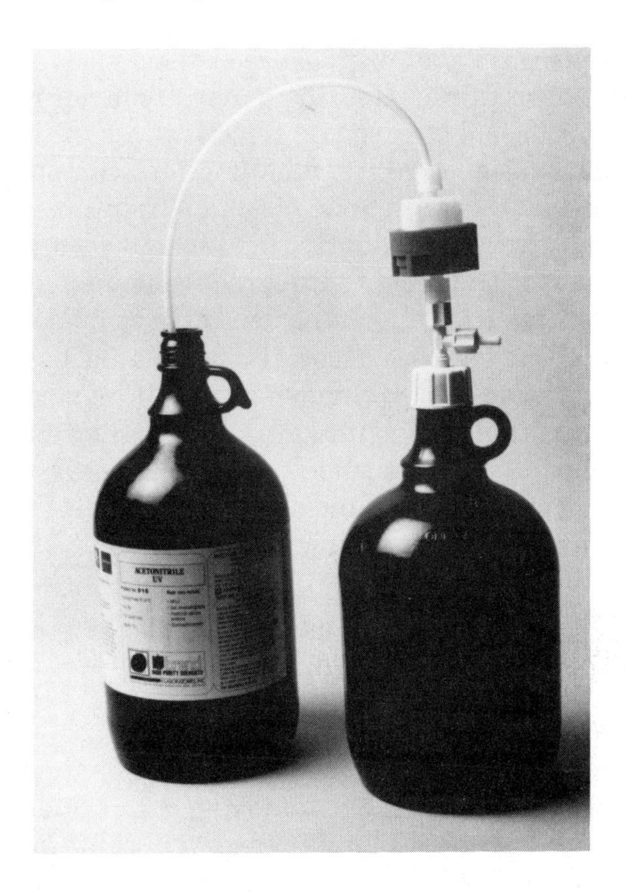

**Fig. 6.9.** Filter-degasser apparatus allows vacuum filtration of mobile phase directly into reservoir. Courtesy of Lazar Research Laboratories.

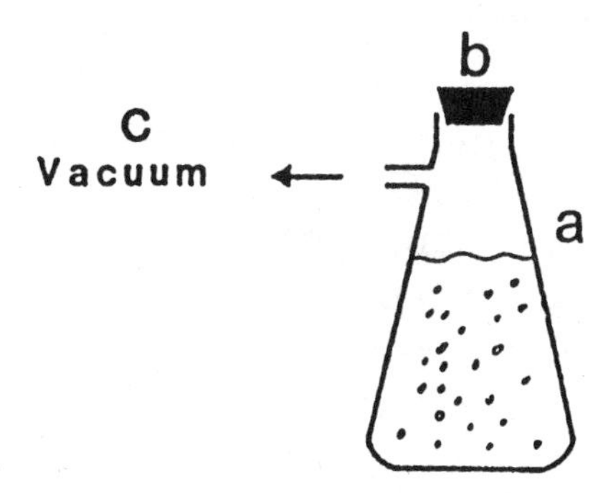


**Fig. 6.10.** Vacuum degassing apparatus: (a) vacuum flask (e.g., 1-L), (b) stopper (e.g., Teflon), (c) vacuum source (e.g., 10 mm Hg). Reprinted from ref. (1a) with permission.

Heating increases the vapor pressure of the dissolved gas, so that it bubbles out of solution. Heating generally is not recommended for solvent mixtures, because the selective loss of the more volatile component will change the mobile phase composition. However, in cases where complete degassing is required (e.g., electrochemical detectors in the oxidative mode) heating often is the best choice because it is the most thorough degassing method (see Fig. 6.7). A reflux-condenser should be used during heating in order to minimize evaporative loss. Special care should be taken to use adequate ventilation and to remove any ignition sources from the vicinity when using this technique. In use, the reservoir can be placed on a warm hot-plate to help minimize redissolution of the air; some LC systems (e.g., Hewlett-Packard 1084) allow heating of the reservoir during degassing.

## 6.2. Reservoirs

### *Reservoir Container*

The reservoir is a glass container that holds the mobile phase before it is pumped into the system (Fig. 6.1). One-liter and 4-L solvent jugs are the most popular reservoir containers; they can be used directly for helium sparging, but should not be used for vacuum degassing because of the danger of collapse under vacuum. Table 6.2 lists several types of reservoirs designed to accommodate specific methods of degassing. When degassing is not required, any reservoir design can be used.

### *Inlet Line and Frit*

The inlet line to the pump (Fig. 6.1) typically is Teflon tubing 1/8 in. od x 1/16 in. id with an inlet frit attached to the reservoir end of the line. Alternatively, some vendors (e.g., Waters) use 1/8 in. od thin-walled Teflon tubing for the inlet line. The inlet frit is connected to the inlet line with either a slip-on connector or a conventional compression fitting. The frit has two functions: (a) to keep particulate matter out of the pump, and (b) to anchor the line in the bottom of the reservoir (thus the frit is often called a "*sinker*"). Frits of 10-µm porosity are preferred over 2-µm frits because of their lower resistance to liquid flow, and therefore reduced likelihood of causing cavitation in the inlet line.

### *Reservoir Caps*

In all cases, a loosely-fitting cap should be used if one is not provided with the system. A solvent-bottle cap with vent- and inlet-line holes, or a piece of aluminum foil wrapped around the neck of the reservoir will exclude laboratory dust. If the cap fits too tightly, a partial vacuum may form as the mobile phase is removed; cavitation ("vapor-lock") in the inlet lines will then disrupt pump delivery. Two specialized caps that facilitate degassing are listed in Table 6.2 (footnotes b and c).

## 6.3. Problem Prevention

### *Solvent Filtration*

Pure HPLC-grade solvents, or mobile phases containing only HPLC-grade solvents do not require filtration. HPLC-grade sol-

Table 6.2
Reservoir Selection

| | Degassing method[a] | | | | |
|---|---|---|---|---|---|
| *Reservoir type* | None | He | Vacuum | Sonic | Heat |
| 1-L or 4-L solvent jug | + | + | – | + | |
| Coated or jacketed jug | | | + | | |
| Lab glassware (e.g., flask) | | | – | | + |
| Vacuum flask | | | + | | |
| *Modified reservoirs* | | | | | |
| Filter-degasser[b] | | | + | | |
| Vacuum-sparger[c] | | + | + | | |
| Sparging cabinet[d] | | + | | | |
| Sparging manifold[e] | | + | | | |

[a](+) recommended, (–) not recommended.
[b]Lazar Research Laboratories
[c]Omnifit Ltd.
[d]Perkin-Elmer
[e]Hewlett-Packard

vents are filtered through an 0.2-μm filter prior to bottling at the factory, and lab filtration can actually contaminate the mobile phase! Other solvents and mobile phases should be filtered through an 0.5-μm filter prior to putting them in the reservoir. Filtration is important especially if any solid reagents (e.g., buffer salts) are used in mobile phase preparation, because particulates can contaminate pump check valves, damage injectors, and block column frits.

The solvent filtration kit shown in Fig. 6.11 uses disposable filters selected to match the solvent type. It is important to match the filter type with the solvent; some filters dissolve in certain solvents, whereas others are impervious.

See ref. (7a–c) for a discussion of the compatibility of different membrane filters with various solvents. Polyvinylidene fluoride filters are the best single choice for mobile phase filtration, because they are compatible with all LC solvents (see Table 6.3). When using aqueous mobile phases (e.g., methanol/water), it is helpful to prewet the filter with 1–2 mL of organic solvent (e.g., methanol) to speed filtration.

**Fig. 6.11.** Solvent filtration kit: Filter is placed on a stainless-steel screen at (a), clamped to the funnel cup (b), and fitted to a vacuum flask via a ground-glass joint. Courtesy of Millipore Corporation.

Table 6.3
Solvent and Filter Compatibility

| Trade name[b] | Porosity, μm | Solvent type[a] Aqueous | Aq/Org | Organic |
|---|---|---|---|---|
| HATF | 0.45 | + | | |
| Alpha Metracel | 0.45 | | | + |
| Durapore | 0.45 | + | + | + |
| FP Vericel | 0.5 | + | + | + |
| Fluoropore | 0.5 | | | + |
| TF-450 | 0.45 | | | + |

Information from Gelman (G) and Millipore (M) product literature; check specific solvents with filter specifications before first use.
[a](+) compatible with this solvent type.
[b]HATF (M) is cellulose ester, Alpha Metracel (G) is regenerated cellulose; Durapore (M) and FP Vericel (G) are polyvinylidene fluoride; Fluoropore (M), TF-450 are PTFE.

## *Cleanliness*

Cleanliness is the key to reservoir maintenance. If solvent jugs are used as reservoirs, they should be discarded occasionally (e.g., once a month). Commercial reservoirs should be cleaned at similar intervals by washing with acid, solvent, or soap and water. Whatever method is chosen, be sure to use HPLC-grade water or solvent as a final rinse, so that no residue is left from the washing process.

In order to avoid problems, the inlet frit should be replaced (a) when it shows signs of blockage (e.g., bubbles appear in the inlet line), or (b) once every 3–6 months. Goldberg et al.[8] found that, under certain ion-pairing conditions, the inlet frit accumulated residues that contributed to retention irreproducibility. When the frit was removed, the problem was corrected. If problems with retention reproducibility are found with ion-pair LC, check for frit problems by removing the inlet frit. If the LC is operated without an inlet frit in place, take *extra care* to filter all mobile phases through a 0.5-μm filter.

## *Solvent Quality*

Many chromatographic problems can be avoided by the selection of high-quality reagents for mobile phase preparation. Reagents that have been specially purified for LC use are labeled "HPLC-grade" in vendor catalogs. These reagents are manufactured with special attention to minimizing particulate matter and UV absorbance. Use only HPLC-grade solvents, including water for mobile-phase preparation. Detailed descriptions of solvent specifications can be found in ref. (2) and (3). In-house generation of HPLC-grade water using a purification system such as the Milli-Q system from Millipore, is much less expensive than buying HPLC-grade water from a solvent vendor. Similarly, use high-purity buffers and salts in mobile phases.

Interferences can be present in HPLC-reagents in the form of stabilizers or contaminants. Some solvents, such as THF, form explosive peroxides upon extended exposure to air. For this reason, THF is available with a peroxide-inhibiting stabilizer (e.g., butylated hydroxytoluene); stabilized THF is not suitable for most LC applications. When stabilized THF is used with UV detection, a large baseline offset is normally observed. When unstabilized solvents are used, (a) discard opened bottles regularly (e.g., once a month), and (b) do not evaporate the solvent to dryness. Although HPLC-grade reagents undergo exacting quality control, out-of-specification batches sometimes reach the field. If you suspect that you have a substandard batch of solvent, check this by switching to another batch number or a similar product from another vendor. In order to be able to trace such problems, be sure to record the solvent batch (lot) number each time the mobile phase is prepared.

In some cases, stabilizers added to HPLC-grade solvents can change the selectivity of the mobile phase. One worker,[9] reported large differences in retention when a normal-phase LC assay was performed with a chloroform/hexane mobile phase containing different stabilizers. When the chloroform was stabilized with ethanol, solute retention was normal. When the mobile phase contained a different grade of chlorform (different part number, same vendor), that was stabilized with amylene (2-methyl-2-butene), no peaks eluted from the column. In this case, the ethanol stabilizer (at 0.5–1.0% levels) was sufficient to modify the strength of the mobile phase so that the assay worked. Thus, "equivalent" mobile phases made up from slightly different starting solvents from the same vendor are not necessarily the same.

Another group[10] reported that column stability problems were encountered with silica columns when different brands and grades of methylene chloride were used in the mobile phase. Upon close examination, it was determined that residual HCl in the solvent was causing the problem. The solvents tested contained from 5 to 781 ppb of residual chloride. The higher levels of chloride were sufficient to damage the silica-based columns that were used.

These examples of problems caused by minor solvent components should serve as reminders to pay careful attention to the specification and use of LC solvents. Solvents from the same manufacturer with the same part number should be equivalent, but solvents with different part numbers or from different manufacturers may not be interchangeable for all assays.

### *Spares*

An empty solvent bottle or two should be kept as backup reservoirs. These can be used to replace old solvent jugs or as temporary substitutes if a commercial reservoir is broken. A spare inlet frit for each line (e.g., three frits for a ternary system) should be stocked, along with sufficient tubing to replace the inlet lines.

## 6.4. Problems and Solutions

There are four main problem areas related to the reservoir or mobile phase: (1) insufficient mobile-phase degassing, (2) an inadequate supply of mobile phase to the pump, (3) contamination of the mobile phase or reservoir, and (4) chromatographic problems. The first three areas are discussed in the following sections and are summarized in Table 6.4 (at end of this chapter). Problems with the chromatogram are covered in Chapters 14 and 15.

### *Insufficient Degassing*

Insufficient degassing is the most likely cause of fluctuating pressures or sharp noise-spikes (due to bubbles) in the chromatogram. Dissolved gas in the mobile phase comes out of solution when the mobile phase warms up, or when mobile-phase components are mixed, and then interferes with pump and/or detector operation. First, try degassing the mobile phase to solve the problem (or use a more effective method, such as helium sparging). With low-pressure mixing, a persistent degassing problem can be confirmed by pumping a premixed and degassed mobile phase. If the problem then disappears, it is likely that an improper degassing technique is the cause. Once degassing is eliminated as a cause of the problem, look elsewhere for problems (e.g., loose fittings, blocked frits).

In severe cases, it may be necessary to operate the system only with premixed and degassed mobile phase. When the system pressure is steady, yet noise spikes are seen in the chromatogram, a post-detector backpressure restrictor (e.g., a column or capillary tube) will often solve the problem (see Chapter 12).

### *Inadequate Mobile Phase Supply*

*Insufficient mobile phase* in the reservoir causes air to enter the inlet line and disrupts the pump flow and pressure. Always fill the reservoir with sufficient mobile phase for all of the samples to be run, and position the inlet-line frit at the bottom of the reservoir. Securing the line is especially important with stiff Teflon lines, because they tend to creep out of the reservoir under the influence of system vibrations. In this case, the reservoir cap can be modified (Fig. 6.12) to hold the line in place. Drill an undersized hole in the cap liner and force the line through the hole to prevent movement. Extend a little more line than is necessary to reach the bottom of the reservoir, so that the tension in the bent line holds the inlet frit at the bottom. Be sure to provide a vent hole in the cap (e.g., 1 mm) so that a vacuum is not formed as mobile phase is pumped out. A helium-sparging line can be held in place in a similar manner with another hole.

A *blocked inlet frit* can cause *cavitation* of the mobile phase in the inlet line. This is analogous to vapor-lock in an automobile. Under the partial vacuum created when the pump tries to draw mobile phase through the blocked frit, the mobile phase vaporizes and/or dissolved gas evolves. When this gas reaches the pump, pressure and flow instability are seen.

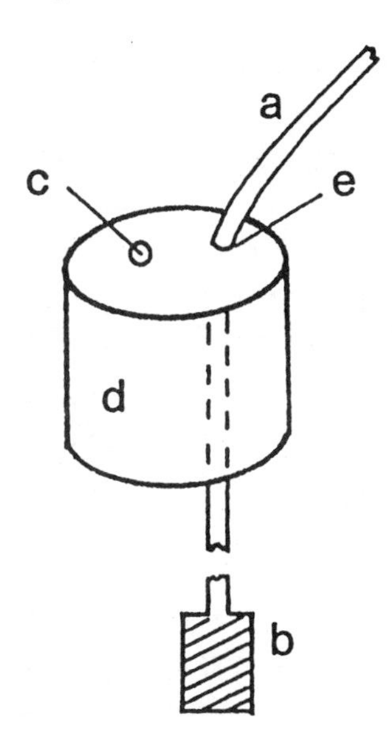


**Fig. 6.12.** Reservoir cap modification to restrict inlet-line movement: (a) inlet line (e.g., 1/8-in. od), (b) frit, (c) vent (1 mm maximum), (d) cap, (e) undersized hole (e.g., 3/32-in.).

Frit blockage usually is the result of particulate contamination of the mobile phase or microbial growth in the reservoir (see below). Confirm the presence of a blocked frit by temporarily removing the inlet frit and operating the system. If the problem disappears, a blocked frit is likely. Replace a blocked inlet frit with a new one; some workers attempt to clean the inlet frit by sonication in dilute nitric acid, but this is only marginally successful. In some cases, it may be helpful to replace a 2-μm porosity frit with a 10-μm frit. If inlet-frit blockage is a recurring problem, review your mobile-phase preparation procedures (e.g., measuring, mixing, filtering, degassing) to locate the source of contamination.

If the *resistance to flow* is too high, cavitation can occur in the inlet lines. Flow resistance is a function of the inlet-frit porosity, the inlet-line id, the flow rate, and the solvent viscosity. Flow resistance (other than a blocked frit) generally is a problem only at higher flow rates (e.g., above 10 mL/min). In this case, one or more of the following changes can help solve the problem: (a) increase the frit porosity (e.g., 10 vs 2 μm), (b) increase the inlet line id (e.g., 1/16-in. id minimum), (c) elevate or pressurize the reservoir, (d) lower the flow rate, (e) operate without an inlet frit.

If the reservoir is *too tightly capped,* a partial vacuum can form in the reservoir as mobile phase is pumped out. Loosen the cap or add a vent hole to solve this problem. If the system uses a helium atmosphere to prevent redissolution of air, it is also possible to form a partial vacuum if the helium supply runs out or is shut off.

A *crimped or blocked inlet line* can also cause pump starvation. Check the line for crimps, twists, or heavy objects placed on it. Replace the line if necessary.

*Related problems.* Pump starvation can be caused by problems elsewhere in the LC system. Possiblities include: loose or leaky fittings, blocked low-pressure-mixer frits, perforated solenoid diaphrams (low-pressure mixer), and pump malfunction (seals, check valves). See Chapters 7 (Pumps) and 9 (Fittings) for a discussion of these problems.

### *Contamination of Reservoir or Mobile Phase*

The symptoms of reservoir or mobile-phase contamination are (a) increased long-term noise, or (b) a high detector baseline. Mobile-phase contaminants usually *do not* show up as extra peaks in the chromatogram; contaminants are pumped into the system at a steady concentration and therefore also elute at a steady concentration. With gradient elution, however, contaminants can collect at the top of the analytical column under weak mobile-phase conditions and later elute when the mobile-phase strength is increased. Contaminants in the mobile phase also can give spurious peaks when a sample is injected (see Chapter 15). Abrupt changes in long-term noise or baseline signal may be seen when the mobile phase is replenished, or after the LC has been idle (e.g., overnight). If the baseline changes when new mobile phase is added, this suggests that either the old or new mobile phase may be contaminated. If a change in the baseline occurs after the LC has been idle, corrosion or microbial growth in the system may be responsible (see below).

A *dirty reservoir* can contaminate an otherwise good batch of mobile phase. Replace dirty solvent-bottle reservoirs or clean specialized reservoirs. It is good practice to dedicate a reservoir to each mobile-phase type and to clean or discard the reservoir monthly in order to prevent contamination.

*Contaminated mobile phase* can result from the use of poor-quality reagents; always use high-purity reagents and HPLC-grade solvents. Mobile phases can also be contaminated by improper handling or use of dirty glassware during preparation. If bad reagents or solvents are suspected, switch to reagents from a different manufacturing lot or another manufacturer; discard all contaminated mobile phase.

*Microbial growth in buffers* can contaminate the mobile phase. This can be a special problem for mobile phases that support microbial growth (e.g., aqueous mobile phases containing acetate buffer). If the mobile phase is left in the system, microbial growth in the lines can occur, contaminating the system with byproducts of the organisms. Prevent this problem by thoroughly flushing the system with nonbuffered mobile phase daily; discard any partially used buffer. If microbial growth is suspected, the inlet line and frit can be cleaned by washing with dilute nitric acid, but often it is easier to replace them.

Stock solutions of buffers can also support microbial growth, so these should be stored under refrigeration and should contain an added growth-inhibitor (e.g., 0.04% sodium azide) if necessary. Daily filtering (0.2-µm biological filter) of buffered mobile phases may be required to prevent system contamination.

Finally, the mobile phase can become *contaminated during filtering or degassing*. Vacuum filtering or degassing can introduce pump oil or other contaminants from the vacuum hose if it is not connected and disconnected properly. Helium sparging with impure helium or a dirty frit (was it placed on a dirty lab bench before use?) can transfer contaminants to the mobile phase.[11] Be sure to use clean and inert stoppers, stir bars, filters, and glassware when preparing and degassing mobile phase. Discard the contaminated mobile phase and eliminate the cause of contamination before proceeding.

Table 6.4
Reservoir-Related Problems and Solutions

| Cause of problem | Symptom | Solution |
|---|---|---|
| Insufficient degassing | Pressure fluctuations, noise, spikes in chromatogram | 1. Degas mobile phase<br>2. If already degassed, use helium sparging<br>3. Premix and degas mobile phase |
| *Inadequate mobile phase supply to pump* | | |
| Insufficient mobile phase | Pressure fluctuations, air in inlet line | 1. Add mobile phase to reservoir<br>2. Position inlet frit below surface of mobile phase<br>3. Position inlet frit so that helium bubbles are not drawn in |
| Blocked inlet frit | cavitation in inlet line, pressure fluctuations | 1. Test by removing inlet frit, if problem disappears, replace frit<br>2. Filter mobile phases to prevent recurring blockage |
| Flow resistance too high | Cavitation in inlet, pressure fluctuations (especially at higher flow rates) | 1. Test by lowering flow rate; if performance improves, go to #2<br>2. Replace inlet frit (if still bad, increase porosity to 10 μm)<br>3. Increase inlet line i.d.<br>4. Elevate or pressurize reservoir<br>5. Lower flow rate<br>6. Operate without inlet frit (filter mobile phase!) |
| Cap too tight on reservoir | Cavitation in inlet line, pressure fluctuations | 1. Loosen or vent reservoir cap<br>2. For closed resevoirs, replace used mobile phase with helium |
| Crimped or blocked inlet line | Cavitation in inlet line pressure fluctuations | 1. Locate and remove obstruction<br>2. Replace inlet line |
| Related problems | (see text) | |
| *Contamination of reservoir or mobile phase* | | |
| Contaminated inlet line, frit or reservoir | Excessive long-term noise, high background signal | 1. Replace frit<br>2. Clean or replace inlet line<br>3. Clean or replace reservoir |

*(continued)*

Table 6.4
Reservoir-Related Problems and Solutions *(continued)*

| Cause of problem | Symptom | Solution |
|---|---|---|
| *Contamination of reservoir or mobile phase* | | |
| Contaminated mobile phase | Abrupt change in background signal when new mobile phase is added | 1. Replace with new mobile phase<br>2. Use solvents from a different lot or manufacturer<br>3. Use higher-grade solvents and reagents<br>4. Prevent contamination when mixing and degassing mobile phase |
| Microbial growth in buffers | Abrupt change in background signal (generally after LC stands idle), gradual change in background when system in continuous use, blockage of column frit with rise in pressure | 1. Flush system and use fresh buffered mobile phase daily<br>2. Store buffer stock solution under refrigeration<br>3. Add growth inhibitor to mobile phase<br>4. Filter mobile phase daily through 0.2-μm filter |
| Contamination during filtering or degassing | Abrupt change in background signal when new batch of mobile phase is used | 1. Use clean glassware<br>2. Replace sparging frit<br>3. Avoid mobile phase contact with rubber hoses and stoppers<br>4. Use new filter for each mobile phase batch<br>5. Be sure filter is compatible with solvent |
| Related problems | (see text) | |

## 6.5. References

[1]Brown, J. N., Hewins, M., van der Linden, J. H. M., and Lynch, J. H. M. (1981) *J. Chromatogr.* **204,** 115.

[1a]Snyder, L. R. and Dolan, J. W. (1985) *Getting Started in HPLC, User's Manual,* LC Resources, Lafayette, CA.

[2]*High Purity Solvent Guide, 2nd ed.* (1984) Burdick & Jackson Laboratories, Muskegon, MI.

[3]*HPLC Solvent Reference Manual* (1985) J. T. Baker Chemical, Phillipsburg, NJ.

[4]Bakalyar, S. R., Bradley, M. P. T., and Honganen, R. (1978) *J. Chromatogr.* **158,** 277.
[5]Snyder, L. R. (1983) *J. Chromatogr. Sci.* **21,** 65.
[6]Kirkland, J. J., Du Pont, personal communication.
[7]Williams, D. D. and Miller, R. R. (1962) *Anal. Chem.* **34,** 657.
[7a]Merrill, J. C. (1984) *A Laboratory Comparison of Popular HPLC Filters and Filter Devices for Extractables,* Gelman Sciences, Ann Arbor, MI.
[7b]Merrill, J. C. (1986) *A Laboratory Comparison of Nylon Syringe Filter Units for Extractables in HPLC,* Gelman Sciences, Ann Arbor, MI
[7c]Merrill, J. C. (October 1987) *Amer. Lab.* **19(10),** 74.
[8]Goldberg, A. P., Nowakowska,E. Antle, P. E., and Snyder, L. R. (1984) *J. Chromatogr.* **316,** 241.
[9]Ciurczak, E. and Dolan, J. W. (1986) *LC/GC* **4,** 894.
[10]Nichols, W. A., Mayer, A. C., and Alexander, M. (1987), quoted from *LC/GC* **5,** 874.

# Chapter 7

# PUMPS

## Introduction

LC pumps draw mobile phase from the reservoir(s) and drive it through the column under pressure. Pumps for use with analytical columns typically have a flow-rate range of 0.1–10.0 mL/min at pressures up to 6000 psi; specialized pumps for microbore and preparative chromatography provide flow rates outside this range. Most LC pumps include pulse-damping modifications in order to reduce pressure pulses.

LC pumps with the same nominal specifications generally can be interchanged without changing system performance. Two problems can arise from mismatching the pump with other components in the system, however. *First,* the new pump may not fit into the mechanical design of the LC, especially when all the modules are mounted in a cabinet and/or are controlled by a system controller. *Second,* the new pump may have an inappropriate output or internal volume. For example, an analytical pump with a minimum flow rate of 0.1 mL/min cannot be substituted for a microbore pump operating in the 0.01–0.05 mL/min range.

Pump problems often show up as abnormalities in the chromatogram. Increased baseline noise and retention-time irreproducibility can result from pump wear, contamination by resulting particulates, or air entrainment in the pump. Because the pump contains more moving parts than any other LC module, mechanical wear is a common source of problems and a key area to address in preventive maintenance. The three most common pump problems are (a) check valve failure, (b) seal leakage, and (c) air bubbles in the pump.[1]

## 7.1. Basic Pump Design

Most LC pumps used today are based on the reciprocating-piston pump, which works like an automobile engine. The five main parts of the reciprocating pump are shown in Fig. 7.1: the *Piston, Cylinder, Pump Seal,* and two *Check Valves.* The assembled cylinder, seal, and two check valves are called the *Pump Head.* A driving *Cam* is attached to the pump motor, and a *Connecting Rod* translates the rotary motion of the cam into the reciprocating piston motion.

### *Pump Operation*

A schematic drawing of pump operation is shown in Fig. 7.2. During the *Intake Stroke* (or *Fill Stroke*), the piston is withdrawn from the cylinder, lowering the pressure upstream from the *Inelt Check-Valve.* The inlet check-valve opens, and liquid flows into the cylinder. The *Outlet Check-Valve* closes, because the system pressure exceeds the cylinder pressure. On the *Delivery Stroke,* the piston moves into the cylinder, the inlet check-valve closes, the outlet check-valve opens, and liquid flows to the column. The delivered volume for pumps used with analytical columns typically is 35–400 μL/stroke.

### *Check Valves*

The check valves control the direction of solvent flow through the pump head. Typically, the inlet check-valve is mounted on the bottom and the outlet check-valve is mounted on the top of the pump head (e.g., Fig. 7.3a); this facilitates bubble-clearing (dis-

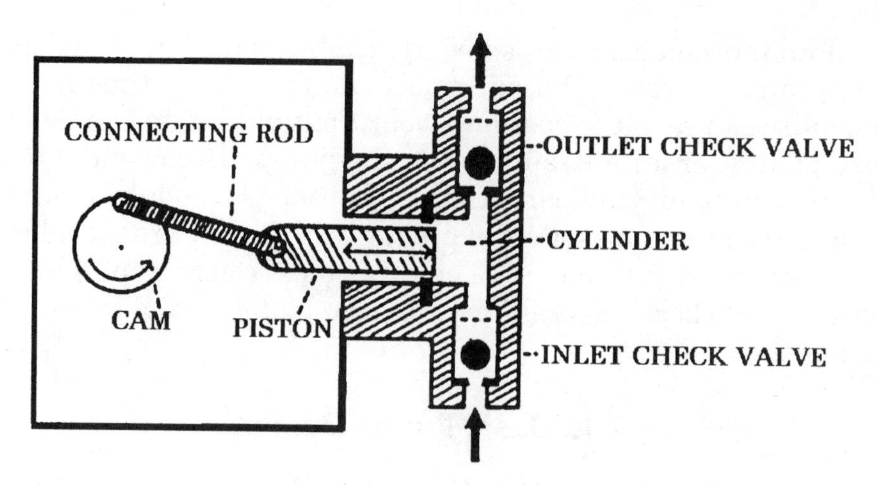


**Fig. 7.1.** Reciprocating-piston pump. Reprinted from ref. (2) with permission.

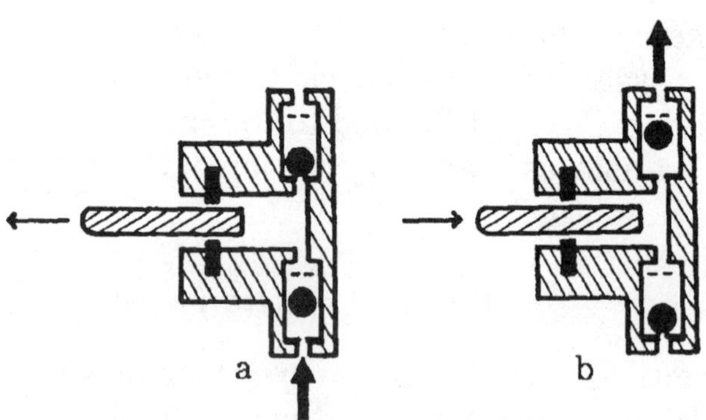


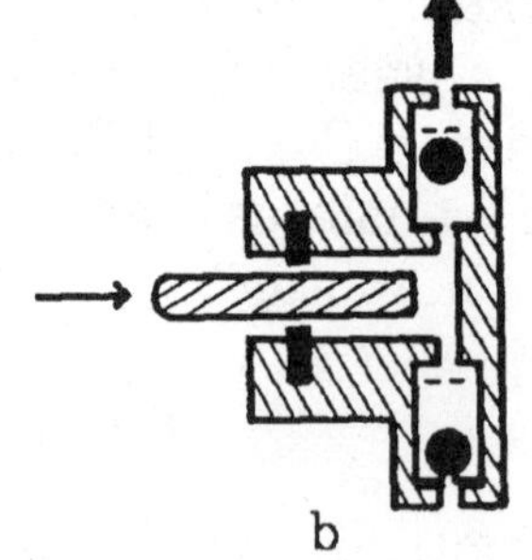


**Fig. 7.2.** Pump operation: (a) intake stroke: piston is withdrawn from cylinder, outlet check-valve closes, and liquid flows in through inlet check-valve; (b) delivery stroke: piston is pushed into cylinder, inlet check-valve closes, and liquid flows out through outlet check-valve. Reprinted from ref. (2) with permission.

cussed below). The parts of the check valve are shown in Fig. 7.3b,c. The housing is stainless steel and contains a small ball (usually made of ruby). A seat (typically of sapphire) is fitted in the inlet end of the housing, and a screen or retainer is fitted in the outlet. When liquid flows into the inlet of the check valve (seat end), the ball is lifted from the seat (and retained by the screen), allowing liquid to flow through the check valve (Fig. 7.3b). When the flow reverses, the ball is forced onto the seat and

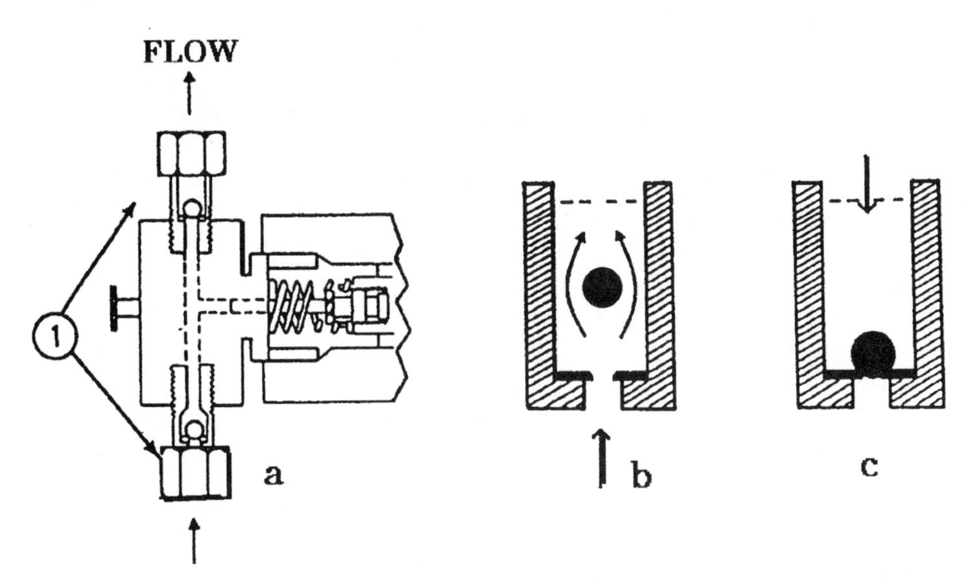


**Fig. 7.3.** Check valves: (a) cross-section of pump head showing check valves (1) and direction of flow; (b) check valve opened to permit liquid flow; (c) check valve closed to stop flow. Courtesy of LDC Milton Roy (a) and ref. (2) (b, c).

blocks flow in the reverse direction (Fig. 7.3c). Some check valves are more complicated in design than the ones shown here. They may have two sets of balls and seats to minimize leakage, or springs to ensure that the ball seals against the seat.

## *Pump Seals*

Pump seals allow the piston to move freely into and out of the pump head, while preventing mobile-phase leakage. All of the pumps discussed in this chapter (except the piston-diaphragm pump) use pump seals. Figure 7.4 is an exploded diagram showing how the pump seal fits into the pump head. The seal is pressed into the head and mechanically held in place when the head is installed on the pump. Figure 7.5 shows a closeup of one pump-seal model. Seals are made of a flexible, yet inert material, such as graphite-filled Teflon. Notice that the two sides of the seal are not equivalent. One side contains a spring, which helps to seal the inner flange around the piston. When installed, this spring *must* face the pump chamber, or the seal will leak. By

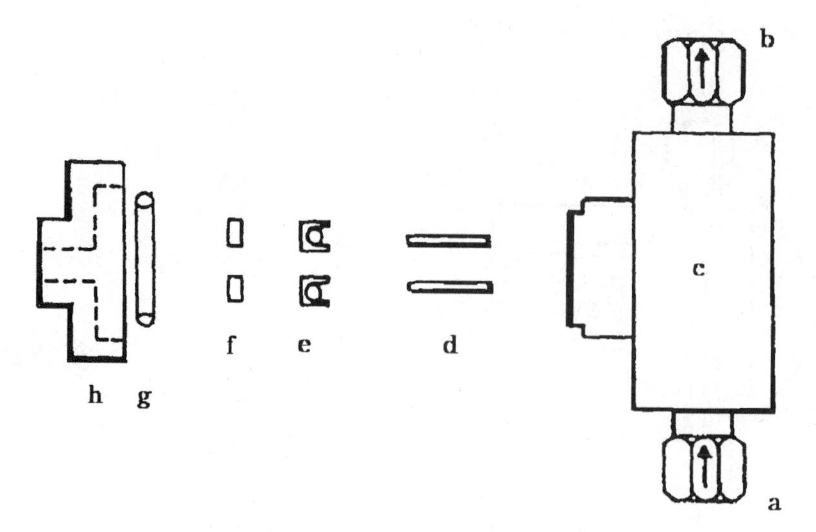


**Fig. 7.4.** Exploded diagram of pump head assembly: (a) inlet check-valve, (b) outlet check-valve, (c) pump head, (d) piston guide, (e) pump seal, (f) seal backup washer, (g) o-ring, (h) seal retaining cap. Courtesy of LDC/Milton Roy.

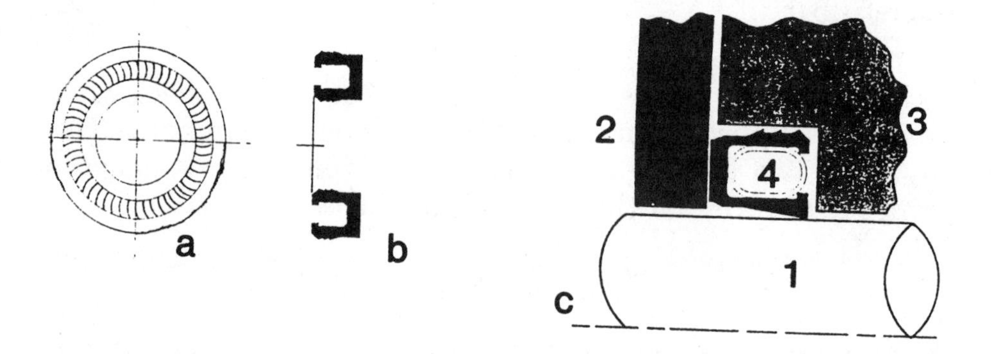


**Fig. 7.5.** Pump seal: (a) face-view, high-pressure side, showing spring; (b) cross-section of (a), rotated 90° with spring not shown; (c) installed seal showing sections of piston (1), backup washer (2), pump head (3), and seal (4). Courtesy of Bal Seal Engineering Co.

design, pump seals are not 100% efficient at preventing leakage, so that the piston can move freely. Thus, the surface of the piston behind the seal generally is damp. It is the evaporation of this mobile phase that causes abrasive crystal buildup when buffers are used. The sapphire piston can be scratched by these crystals, and mobile-phase leakage increases. Buffer crystals also can

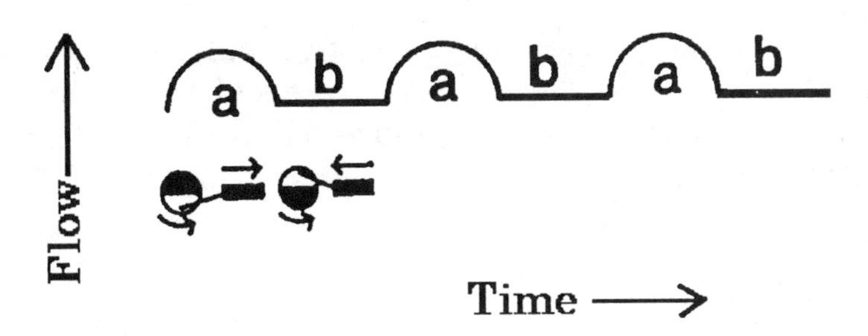


**Fig. 7.6.** Single-piston pump output: (a) delivery stroke; (b) intake stroke; insets show cam rotation and direction of piston travel. Reprinted from ref. (2) with permission.

accelerate seal wear, so it is important to flush buffered mobile phases from the pump when it is not in use. As the pump seal wears, it can cause secondary damage to other parts: (a) mobile phase can leak into the pump mechanism, corroding bearings or electronic parts, and (b) particulates given off as the seal wears can cause frit blockage or column damage downstream from the pump.

### *Pump Output*

A graph of the flow (or pressure) from a single-piston reciprocating pump is seen in Fig. 7.6. During a pumping cycle, the pump delivers mobile phase half of the time; the cylinder fills during the other half. As a result, the flow rate varies widely, causing undesirable flow and pressure pulses. Pressure pulses can damage LC columns and, with many detectors, cause baseline noise (oscillations) in the chromatogram. Mechanical and/or electronic *Pulse Suppression* is used in most pumps to smooth out pump pulsations.

In order to smooth the flow of solvent, many pumps have electronic sensors that adjust the flow rate in response to pump pressure. Other systems measure the pump delivery and adjust the motor speed as necessary. To improve the accuracy of pump delivery, many pumps use manual or automatic flow-rate adjustment to compensate for mobile phase compressibility.

Many LC systems also mix the solvents or mobile-phase components on-line to form the final mobile phase. Pumps with

*Low-Pressure Mixing* use timed solenoid valves and a mixing chamber to combine the desired proportions of components before they reach the pump. In most systems using *High-Pressure Mixing,* a separate pump for each mobile-phase component feeds solvent to a high-pressure mixer. These mixing techniques are discussed further in Section 7.3.

A *Purge Valve* mounted between the pump and injector facilitates mobile phase changeover and bubble removal. When opened, this valve releases the system pressure so that the pump can be operated at high flow rates (e.g., 10 mL/min) for rapid washout during a change in solvent.

## 7.2. Common Pump Variations

Four variations of the reciprocating-piston pump are commonly used in LC: (a) single-piston pumps, (b) pumps that use two or three pistons in parallel, (c) tandem-piston pumps, and (d) piston-diaphragm pumps. Syringe pumps are a fifth common pump design. Various pulse-suppression and mixing techniques are used with different pumps.

### *Single-Piston Pumps*

The action of single-piston reciprocating pumps was described earlier (Sections. 7.1). These pumps usually are modified to minimize flow pulsations by (a) altering the cam shape or (b) varying the pump-motor speed throughout the pumping cycle (sometimes referred to as an "electronic cam"). Figure 7.7 compares the output of an unmodified pump to that of a pump with a modi-fied cam. In this case, the fill-time is shortened by rapidly withdrawing the piston, then the piston moves forward quickly until the desired delivery rate is reached. The piston speed then is adjusted to deliver a constant flow and finally the next rapid-refill cycle begins. This modified pumping cycle smooths the flow and reduces pulses. A shaped cam and/or modified motor speed commonly are used in other pump variations.

A pulse-suppressor often is used with LC pumps to further dampen pump pulsations. One pulse-suppression design is a flattened, thin-walled piece of tubing coiled in a large spiral (e.g., 100 cm of tubing wrapped in a 7.5-cm diameter spiral), as shown

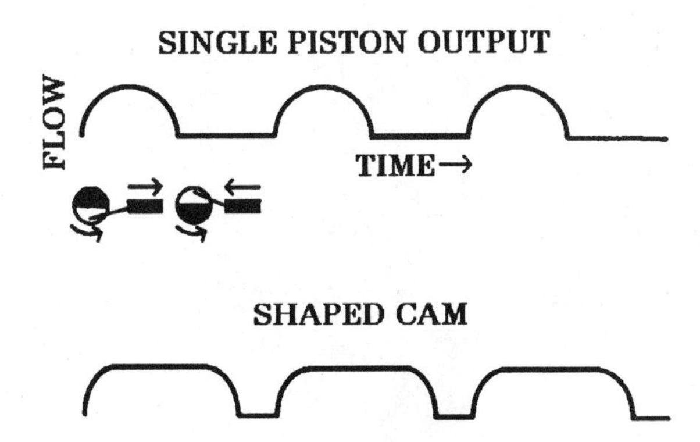


**Fig. 7.7.** Normal vs shaped-cam output of a single-piston (flow vs time, as in Fig. 7.6. Altered cam shape shortens intake stroke and lengthens delivery stroke. Reprinted from ref. (2) with permission.

in Fig. 7.8. The coil stretches a little when the pressure rises and contracts when the pressure drops, which changes its volume. This reduces flow and pressure pulsations. These pulse dampers can be purchased separately (e.g., Handy & Harmon) for addition to existing systems; they are mounted between the pump and the injection valve.

## *Dual-Piston Pumps*

In a dual-piston pump, a second piston, operated 180° out-of-phase is added to the single-piston pump. Figure 7.9 shows how the complementary overlap of the delivery strokes results in a significant pulse reduction. A modified cam and/or pulse damper further reduce pump pulsation. In practice, the two pistons are configured to operate in one of two modes: (a) the pistons are mounted parallel to each other and driven by two cams, which in turn are driven by a single shaft, as seen in Fig. 7.10; or (b) the pistons are 180° opposed, operated by the same cam, in a manner similar to the three-piston configuration shown later (Fig. 7.12).

## *Three-Piston Pumps*

A third piston can be added to the dual-piston pump in order to further reduce pump pulsations, as shown in Fig. 7.11. The

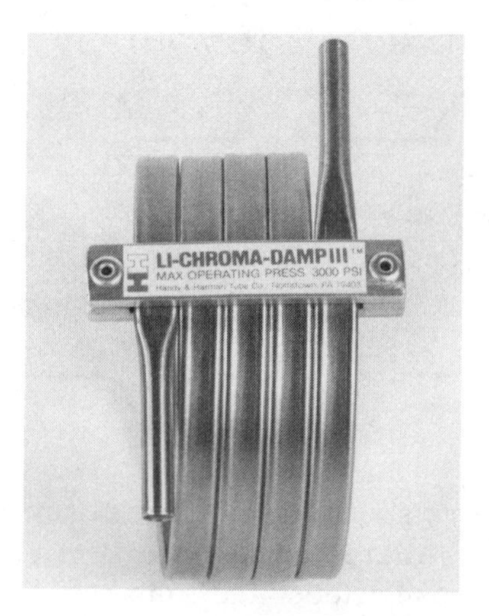


**Fig. 7.8.** Pulse-suppressor: 3-in. diameter coil, 3000 psi maximum operating pressure, <5 mL internal volume. This model is designed to add on to existing LC systems. Courtesy of Handy & Harman Tube Co., Inc.

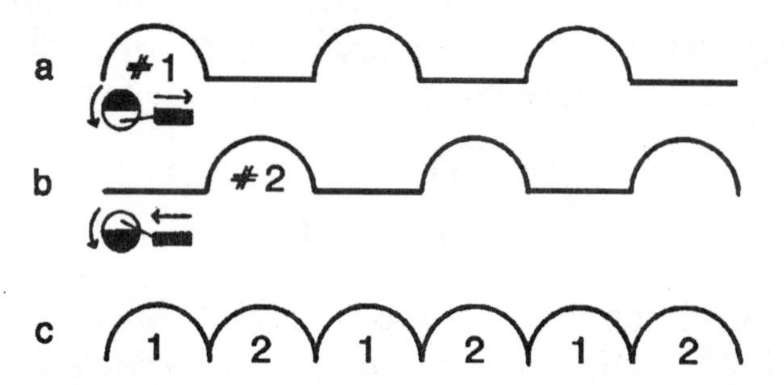


**Fig. 7.9.** Dual-piston overlap: (a) piston #1 output; (b) piston #2 output; (c) combined output of pistons #1 and #2 (flow vs time, as in Fig. 7.6). Reprinted from ref. (2) with permission.

pistons can be mounted in parallel and driven by three cams on the same shaft (e.g., DuPont) or by a swash-plate (e.g., Jasco). Alternatively, the pistons can be mounted in the same plane, 120° opposed, and driven from the same cam (e.g., Nicolet Analytical Division; see Fig. 7.12). Shaped cams and/or electronic pulse reduction can be used to further reduce pulsation.

Pumps containing more than three pistons are feasible but impractical; the increase of cost, complexity, and maintenance is too great for the small improvement in performance.

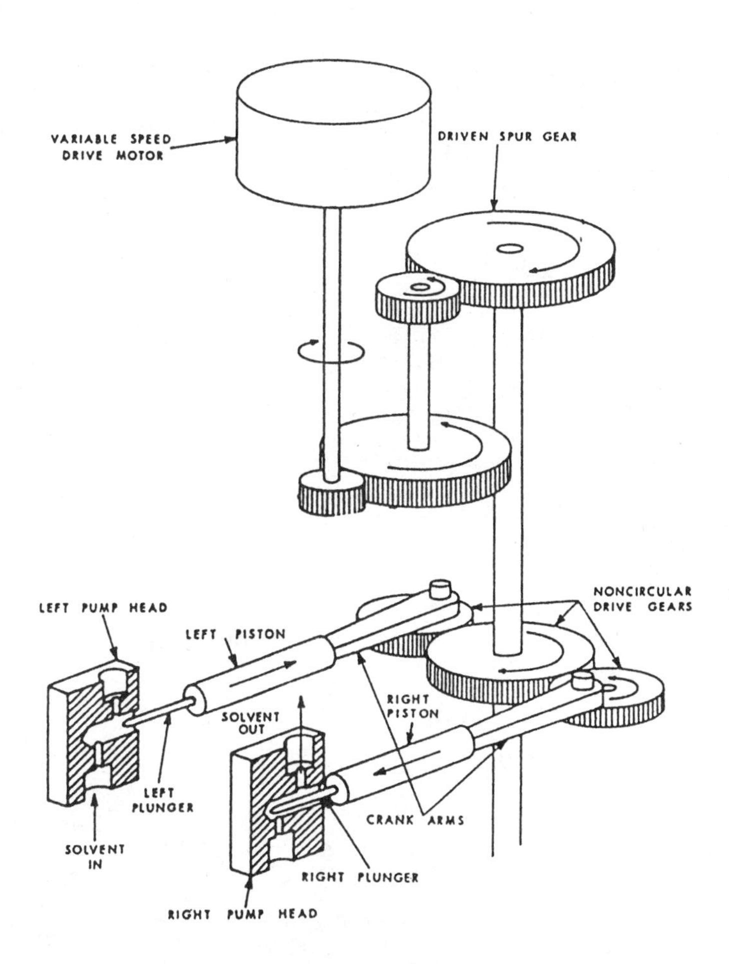


**Fig. 7.10.** Dual-piston pump configuration; parallel pistons are driven by two cams. Courtesy of Waters Chromatography Division.

## *Tandem-Piston Pumps*

Tandem-piston pumps use two pistons operating in series to minimize pulsations. A diagram of the pump operation is seen in Fig. 7.13. In this case, one piston delivers mobile phase at twice the rate of the second. Figure 7.13 shows this as a difference in piston diameters; alternatively, one piston can travel twice as fast as a second piston of the same diameter. During one portion of the pump cycle (Fig. 7.13a), one piston delivers mobile phase and the other fills, just as in the dual-piston configuration. After the top piston has finished its delivery stroke, it begins the fill stroke

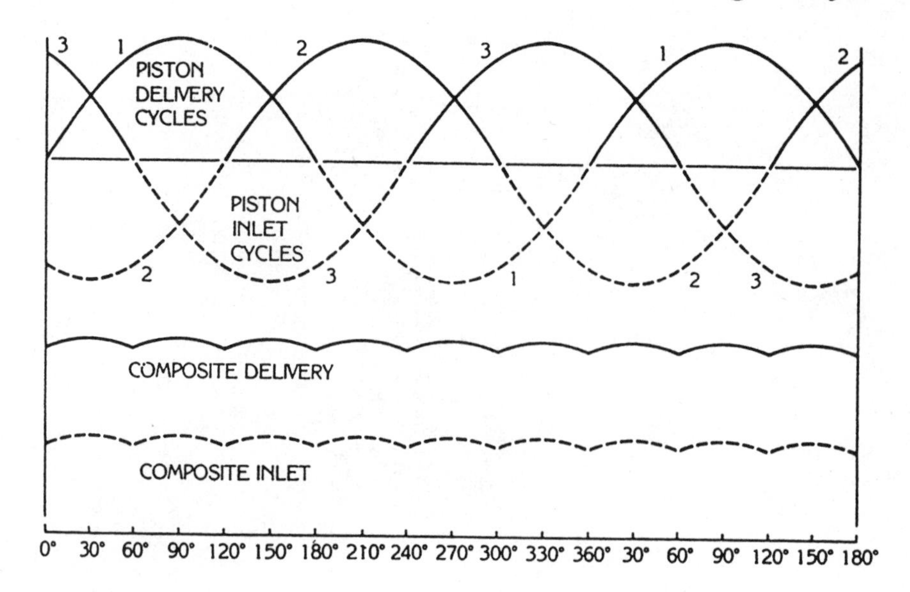


**Fig. 7.11.** Piston overlap for a three-piston pump. Reprinted from ref. (3) with permission.

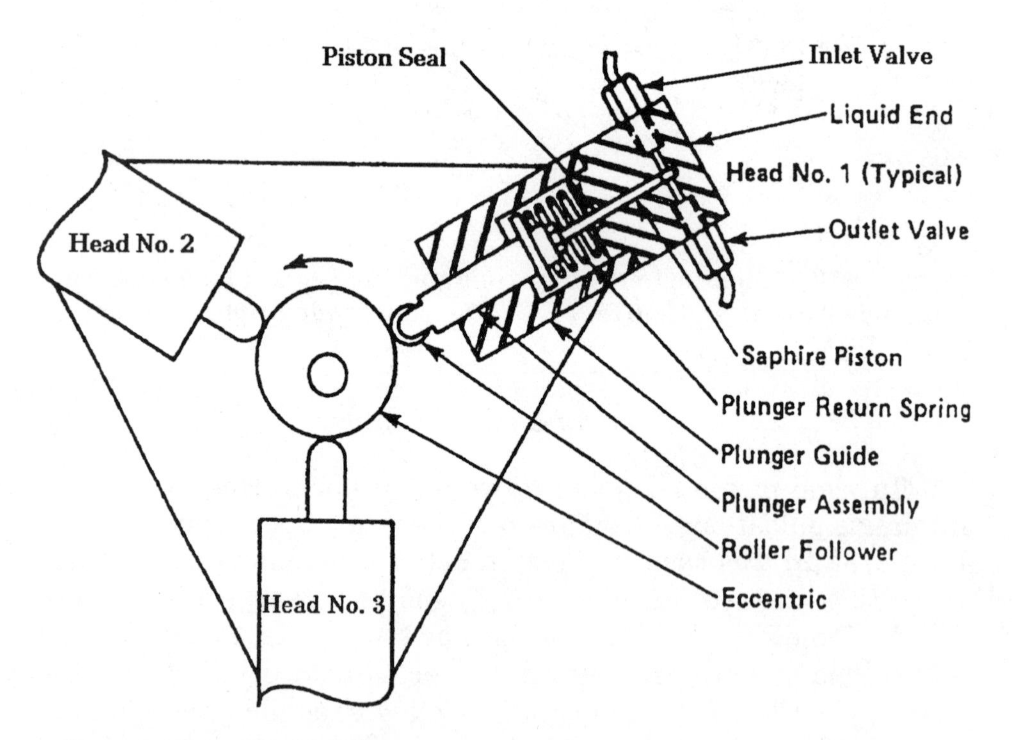


**Fig. 7.12.** Three-piston pump: pistons mounted in a plane and driven by a single cam. Courtesy of Nicolet Analytical Division.

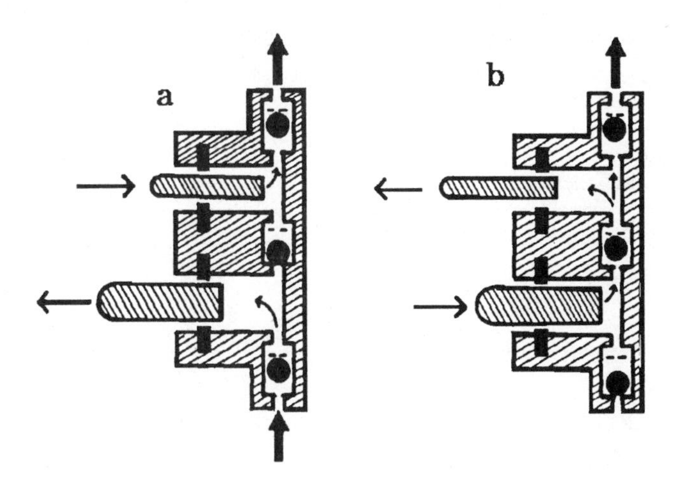


**Fig. 7.13.** Tandem-piston pump: (a) top chamber delivers mobile phase to column while bottom chamber refills at twice the rate; (b) bottom chamber delivers mobile phase to column and simultaneously refills top chamber. Reprinted from ref. (2) with permission.

(Fig. 7.13b). Now the bottom piston delivers at twice the rate of the first, so it (a) fills the top chamber as the top piston is withdrawn, and (b) simultaneously delivers mobile phase to the column. Then the cycle continues with the top piston delivering and the bottom one filling at twice the rate. The tandem-piston pump uses three check valves vs four for the dual-piston pump, thus increasing reliability and ease of maintenance by reducing the number of failure-prone components.

A diagram of a commercial tandem-piston pump (Perkin-Elmer) is seen in Fig. 7.14. In the upper figure, the metering pump is filled with up to four solvents while the high-pressure pump delivers mobile phase to the column. In the lower figure, solvent in the metering pump is pumped to the high-pressure pump as well as to the column (i.e., through the high-pressure-pump head). The spring between the two pistons reduces pulses and accommodates solvent compressibility. Sufficient mixing takes place as the mobile phase travels between the two pump heads so that a separate mixer is not required (although mixing does take place in the pulse damper). A pulse damper downstream of the high-pressure pump smooths any residual pump pulsations.

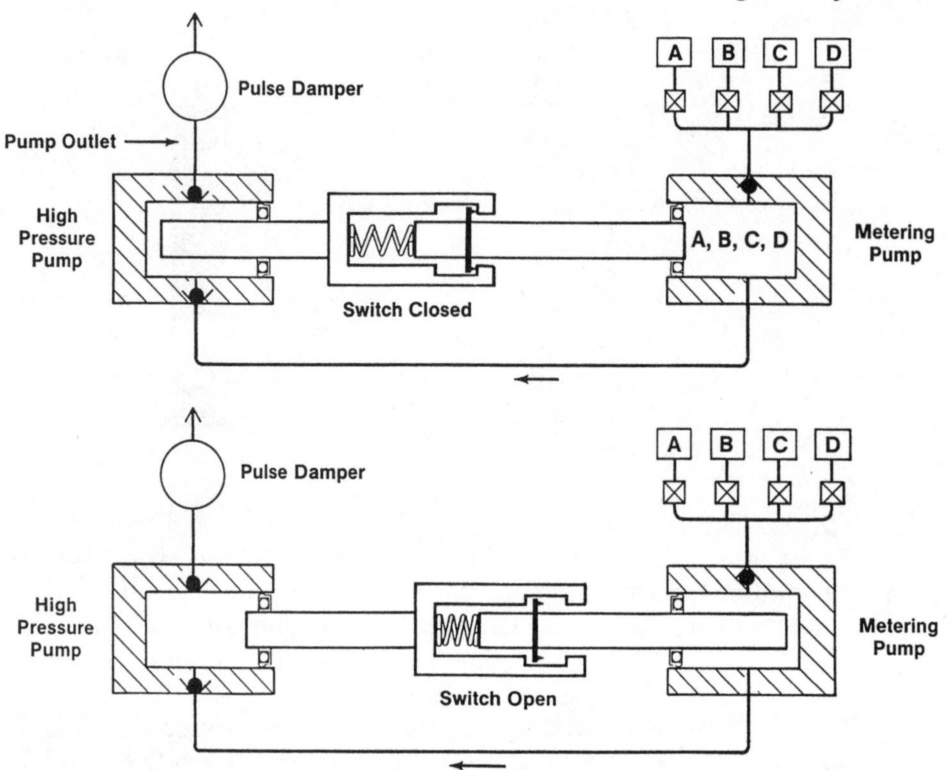


**Fig. 7.14.** Tandem-piston pump. Top, metering pump fills while high-pressure pump delivers mobile phase to column. Bottom, metering pump delivers to high-pressure pump and to column (through high-pressure-pump head). Courtesy of Perkin-Elmer.

## *Piston-Diaphragm Pumps*

Another variation of the reciprocating-piston pump is the piston-diaphragm pump (often called a diaphram pump). In this case, one side of the pumping chamber consists of a flexible diaphram (Fig. 7.15). During the intake cycle, low-pressure metering pumps are used to fill the pump chamber, because the diaphram pump has no suction. To deliver solvent, the delivery piston pumps oil against the back of the diaphram, causing it to deflect and force mobile phase out of the pumping chamber into the column. At the end of the delivery cycle, the piston withdraws, reducing the pressure on the back of the diaphram and the metering pump fills the chamber. The inlet and outlet check-valves work in the conventional manner. An advantage of this

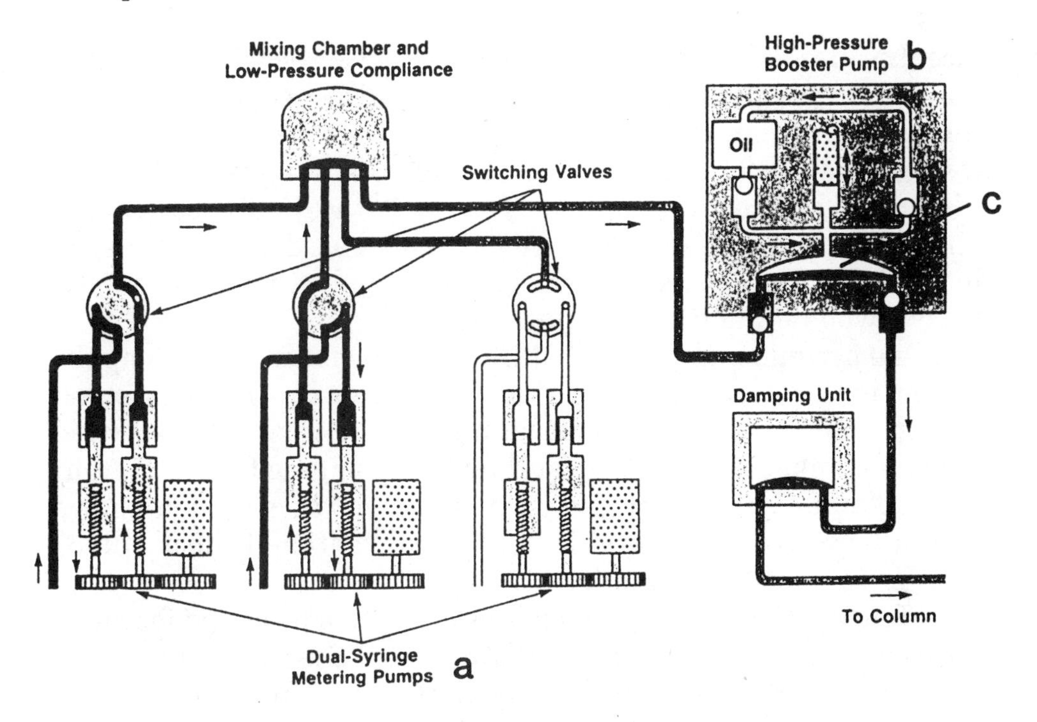


Fig. 7.15. Piston-diaphragm pump: (a) low-pressure metering pumps feed solvents to main pump (b) during intake stroke; main pump (b) forces oil against diaphram (c), and solvent is delivered to column during delivery stroke. Courtesy of Hewlett Packard.

pump is that there are no pump seals to wear or leak. On the other hand, diaphragm failure (which is rare) can contaminate the system with pump oil and ruin the LC column.

## *Syringe Pumps*

Syringe pumps are single-piston pumps with much slower total pump-cycle times than the reciprocating-piston pumps discussed above. The pump output is usable only for one piston-volume, because the flow is stopped for a relatively long time (e.g., several seconds) while the pump refills. This means that the chromatogram's elution volume must be less than the pump delivery volume. For example, if the pump can deliver 10 mL of mobile phase, the last peak must be off the column in 5 min if a 2 mL/min flow rate is used. Syringe pumps have the advantage of providing near pulse-free delivery. Because of this, syringe pumps were once the pump-of-choice for LC. However, with the advent of more stable column packings plus reliable, pulse-free reciprocating pumps, syringe pumps became less popular. Now syringe pumps are used mainly for microbore LC (1–2 mm id columns) and supercritical-fluid chromatography. They are the preferred pumps when flow-sensitive detectors (e.g., flame-based and mass-spec detectors) are employed. They also have applications for routine assays, when elution volumes are low.

One syringe pump (Applied Biosystems) for microbore applications is shown in Fig. 7.16. This pump uses two 10-mL pistons that operate in parallel, so that high-pressure mixing for isocratic or gradient elution is possible at flow rates ranging from 1 μL/min to 1 mL/min. This pump also uses an active valve (which allows for refill or solvent delivery) instead of conventional check valves in order to improve reliability.

Figure 7.17a shows a syringe-pump system designed for automated quality-control analysis (Waters QA-1 System). This pump offers a less-expensive alternative to using reciprocating-piston pumps for routine assays. An autosampler, pump, detector, and data system are contained in a single unit. The pump refills before each sample is run, so separations must be designed to elute the last peak in less than the 40-mL pump delivery volume. Thus, bands of $k' < 15$ can be eluted for columns with void volumes of 2.5 mL (e.g., 25 x 0.46-cm) in a single piston stroke. A pump such as this has the potential of a significant delivery error caused by the large volume of mobile phase that must be compressed prior to delivery. This problem is overcome by fully compressing the mobile phase prior to sample injection (see Fig. 7.17b). Another feature of this system reduces the problem of bubbles in the pump: the pump is mounted at a slight angle, so that any trapped bubbles will tend to float to the outlet check valve and be swept from the pump head at the beginning of the pump stroke (see Fig. 7.17c).

## 7.3 . Common Mixing Variations

Manual mixing of mobile phases is time-consuming, error-prone, and can be wasteful if excess mobile phase is mixed and then discarded. On-line mixing, either on the high- or low-pressure side of the pump, eliminates these three problems and facilitates automation of the LC system.

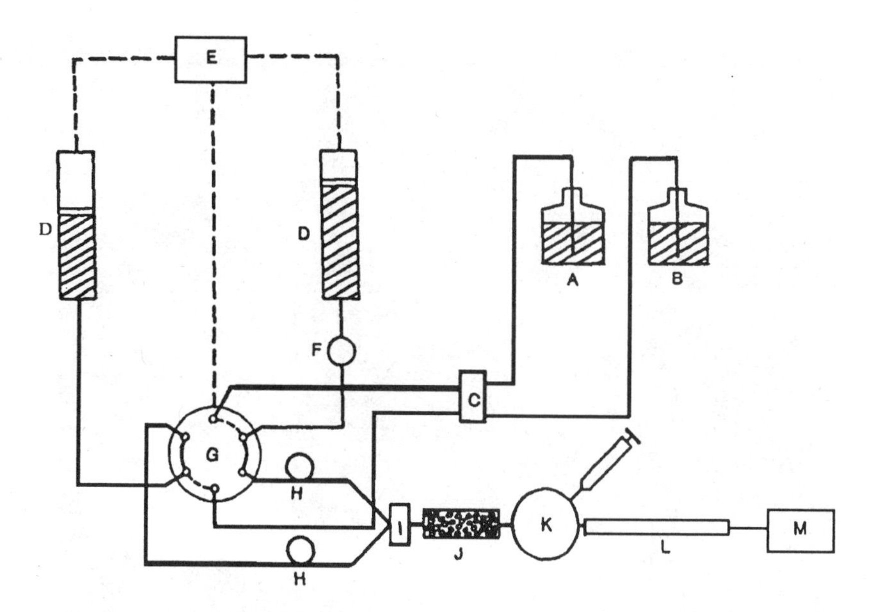


**Fig. 7.16.** Applied Biosystems (formally Brownlee Labs) syringe pump and schematic flow diagram. (A,B) reservoirs; (C) inlet manifold; (D) pump cylinder; (E) microprocessor; (F) pressure transducer; (G) switching valve; (H) tubing to prevent backflow; (I) mixing tee; (J) mixing chamber; (K) injector; (L) microbore column; (M) detector. Dotted lines represent electronic control connections; solid lines are the flow path. Courtesy of Applied Biosystems.

## *Low-Pressure Mixing*

Low-pressure mixing uses a set of solenoid-controlled valves (Fig. 7.18) to meter the proper proportion of mobile phase components into a low-pressure mixing chamber, as seen in Fig. 7.19. Two, three, or four valves may be used for the formation of binary, ternary, or quaternary mobile phases. For example, to make a 35/25/40 methanol/acetonitrile/water mobile phase, the methanol valve would be open 35% of the time, the acetonitrile valve 25% of the time, and the water valve 40% of the time during each valve cycle. Manufacturers "optimize" the total valve cycle-time (ranging from milliseconds to seconds) for each model of LC pump. Shorter cycle-times usually result in more homogeneous mobile phases, but give less accurate proportioning of mobile phases that contain less than 5% of a component. Longer cycle-times give more accurate proportioning, but may require extra mixing to en-

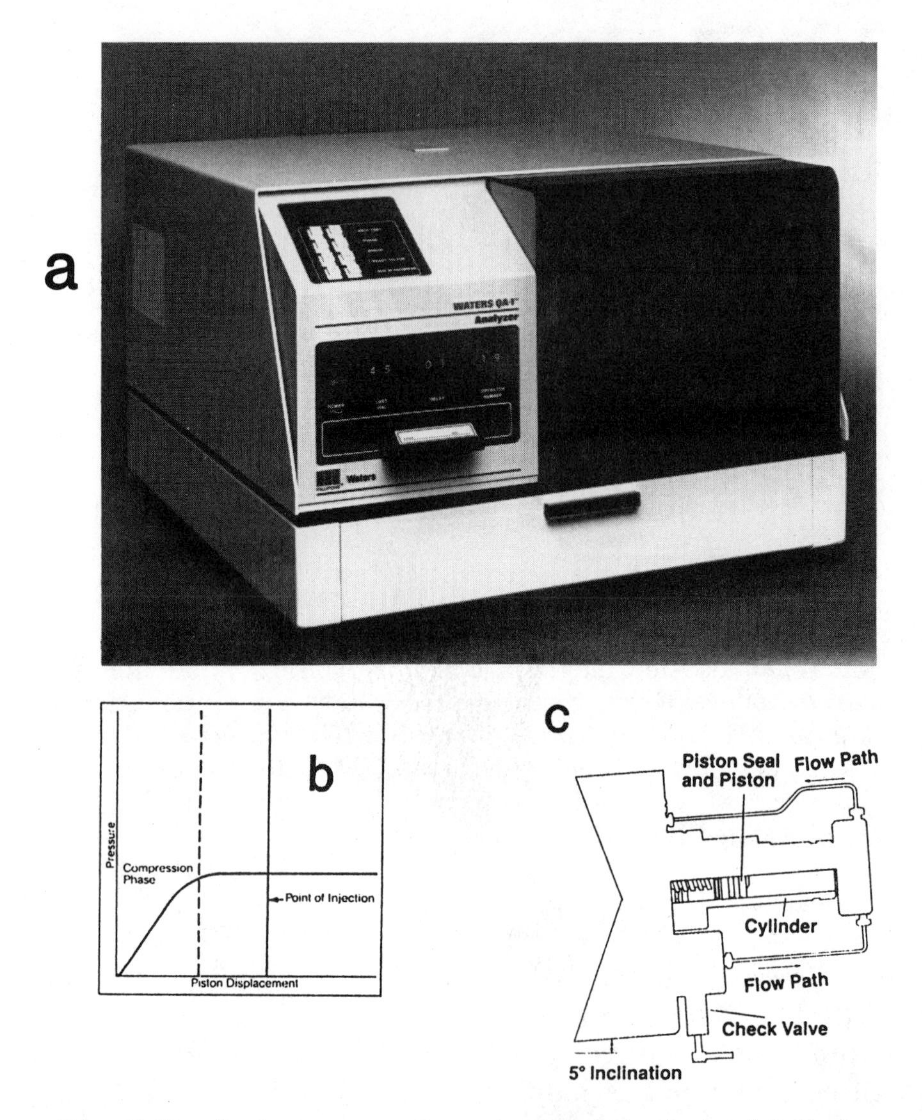


**Fig. 7.17.** Syringe pump for automated analysis. (a) System unit containing sample handler, pump, injector, column, and detector; (b) compression of mobile phase prior to sample injection; (c) pump head mounted at an angle to facilitate bubble clearing. Courtesy of Waters Chromatography Division.

sure homogeneity. Mixing chambers use either (a) static mixers (fritted disks or columns packed with glass beads), or (b) dynamic (stirred) mixers to ensure mobile-phase homogeneity. The turbulence created by passing the mobile phase through the check

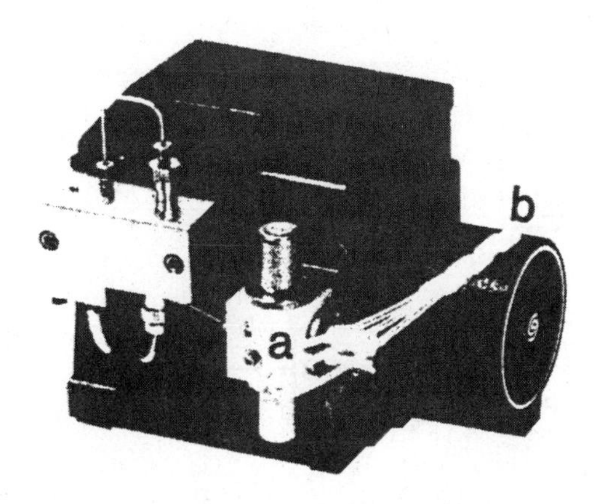


**Fig. 7.18.** Solenoid-controlled proportioning valves: (a) valves mounted on Teflon manifold; (b) mobile-phase inlet lines. Courtesy of Spectra-Physics.

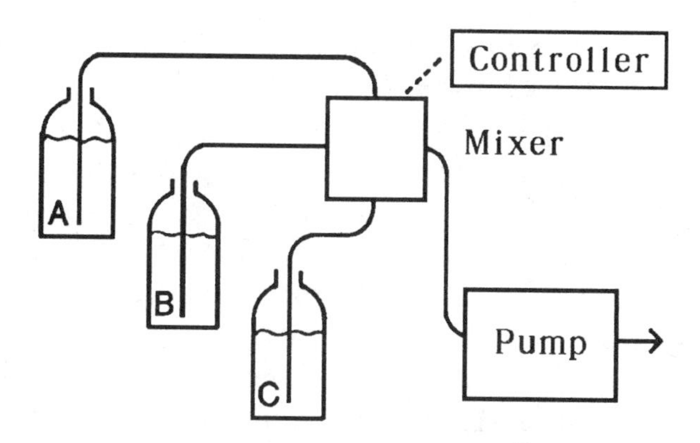


**Fig. 7.19.** Low-pressure mixing: up to three solvents (A, B, C) can be proportioned into mixing chamber on low-pressure side of pump. Reprinted from (2) with permission.

valves and pump chamber is the basis for mixing in some pumps (e.g., Perkin Elmer Series 4, Fig. 7.14). In general, low-pressure mixing is quite susceptible to mobile-phase outgassing, so mobile-phase degassing is critical. Low-pressure mixing requires only one pump, and thus is less expensive than comparable high-pressure mixing systems when three or more solvents (and pumps) are used. Low-pressure mixing systems also can be purchased at moderate cost (e.g., AutoChrom) for use with existing pumps.

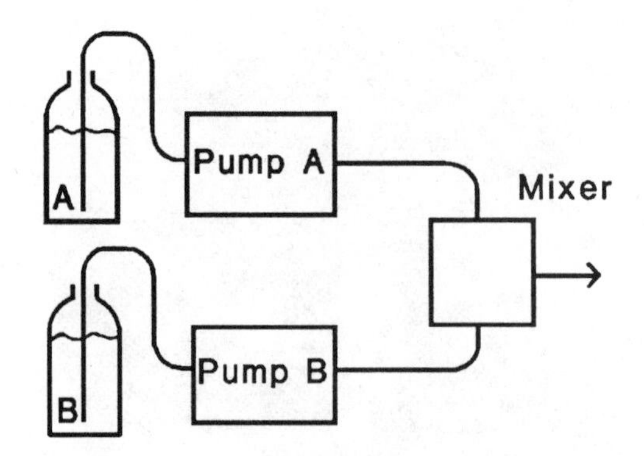


**Fig. 7.20.** High-pressure mixing: solvents A and B are independently pumped (pumps A, B) into a mixing chamber on the high-pressure side of the pumps. Reprinted from ref. (2) with permission.

## *High-Pressure Mixing*

In high-pressure mixing, the solvent flow from two or more LC pumps is combined and mixed to form the mobile phase (Fig. 7.20). This technique relies on the relative flow rates of each pump to proportion the mobile-phase components. For example, a 1 mL/min flow of 30/70 THF/water requires that one pump must deliver 0.3 mL/m THF and the other pump must deliver 0.7 mL/min water. Proportioning small amounts of a mobile-phase component at low flow rates is limited by the pump's ability to deliver accurately under these conditions. For example, if a 2% blend of A in B at 0.5 mL/m is desired, the A-pump must be able to deliver accurately (±5–10%) at 0.01 mL/min. Mixing usually takes place in a stirred (dynamic) mixer or in a packed-bed (static) mixer. High-pressure mixing is less prone to bubble problems than low-pressure mixing, so solvent degassing may not be necessary. Because high-pressure mixing requires a pump for each mobile-phase component, high-pressure mixing is not very economical if more than two components must be mixed.

## *Other Mixing Variations*

Two other mixing variations are encountered commonly. One uses low-pressure proportioning plus high-pressure mixing (Fig. 7.21, Varian). This gives the advantage of requiring only a single

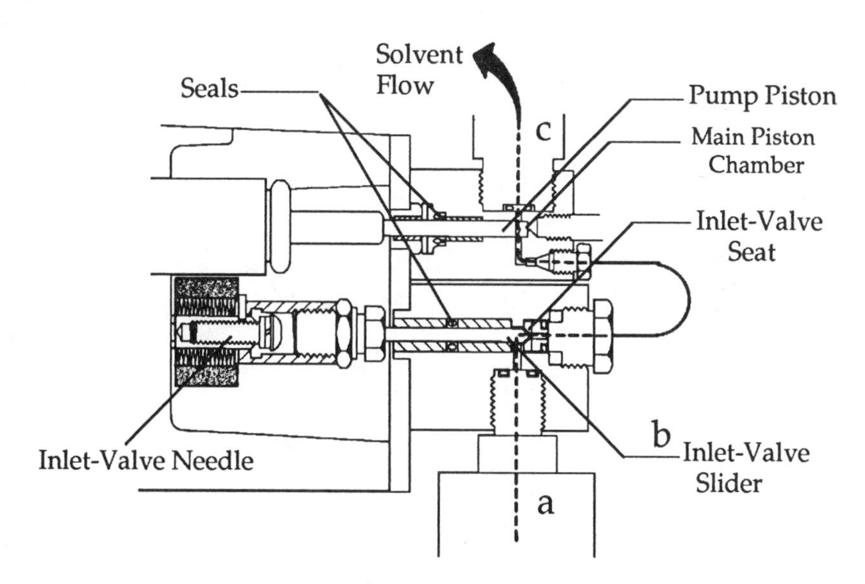


**Fig. 7.21.** Pump using low-pressure proportioning and high-pressure mixing: (a) solenoid-operated proportioning valves (one of three shown) feed solvent directly into pumping chamber; (b) mechanically operated inlet check-valve prevents bubble problems; (c) proportioned, but unmixed solvents are delivered to high-pressure mixing chamber (not shown). Courtesy of Varian Associates.

pump, yet avoids the need to degas the mobile-phase components. The second mixing technique uses three syringe pumps to meter solvents to a low-pressure mixer and then feeds the mobile phase into a piston-diaphragm pump (Fig. 7.15, Hewlett-Packard).

If the mixing characteristics of your system are unsatisfactory for your applications, you can purchase a mixing chamber to add to the LC. The mixer shown in Fig. 7.22 is designed for micro-LC applications. Several LC companies also sell stand-alone mixing static and/or dynamic mixing chambers with various volumes for conventional LC applications.

## *Gradient Elution*

Gradient elution requires a change in composition of the mobile phase during the chromatographic run (see Chap. 3). LC systems incorporating either low- or high-pressure mixing can be used for gradient elution, if the pump controller can provide the desired gradients. For isocratic applications, the volumes contrib-

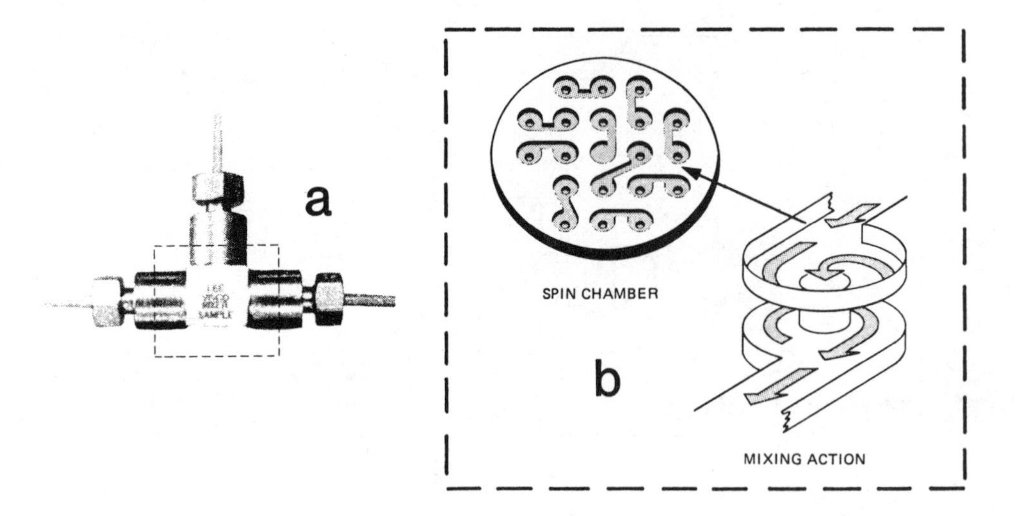


**Fig. 7.22.** Supplemental mixing chamber, 9 μL internal volume: (a) external view; (b) internal view of dashed box in (a). Courtesy of The Lee Company.

uted by the mixer, feed lines, and pump are not very important (except for solvent changeover). In gradient elution, however, the volumes of these components should be minimized, for a well-defined gradient and a rapid return to the starting conditions.

## 7.4. Summary

There are no "good" or "bad" designs of the pump or mixer; each kind has tradeoffs that make it more or less suited for particular applications. Pump and mixer features are summarized in Tables 7.1 and 7.2, which are meant as an overview—individual models have many features not shown.

## 7.5. Problem Prevention

### *Procedures*

Reliable pump operation requires that proper attention be paid to system cleanliness, solvent and reagent quality, mobile-phase filtration, and mobile-phase degassing. These items are summarized in Table 7.3 and discussed in Chapter 6. The pump

Table 7.1
Pump Features

| Feature | Pump type[a] | | | | | |
|---|---|---|---|---|---|---|
| | Single-piston | Dual-piston | Triple-piston | Tandem-piston | Piston-diaphragm | Syringe (modern) |
| Complexity | + | ++ | +++ | ++ | +++ | + |
| Smooth flow | + | ++ | +++ | ++ | ++ | +++ |
| Check valves | 2 | 4 | 6 | 3 | 2 | 0–2 |
| Pump seals | 1 | 2 | 3 | 2 | 0 | 1 |
| Cost[b] | + | ++ | +++ | ++ | +++ | +++ |

[a](+) low, (++) medium, (+++) high.
[b]Higher-priced units may have added features not present on less-expensive models

Table 7.2
Mixing Tradeoffs

| Type | Advantages | Disadvantages |
|---|---|---|
| Manual | –Inexpensive equipment | –Wasteful<br>–Inconvenient |
| Low Pressure | –Low cost<br>–No waste<br>–Convenient | –Bubbles |
| High Pressure | –No waste<br>–Convenient | –Equipment Cost |

seals should be replaced every three months for maximum reliablity. Certain pump models may require occasional lubrication; check the operator's manual for recommendations.

Flush all buffers from the pump at the end of each day's use to prevent buildup of buffer crystals. Some workers find that seal life can be extended and pump reliability can be increased if the piston is flushed daily behind the seal. Many commercially available pumps have vent and drain holes behind the pump head which can be used for this purpose. To clean behind the pump

Table 7.3
Preventing Pump Problems

| |
|---|
| Use high-quality reagents and HPLC-grade solvents |
| Filter mobile phases and solvents |
| Low-pressure mixing<br>-degas mobile phase |
| High-pressure mixing<br>-degassing may be helpful |
| Flush stagnant mobile phase and bubbles from pump at beginning of each day |
| Flush buffer from system daily |
| Avoid leaving pump in pure water or corrosive solvents |
| Replace pump seals every 3 months |
| Lubricate if necessary |
| Check pump operator's manual for additional suggestions |

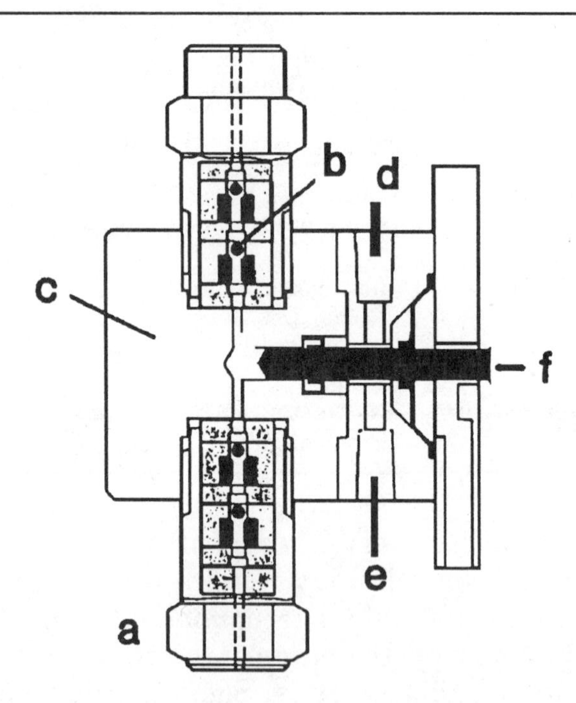


**Fig. 7.23.** Flushing behind the pump seal to minimize salt buildup: (a) inlet check-valve; (b) outlet check-valve; (c) pump head; (d) inlet to flushing channel; (e) drain from flushing channel; (f) piston. Courtesy of LKB Instruments, Inc.

Table 7.4
Spare Pump Parts

| Spare Pump Parts |
|---|
| Pump seals and insertion tool |
| Check valves (inlet and outlet) |
| Assembled pump head (optional) |
| Miscellaneous fittings |
| Mixer spares (e.g., frits, seals) |
| Fuses |

head, use a syringe to flush about 10 mL of water through the vent hole into the cavity as shown in Fig. 7.23; place a beaker under the drain hole to catch the waste. Then flush about 10 mL of methanol or propanol through the vent hole to remove any water residue. Caution: before this procedure is first used, be sure to remove the pump head and check the cavity behind the pump head to be sure that no parts are exposed that will be damaged by the water.

When the pump is operating properly, the baseline of the chromatogram should be smooth and even, and retention times should be reproducible. The pump pressure should be steady, with fluctuations no greater than about ±2% of the average (during isocratic operation). During gradient runs, the pressure change should be gradual and smooth.

## *Spares*

A suggested list of spare pump parts is given in Table 7.4; remember to reorder any parts as soon as they are removed from your spares kit. An extra pump seal for each pump head should be stocked, plus any bushings or guides that the manufacturer recommends replacing at the same time as the seal. A pump-seal insertion tool should be part of your tool kit if it is available for your particular model of pump. One pair of check valves (inlet and outlet) usually is sufficient, or stock an assembled pump head (complete with check valves and a new pump seal). Stock a single piston-assembly (i.e., piston and connecting rod) or rely on overnight shipment from the vendor when a piston is required. A spare piston seldom is needed if care is taken when removing and replacing the pump head. Your fittings kit should contain the necessary connectors for mobile-phase inlet and outlet lines. If your pump uses a mixer that requires regular maintenance, keep the necessary spares.

## 7.6. Problems and Solutions

A summary of pump problems and solutions is given in Table 7.6 at the end of this chapter.

### *Check-Valve Problems*

Check-valve leakage caused by (a) contamination by particulates, or (b) air bubble entrapment is the most common pump problem. Steady pressure and flow are essential for reliable LC operation, so the check valves must seal properly, or deviations in flow and pressure will result. This is one of the primary reasons for filtering mobile phases and using solvent-line inlet frit—it takes only a tiny speck of dust on the check-valve seal to cause leakage. Buffers left in an unused pump can crystalize and also cause check-valve leakage.

The best insurance against *particulates in check valves* is to (a) filter all mobile phases (or use only HPLC-grade solvents), (b) use a solvent-line inlet frit, and (c) flush the LC system daily with nonbuffered mobile phase. Contamination by particulates is likely if the pump will not successfully deliver a degassed solvent such as methanol (see below). Sometimes particulates can be washed out by using a series of solvents of different polarity. For example, pump 25 mL each of water, methanol, isopropanol, and methylene chloride (open the purge valve to waste during flushing); then return through the same series to a solvent compatable with your column. If the problem persists, remove the check valve(s) and clean them by drawing solvent through them in the normal direction of flow, using a vacuum aspirator. If flushing does not clear the contaminant, it is best to replace the check valve with a new (or factory-rebuilt) valve. Pump methanol or isopropanol through the new check valve to help remove the residual air. Check-valve rebuilding kits are available, but these require considerable skill and cleanliness to use successfully. During manufacturing, check valves are assembled in dust-free clean rooms—conditions that are hard to duplicate in most labs.

A second common cause of check-valve failure is *air entrapment.* Air bubbles can be caught inside the check valve body; this then prevents the ball from returning to the seat when the flow is reversed. Check-valve cleanliness and mobile-phase degassing will minimize this problem. It is difficult to distinguish whether air is trapped in (a) the check valves or (b) elsewhere in the pump head. However, an exact diagnosis is not important, because the remedy is the same for both cases. Air entrapment is indicated when pressure and flow vary widely during a single pump cycle. When bubble problems are encountered, the bubbles usually can be removed by (a) bleeding the check valves and/or (b) flushing with degassed methanol.

The easiest way to remove a bubble is to open the purge valve and run the pump for 10–15 s at 5–10 mL/min. Tapping the pump heads sharply with a screwdriver handle (while the pump is being flushed) sometimes dislodges the bubble. Then reduce the flow and close the purge valve. If the bubble persists, change the solvent to degassed methanol and (a) flush the pump again, or (b) bleed the check valves (see below).

Air-bubble removal is facilitated by changing from the usual mobile phase to degassed methanol. Methanol thoroughly wets all the internal pump surfaces, making the bubbles less likely to stick. Using degassed solvent encourages the bubbles to redissolve and be washed out. Flushing with degassed methanol also is effective for removing bubbles trapped in detector flow cells and for initially priming and flushing a newly-rebuilt pump head.

Trapped bubbles are less likely, and bubble removal is easier, if there is a positive head-pressure on the reservoir. This can be accomplished by (a) keeping the reservoir higher than the pump, or (b) pressurizing (e.g., 5 psi) the reservoir with helium. Additionally, some pumps (e.g., Waters) have pump priming valves that allow you to force mobile phase through the pump head to help remove air bubbles.

With multiple-piston pumps, you will often see a drop in pressure associated with a single head. You need to identify the problem head before you can bleed the air from it. Observe the pressure meter and pump action for several cycles to determine when the fill and delivery strokes take place. Now you can correlate the pressure drop with the delivery stroke of a single pump head.

To *bleed the check valves,* proceed as follows (after ensuring there is adequate mobile phase in the reservoir): *First,* change to a degassed, nonbuffered mobile phase (e.g., methanol), because the bleeding procedure is rather messy and will spill mobile phase onto the outside of the pump head. (If the failure occurs after the pump has been operating reliably, it is likely that the bubble can be quickly removed, so you may want to bleed the check valves with the usual mobile phase; be aware of potential problems from buffer-leakage.) *Second,* steady the outlet check-valve of the suspect head with an open-end wrench and loosen the compression fitting for the outlet tubing (about 1/3-turn) just as the delivery stroke starts for that pump head. You will see mobile phase and air bubbles leak out of the fitting during the delivery stroke. A paper towel wrapped around the base of the check valve will absorb the leaking mobile phase. (If the ferrule sticks in the fitting, loosen the nut one turn and wiggle the outlet tubing until you feel the tubing and ferrule loosen, then proceed.) Tighten the fitting when the end of the delivery stroke is reached. *Third,* observe the pressure meter again for a steady reading. You may have to repeat the third step several times before the bubbles are removed. *Fourth,* if the above steps do not remove the bubble, repeat the process by loosening the check valve where it screws into the pump head. *Finally,* rinse any spilled buffer from the pump surfaces and wipe up any spills. If no bubbles are seen escaping from any of the fittings during bleeding, the problem is likely to be a result of particulates in the check valves, a bad pump seal, or one of the "Other Problems" noted below.

*Check-Valve Replacement.* A faulty check valve is replaced by removing the inlet (or outlet) line, unscrewing this valve from the pump head, and screwing a new check valve into the head. Care should be taken to steady the pump head when the check valves are loosened or tightened, so that it does not twist and break the piston. Some pumps have a seal where the base of the check valve contacts the pump head; take care not to damage this seal by overtightening the check valve. The pump head should be thoroughly flushed, using the procedure noted below in the final seal-replacement step, before proceeding with the next LC separation.

*Other Problems.* Symptoms similar to check-valve failure can also be seen when problems of pump starvation, frit blockage, poorly degassed mobile phases, and so on, are encountered (see Chapter 6).

### *Pump-Seal Problems*

The two most common pump-seal problems are (a) seal leakage, and (b) downstream contamination from seal particles. Unfortunately, by the time any symptoms of these problems appear, the seal is long overdue for replacement. For this reason, it is cost-effective to replace the pump seals before they fail (e.g., every three months), rather than to spend the time and money correcting secondary problems created by seal failure. Another pump-seal problem is contamination of the system, if the seal material is not compatible with the solvents being used.

*Seal Wear and Failure.* Because the seal contacts the moving piston, it is the part of the LC system most subject to abrasion. Using buffers or other salts in the mobile phase will accelerate seal wear. Seal wear cannot be prevented; it can only be minimized by proper pump care.

Pump-seal failure shows up as (a) an inability to pump at high pressures, and (b) mobile-phase leakage behind the pump head. Sometimes a change in sample retention is noticed. When a pump that would previously pump to high pressures (e.g., 5000 psi), is unable to produce those pressures, seal failure is a probable cause. (Be sure that the upper-limit pressure switch has not been lowered inadvertently.) A leaky seal acts like a pressure-relief valve, because it will no longer seal properly at higher pressures. Part of the flow leaks past the seal, so the flow rate to the column is lower than the controller setting. Most pumps have a drip-hole below and behind the pump head; when liquid drips from this hole, seal failure has occurred. If failure is suspected, replace the seal.

*Seal Incompatibility.* For the most part, pump seals are compatible with all mobile phases. Some seals, however, have been developed for extended life in aqueous reversed-phase solvent systems (e.g., methanol, water, and acetonitrile). These seals may rapidly degrade in certain normal-phase solvents. In one case, a manufacturer inadvertently shipped LC units with these (reversed-phase) seals installed; customers who pumped pure THF through them observed chromatographic interferences and

Table 7.5
Pump Seal Replacement

1. Preparation
   a. Collect necessary parts and tools:
      -wrenches
      -large screw (for seal removal)
      -sonic bath
      -new seal
      -seal insertion tool
   b. Flush pump with methanol
   c. Remove inlet and outlet lines
   d. Mark direction of flow on head and check valves
2. Remove pump head
   a. Use care to prevent piston breakage
3. Disassemble head
   a. Note order of assembly
   b. Remove check valves, store in dust-free container
   c. Remove old seal; discard
4. Clean all parts
   a. Clean pump head and associated parts
   b. Inspect and clean piston
   c. Clean check valves (optional; not recommended)
5. Reassemble head
   a. Insert new seal (note proper direction)
   b. Assemble parts (note proper order)
6. Reinstall head
   a. Lubricate piston with methanol
   b. Tighten evenly to prevent piston breakage
7. Flush head to remove air

column failure, as a result of materials leached from the seals by THF. If there is any question about seal compatibility, check with the manufacturer of your system. Alternatively, soak a seal overnight in the desired solvent. Check the seal for any physical changes (e.g., softness, swelling, color changes) and run a UV scan on the solvent to see if any UV-absorbing material has dissolved.

*Pump-seal replacement* is a procedure that should be learned by all LC operators. Specific directions for replacement of the seals in your pump should be reviewed in the operator's or service manual for the pump before proceeding. Because each pump is slightly different, only general guidelines are given here (see Table 7.5).

*First,* prepare for replacement: (a) check to be sure that you have the necessary parts and tools, (b) flush the pump with non-buffered mobile phase (methanol is recommended), and (c) remove the fittings connecting the inlet and outlet lines to the check valves (leave the check valves on the head in most cases). In order to prevent possible confusion during reassembly, it is useful to use a diamond-tipped scribe to mark arrows on the pump head and two check valves, indicating the direction of flow.

*Second,* loosen the pump-head retaining-screws. Most pumps use two allen-head screws or finger-tightened screws to hold the pump head in place. Loosen the screws carefully and evenly to ensure that the pump head is not twisted (the piston can break!). Pull the pump head from the piston, which generally will stay in the pump body. (Some models have piston-head assemblies that are removed as a unit.)

*Third,* disassemble the head. Take the parts off according to the directions in the manual. If you are afraid of getting the parts mixed up, lay them on a paper towel in the order that you remove them. The check valves should be removed at this time and placed in a dust-free place (e.g., a covered beaker). The seal itself can be removed easily with the aid of a large screw (e.g., a 1-in. long, #10 wood screw). Turn the screw into the piston hole in the center of the seal until a few threads grip it securely (Fig. 7.24); then wiggle the screw from side to side and pull it out along with the seal. (Be careful not to scratch the pump head.) Discard the old seal.

*Fourth,* thoroughly clean the parts. The pump head and any washers or retainers should be cleaned for a few minutes in methanol in a sonic bath. This is a convenient time for lubrication of the cam, if your pump requires it. The piston should be rinsed with water and wiped with a lint-free wiper to remove any buffer residues. Inspect the piston for scratches; replace it if it is scratched. Rinse all the parts with methanol and allow them to air-dry. Some workers sonicate the check valves in methanol at this stage; we do not recommend this if the check valves have been trouble-free (they can be inadvertently contaminated during "cleaning").

*Fifth,* insert the new seal and reassemble the head. Use a seal-insertion tool (if one is available for your model of pump) to

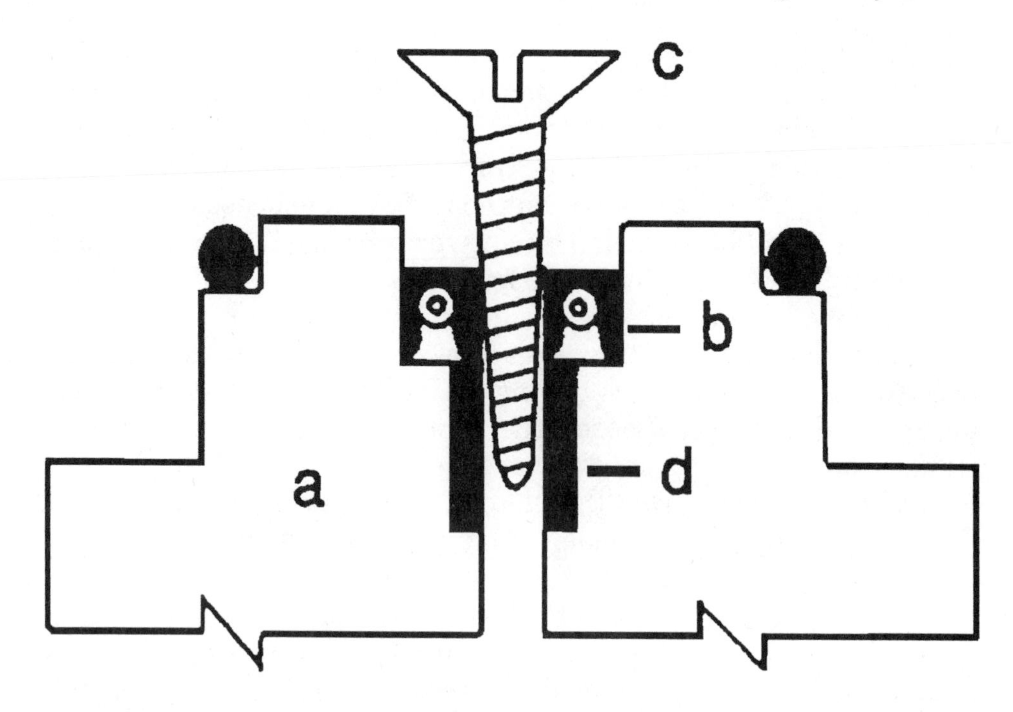


**Fig. 7.24.** Removing the pump seal: (a) pump head (cutaway); (b) pump seal; (c) screw (e.g., a 1-in. long, #10 wood screw); (d) piston guide. See text for procedure.

prevent seal damage during assembly. Figure 7.25 shows the use of a seal-insertion tool. *Be sure* that the spring-side of the pump seal faces the pump chamber (high-pressure side), not the pump motor (low-pressure side), or the seal will leak. Follow an exploded diagram (from the manual) to be sure all the washers and retainers are assembled in the proper order.

*Sixth,* reinstall the head. Place a drop of methanol on the tip of the piston to lubricate it for easier assembly, and slide the pump head into place. Check the direction-of-flow arrows (step 1 above) to be sure that the head is not 180° out-of-line. Tighten the head-retaining screws evenly in order to avoid piston breakage. Attach the inlet line to the inlet check-valve.

*Finally,* flush the head. Pump degassed methanol to prime the pump, using a paper towel to absorb leakage from the outlet check-valve. When methanol freely flows from the outlet check-valve (i.e., few or no bubbles), turn the pump off, attach the outlet line, open the purge valve and pump about 20 mL of methanol

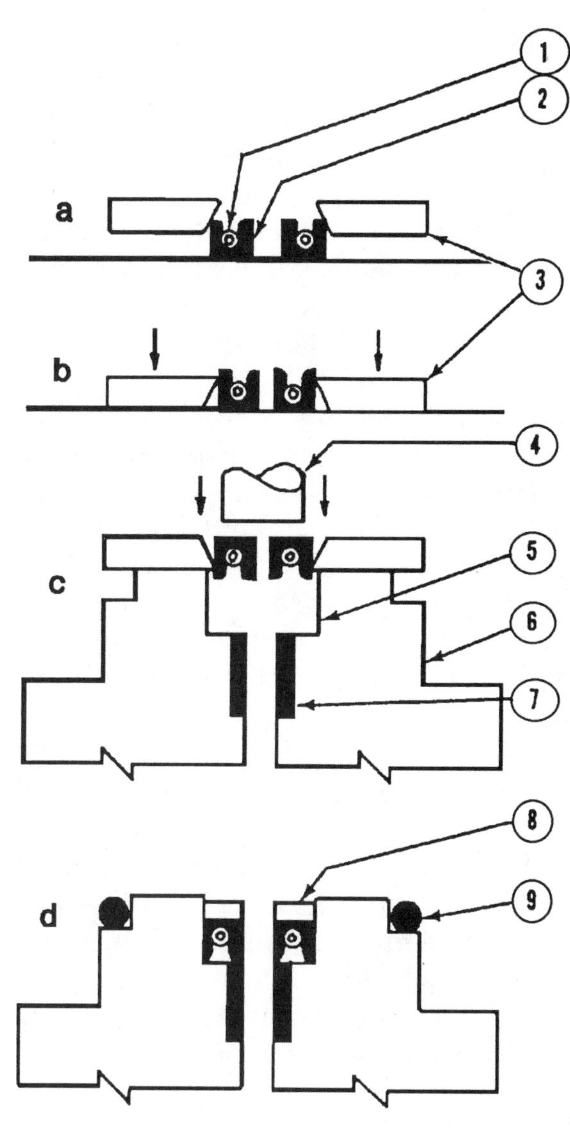


Fig. 7.25. Using a pump-seal insertion-tool: (a) place pump seal (2) on a hard surface with the spring side (1) up, position tapered part of seal tool (3) as shown; (b) press seal tool over pump seal; (c) invert seal and tool and place on pump head (6) as shown, use insertion-tool plunger (4) to push seal out of tool and into seal cavity (5) in pump head so that it rests directly above the piston guide (7); (d) place backup ring (8) and O-ring (9) as shown and reassemble pump head. Courtesy of LDC/Milton Roy.

through the pump to remove any remaining trapped air. Change to mobile phase and proceed with your assay.

## *Piston Problems*

Three types of piston problems can occur: (a) piston breakage from mishandling, (b) piston scoring from abrasion, and (c) piston seizing from poor operational practice. Reasonable care and system maintenance should eliminate these problems completely.

*Piston Breakage.* A broken piston is rare, unless it is broken accidentally when the pump head is being serviced. It is obvious when the piston breaks during disassembly, because the remaining base of the piston is jagged and the tip is left in the seal. Piston breakage on reassembly of the head or during operation gives a no-flow condition that cannot be corrected by the air-removal procedures discussed in the check-valve section above. Some pumps (e.g., Waters) have piston-movement indicators that alert you if the piston is not moving. Follow the replacement procedure noted below.

*Piston Scoring.* The piston can become scored if abrasive materials, such as buffer salts, are trapped under or next to the pump seal. The symptom of scoring is seal leakage that is not corrected when a new seal is installed. (A new seal that is installed backward will also leak.) Inspect the piston with a magnifying glass; tiny scratches parallel to the axis of the piston show up under side-lighting if scoring has occurred. Replace a scored piston as noted below; washout procedures should be implemented so that scoring does not occur again.

*Piston Seizing.* A piston can seize in the pump head, if the pump is left standing for a sufficient time (e.g., several days) filled with buffer or other corrosive mobile phase. A seized piston cannot move, so no pumping can occur. If the pump is started under these conditions, (a) the motor can overheat, (b) a fuse can fail, (c) a clutch can slip, or (d) some other secondary failure can occur. In some cases, a seized pump head can be removed in the usual way, but sometimes the piston must be broken in order to release the head. The piston, and often the pump head, must then be replaced.

*Piston Replacement.* To replace the piston, remove the pump head and inspect it for damage. Generally, the broken tip of the piston can be removed using a pair of pliers; clean the head and replace the seal as discussed earlier. Consult the operator's manual for the piston-removal procedure, then remove the broken piston and driver assembly and replace it with a new one. The pistons in most pumps are sapphire rods of the same diameter (e.g., 1/8-in. od), but the piston length and the drive mechanism vary for different pumps, so only use the specified replacement parts.

Table 7.6
Pump-Related Problems and Solutions

| Cause of problem | Symptom | Solution |
|---|---|---|
| Check-valve contamination | Pressure and flow instability | 1. Flush with solvent series<br>2. Replace with new/rebuilt valve |
| Air entrapment | Pressure and flow instability | 1. Open purge valve, pump at high flow<br>2. Bleed check valves (See text)<br>3. Pump degassed methanol |
| Seal wear and incompatibility | Inability to pump at high pressure, leakage behind pump head | 1. Replace pump seal (See text) |
| Seal incompatibility | Excessive baseline noise, mobile phase appears to be contaminated, short column life | 1. Check with manufacturer for compatibility<br>2. Test compatibility by soaking<br>3. Replace seal with compatible seal<br>4. Wash contaminants from system with strong solvent |
| Piston breakage | Visual check of broken piston, no flow, no movement of piston indicators | 1. Replace piston |
| Piston scoring | Seal leakage not corrected by seal replacement, scratches seen under magnification | 1. Replace piston |
| Piston seizing | No flow, no movement of piston indicators, pump overheating | 1. Remove piston and head<br>2. Replace piston<br>3. Inspect head for damage, replace if necessary |
| Other problems | Pressure meter failure, leaks, electronic problems, etc. | (See text) |

## *Other Problems*

A rare problem is the rupture of the membrane used in some oil-filled pulse dampers (and the pump in the Hewlett-Packard LC Systems). These units use a thin membrane that expands against a viscous fluid (oil) to minimize pulsation while maintaining a low volume. Damage to the membrane can result in leakage of the oil into the mobile phase. This can change selectivity and retention with or without pressure changes. Some manufacturers add a dye to the oil so that a major detector offset will be noticed if leakage occurs. If this problem occurs, you probably will have to discard the column and flush the system with strong solvent (e.g., methylene chloride) to remove the oily residue. Consult the operator's manual for replacement procedures.

Other possible pump problems include: pressure-meter failure, purge-valve problems, leaks (other than seals and check valves), mixer problems, and electronic problems. Consult the operator's manual for corrective measures for your pump. Additional information can also be found in Chapter 9 (leaks), and Chapter 2 (troubleshooting trees).

## 7.7. References

[1]*LC User Survey VI,* (1985) Astor Publishing, Springfield, OR.

[2]Snyder, L. R. and Dolan, J. W. (1985) *Getting Started in HPLC, User's Manual,* LC Resources, Lafayette, CA.

[3]Snyder, L. R. and Kirkland, J. J. (1979) *Introduction to Modern Liquid Chromatography,* 2nd ed, Wiley-Interscience, New York.

# Chapter 8

# TUBING

## Introduction

Tubing connects the various modules in the LC system, providing a path for the mobile phase to flow from the reservoir, through the LC, and finally to waste. Different types of tubing are used, depending on the pressure requirements of the particular application. Stainless-steel and Teflon are the most common materials used for LC tubing. The tubing length and internal-diameter must be selected with care, so that system performance is not degraded. Incorrect tubing choices can cause band broadening or high pressure (flow resistance). For the most part, tubing that contacts the sample should be as short as possible and of small internal-diameter (e.g., 0.010-in. id) for the best chromatographic results. Tubing does not "wear out", so no special maintenance is required other than occasional replacement if it becomes blocked or damaged where it mates with a connecting fitting.

In the United States, tubing is manufactured to English-unit specifications, and most LC suppliers sell tubing in English sizes. Because of this, most references to tubing diameter in this book are in English units. Table 8.1 lists the nominal metric-equivalent sizes.

Table 8.1
Tubing Conversions

| Internal diameter | | Volume, |
|---|---|---|
| in. | mm.[a] | µL/cm |
| 0.005 | 0.13 | 0.13 |
| 0.007 | 0.18 | 0.25 |
| 0.010 | 0.25 | 0.51 |
| 0.020 | 0.50 | 2.03 |
| 0.030 | 0.75 | 4.56 |
| 0.040 | 1.00 | 8.11 |
| 0.046 | 1.20 | 10.72 |

[a]Nominal value.

Table 8.2
Tubing Choices

| Tubing Choices |
|---|
| Stainless-steel |
| Type 316 |
| Type 304 |
| Polymeric |
| Fluorocarbon (e.g., Teflon) |
| Polyethylene |
| Polypropylene |
| PEEK |

## 8.1. Types of Tubing

A variety of tubing types are available for LC applications, as seen in Table 8.2. Teflon and stainless-steel are most commonly used, although polyethylene and polypropylene can be used for certain applications.

### *Stainless-Steel Tubing*

Most connecting tubing in LC systems must withstand high pressures, so stainless-steel tubing is used in these cases. Stainless-steel tubing provides chemically inert connections for pressures above about 100 psi. Thus, stainless-steel is used for all high-pressure lines, beginning at the pump outlet and ending at the detector inlet. Stainless-steel tubing is most commonly available as type 316 or type 304. Type 316 is used more commonly and is more corrosion-resistant than 304, primarily because it contains Mo as part of the alloy. For most LC applications, either type of tubing is satisfactory.

Stainless-steel tubing is often classified as (a) LC-grade, or (b) industrial-grade. LC- and industrial-grade tubing have similar dimensional specifications (e.g., ± 0.002-in. od; ± 0.001-in. id), but tubing selected for LC applications has better concentricity.[2] Concentricity, which describes how well the tubing bore is centered, is especially important in small-diameter tubing, because the tubing bore needs to align properly with the matching hole in connecting fittings.

Typically, LC-grade tubing is washed in an organic solvent (e.g., methylene chloride) to remove residual oils from manufacturing, and then passivated with a nitric acid solution to reduce its chemical reactivity.[2] This tubing is widely available from LC supply houses, and instrument and fittings manufacturers. LC-grade tubing is clean and ready to use for most LC applications.

Industrial-grade tubing can be purchased from local tubing suppliers (see "Tubing" in the Yellow Pages). If you are willing to wash and passivate the tubing, this tubing is a less-expensive alternative to LC-grade tubing for most applications.

Stainless-steel tubing can be purchased (a) in precut lengths (e.g., 10cm), or more economically (b) in longer coils (e.g., 10ft). Tube cutting and the use of precut tubing are discussed later in this chapter.

### *Polymeric Tubing*

Polymeric tubing is commonly used when stainless steel is not required. Because it does not have as high a pressure resistance as stainless-steel, polymeric tubing is limited in its use to (a) supply lines connecting the mobile-phase reservoirs and the pump, (b) the outlet side of the detector, and (c) other low-pressure lines such as the injector waste-line. Fluorocarbon (Teflon) tubing usually is the best choice in plastic tubing, because it is inert to the reagents used in LC. Polyethylene or polypropylene tubing is considerably less expensive than Teflon, and can be used for waste lines. These polymers normally should not be used for pump inlet lines or sample transfer lines. Polymeric tubing is easy to cut, and its flexibility facilitates the routing of liquids to waste containers or collection vessels. Polymeric tubing can be purchased in 10–20 ft rolls and cut to length when needed.

*PEEK* (poly-ether-ether-ketone) tubing has recently been introduced as a substitute for stainless-steel tubing for LC applications. It is available in the popular 1/16-in. od x 0.010-in. id size, has pressure limits in excess of 5000 psi, and is more inert to biomolecules than stainless-steel tubing. Its flexibility, inertness, and ease of handling make PEEK tubing a viable alternative to stainless-steel for many applications.

## 8.2. Tubing Sizes

Stainless-steel connecting tubing (1/16-in. od) for LC is available in several diameters, as seen in Table 8.1. Polymeric tubing is available in similar sizes to stainless-steel, as well as the larger sizes noted in the discussion below. Because various tubing dimensions are used in different parts of the LC system, the following discussion is organized by application rather than by tubing diameter. Tubing length is discussed in the next section.

### *Reservoir to Pump*

The inlet lines from the reservoir to the pump are low-pressure suction lines, and must have low flow-resistance; the flexibility of polymeric tubing enhances the convenience of handling these lines. Teflon tubing, 1/8-in. od x 1/16-in. (or larger) id is used most commonly for the inlet lines. Connections can be made with (a) 1/8-in. compression fittings, (b) plastic tube fittings, or, (c), in the case of inlet frits, the tubing can be pushed over a stainless-steel nipple. In cases where oxygen-free mobile phases are necessary (e.g., certain electrochemical applications), the inlet line needs to be stainless-steel, because Teflon is permeable to oxygen.

### *Pump to Injector*

Stainless-steel tubing is used between the pump and injector to accommodate pressures up to 6000 psi. Because there is no contact between the sample and this tubing, the internal diameter and length of the connecting tubing are relatively unimportant. Recommended internal diameters are 0.020-, 0.030-, or 0.040-in.; smaller id tubing is not recommended because (a) it is more prone to blockage, and (b) no chromatographic advantage is gained by further reducing the dead volume in this part of the system. Long lengths of wider-diameter tubing, however, do require more washout time when changing solvents.

### *Injector*

Polymeric tubing (e.g., 0.020-in. id) is used for the injector waste-line, because of the convenience of routing a flexible line. Sample loops can be constructed of any of the stainless-steel tubing sizes listed in Table 8.1; use larger id tubing for larger-volume loops. Sample-loop construction is discussed in detail in Section 10.1.

### *Injector to Column*

The dead volume of this section of connecting tubing needs to be kept to a minimum, so 0.010- or 0.007-in. id stainless-steel tubing should be selected. In applications where small peak-volumes are encountered (e.g., microbore; or 3-cm, 3-µm columns), it may be necessary to use 0.005-in. id tubing. For most applications, however, 0.005-in. tubing is not recommended, because it is prone to blockage.

### *Column to Detector*

The dead volume in this part of the LC system also needs to be minimized; the guidelines immediately above should be followed. In cases where Teflon tubing is used by the manufacturer as the inlet line to the detector, Teflon of the same diameter (e.g., 0.010-in. id) can be used to connect the column and detector. PEEK tubing (which has a higher pressure limit than Teflon) is a good alternative.

### *Detector Waste Line*

The detector waste-line is a low-pressure line, and polymeric tubing usually is selected because of its flexibility and convenience. When a restrictor is required after the detector (see Chapter 12), a 1-m piece of 0.010-in. id polymeric tubing will often do double-duty as a waste line and restrictor combined.[a]

## 8.3. How Much Tubing to Use

As was discussed in Chapter 3, the connecting-tubing contribution to extra-column effects is of most concern between (a) the injector and column, and (b) the column and detector. Elsewhere in the system, tubing length and diameter are less important; lengths should be kept reasonably short for neatness and rapid solvent changeover. Table 8.3 can be used as a guide for choosing the diameter and length of tubing to select for column connections. Several of the more common column dimensions are listed, along with the maximum tubing length for an increase in band

[a] 1 m of 0.010-in. id tubing creates a backpressure of about 25 psi when water is pumped at 1 mL/min; 0.007-in. id gives about 100 psi/meter backpressure.

Table 8.3
Guide to Tubing Length[a]

| Column characteristics | | | | Maximum length (cm) for 5% increase in bandwidth | | |
|---|---|---|---|---|---|---|
| $L$ (mm) | $d_c$ (mm) | $d_p$(μm) | $N$ | 0.007 in. | 0.010 in. | 0.020 in. |
| 33 | 4.6 | 3 | 4400 | 22 | 9 | • |
| 50 | 4.6 | 3 | 6677 | 33 | 14 | • |
| 100 | 4.6 | 3 | 13333 | 67 | 27 | • |
| 150 | 4.6 | 5 | 12000 | 167 | 68 | • |
| 250 | 4.6 | 10 | 10000 | 556 | 228 | 14 |
| 250 | 4.6 | 5 | 20000 | 278 | 114 | • |
| 250† | 2.0 | 5 | 20000 | 50 | 20 | • |
| 250‡ | 1.0 | 5 | 20000 | 12 | • | • |

[a]Reprinted from ref. (1) with permission. $L$ = column length, $d_c$ = column internal diameter, $d_p$ = particle diameter, $N$ = column plate number; flow rate = 1 mL/min, except: † = 0.2 mL/min, ‡0.05 mL/min. (•)less than 8 cm.

width of 5%. For example, Table 8.3 shows that up to 114 cm of 0.010-in. id tubing can be used for connections if a 250 x 4.6 mm column packed with 5-μm particles is used. On the other hand, for the same column any usable length of 0.020-in. id tubing will cause an increase in bandwidth greater than 5%. (A "usable" minimum of 8-cm of tubing is needed to install a column with conventional fittings, assuming 2 cm/fitting and two fittings per end.)

Except where noted, a flow rate of 1 mL/min was used in the calculation of the values in Table 8.3. As shown in Eq. (8.1) the extra-column contribution to band broadening is inversely affected by flow rate ($L$ is column length, cm; $V_R$ is retention volume, mL; $D_M$ is solute diffusivity, cm²/s; $F$ is flow rate, mL/s; $d$ is column diameter, cm; $N$ is column plate number).

$$L = \frac{40\, V_R^2\, D_M}{\pi\, F\, d^4\, N} \tag{8.1}$$

However, for most columns flow rate changes affect the band width, as well. For this reason, in practical terms, flow rate is not a very important factor in extra-column effects. If larger id columns are used, even at increased flow rates, more connecting tubing can be used because larger peak volumes are generated. This effect can be seen by comparing the entries in Table 8.3 for the 250-mm columns of 1.0-, 2.0-, and 4.6-mm id. See ref. (3) for further discussion of tubing length calculations.

What does all of this mean in terms of practical chromatographic performance? If the important peaks in the analysis are baseline separated, and each has $k' > 1$, then a 5–10% increase in bandwidth generally will be insignificant. In this case, you can take the guidelines in Table 8.3 with a grain of salt. However, if peaks of interest elute before $k' = 1$, and $Rs < 1.2$, you will see a significant degradation in the separation when a 10% increase in band width is encountered. If narrow-bore columns or short, fast columns (e.g., 3-cm long x 0.46-cm id) are used, band widths will be narrower than for conventional columns. In this case, the tubing dimensions plus detector cell-volume must be matched to the column in order to avoid excessive band broadening.

## 8.4. Problem Prevention

### *Labeling*

It is difficult to distinguish visually between 0.007-, 0.010-, and even 0.020-in. id tubing; labeling is important in order to prevent confusion. The most important tubing dimension to note on the label is the internal diameter. One supplier of precut tubing (Upchurch Scientific) codes each piece of tubing with a colored band so that the sizes will not be confused. If you need to know the exact tubing length, or if you easily confuse the different fitting types, write this information on the label, too.

### *Tube Cutting*

*Polymeric tubing* is conveniently purchased in coils of 10 ft or more, and is cut to the desired length with a razor blade or sharp knife. Try to make the cut as square as possible for the best tubing connections.

*Precut stainless-steel tubing.* Stainless-steel tubing can be purchased either in precut lengths or in coils from 5 to 100 ft long.

Buying precut tubing may seem like a luxury, but the savings in time, quality, and convenience can be worth the extra expense. Kits are available, with precut lengths of tubing (e.g., 5, 10, and 20 cm) that are convenient for connecting the column to the rest of the LC system. Precut tubing has a very good finish, so you are assured of a squarely-cut end that has been carefully deburred and cleaned. Because proprietary electropolishing techniques are often used in preparing precut tubing, you may be unable to match its quality when you cut tubing in the lab. When fittings are made up with precut tubing, there is little question that the tube end properly seats in the fitting.

The tube end should be cut as squarely as possible, so that it will seat well in the fitting body. Some workers have found, however, that the angle of the cut can be off as much as 30° with no detrimental chromatographic effects.[2] It appears that the tube end acts much like a fire-hose nozzle; the stream is "squirted" into the hole in the fitting without disturbing the surrounding fluid. This suggests that, as long as the tube ends are deburred, any cutting method that leaves the ends reasonably square is adequate for most LC applications.

*Lab-cut stainless-steel tubing.* Precut tubing (a) is several times more expensive than bulk tubing, (b) is not available (except by custom order) in every length that may be required, and (c) may not be available for an immediate need. For these reasons, nearly all chromatographers will need to cut tubing at some time. Three general techniques are used for cutting small-bore tubing: (a) the abrasive-wheel cutoff machine, (b) the rotary cutter, and (c) the file-and-snap method. The most difficult step to master in all three methods is to prevent the hole in the center from closing, while at the same time cutting the end squarely. The tubing bore must be free of burrs or it can easily become blocked; the burr can also break loose and cause downstream blockage. If pliers or a vise are used, be careful not to damage the tube surface where the ferrule seats, or the assembled fitting may leak.

*An abrasive-wheel cutoff machine,* such as the one shown in Fig. 8.1, will produce squarely cut tubing every time. To use this tool, clamp the tubing in the guide (Fig. 8.1), then rotate the guide, and cut the tubing with the abrasive cutting- wheel. Next,

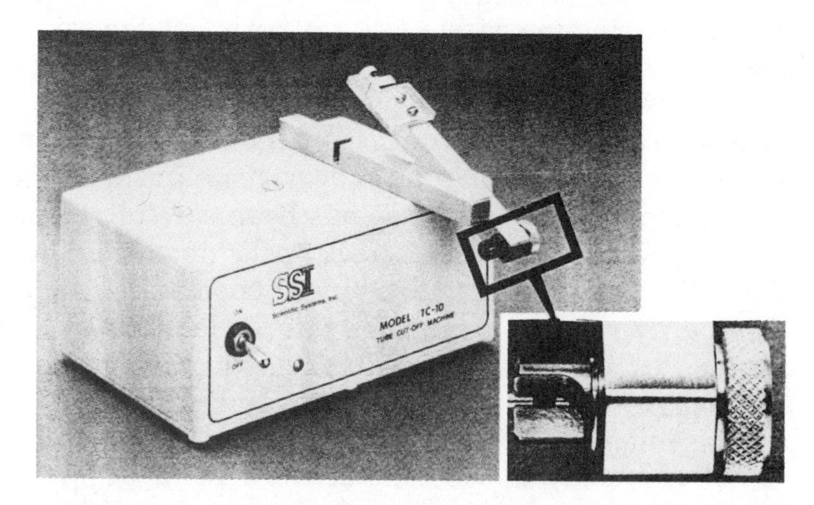


**Fig. 8.1.** Abrasive-wheel tube cutter. Inset shows detail of deburring tool. Courtesy of Scientific Systems Inc.

remove the burr from the inside and outside edges of the tubing with the deburring tool (Fig. 8.1 inset). When used properly, this system gives very good results. Care is required to prevent breaking off the tip of the deburring tool when using small id tubing. The cutoff tool is several times more expensive than the other tube cutting options discussed here, but it is fast and reliable.

A *rotary tube-cutter* (Fig. 8.2) is a second option for cutting stainless-steel tubing. This cutter does not give as smooth a finish on the tube end as the abrasive wheel does, but the tube bore is left fully open and burr-free every time. This tool is also easy to use, is less expensive, and can be carried conveniently in a lab-coat pocket. The rotary cutter is a miniature version of the C-clamp cutter used by plumbers for copper tubing. The tubing is clipped into the guide notch, and as the cutter is rotated, a blade scores the tubing. Next the tubing is removed, held on either side of the score line with a pair of pliers, and then snapped in two by twisting the pliers. This can leave the tubing with a small burr on the outer circumference, but this is removed quickly with a file so that a ferrule will freely slide over the tube. The most difficult part of using this tool is adjusting the cutting wheel for the proper depth of scoring, but once it is adjusted it works well.

The final option for cutting stainless-steel tubing is to use the *file-and-snap* method. A triangle- or knife-edge file is used to file

**Fig. 8.2.** Rotary tubing cutter (Terry tool). Courtesy of Scientific Marketing Inc.

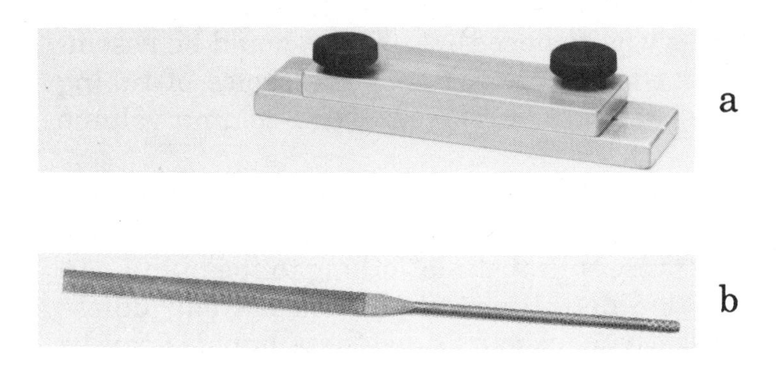


**Fig. 8.3.** Tubing holder (a) and file (b). Courtesy of Upchurch Scientific, Inc.

a groove about one-third of the way through the tubing. Holders, such as the one shown in Fig. 8.3, facilitate the filing operation and are available from several vendors. Alternatively, the tubing can be held in a pair of pliers, a vise, or with your fingers. After the groove is made, the tubing is held on each side of the cut with a pair of pliers and snapped apart. Filing the outside edge to deburr the tube is usually necessary, and the end of the tube seldom is cut squarely; however, the center bore usually is open and burr-free.

Most workers find that the rotory tube-cutter gives the most satisfactory results because it is convenient and works every time.

The abrasive cutter works well for tubing larger than 0.010-in. id, but requires considerable skill to obtain satisfactory ends on smaller tubing. As mentioned above, file cutting gives usable results, but they are not usually pleasing.

For all cutting methods, be sure to clean the cut tubing before using it. Burrs or filings may not cause immediate problems, but they can dislodge later to cause problems elsewhere in the system. Generally, it is sufficient to flush several milliliters of methanol through the tubing to clean it (using a syringe or an LC pump).

### *Normal Operation*

Whenever a piece of tubing is added to an LC system, there should be no significant change in system pressure (unless a blocked piece of tubing was replaced). Fittings should be assembled properly, so that they are leak-free. The choice of tubing diameters and lengths should not increase extra-column volume excessively.

Where translucent polymeric tubing is used in the LC, it may be possible to see air bubbles if they are present in the lines. No air bubbles should be present in the inlet tubing to the pump. An occasional bubble in the detector waste-line is normal; unless spikes are observed in the chromatogram, these bubbles can be ignored.

### *Spares*

A list of recommended tubing spares and tools is given in Table 8.4. A convenient way to stock the most common stainless-steel tubing spares is to purchase a kit of precut tubing, such as is shown in Fig. 8.4. In addition to a selection of precut pieces, a length of 0.010-in. id tubing (e.g., 10 ft. long) should be stocked for cutting into custom lengths (e.g., connecting an autosampler with the rest of the LC system).

Two sizes of polymeric tubing should be kept on hand, as noted in Table 8.4. Solvent inlet-lines and helium sparging-lines require 1/8-in. od x 1/16-in. id Teflon (larger id for some pumps). Because polymeric connecting tubing generally is not used where sample integrity is important, a single diameter usually suffices. The 1/16-in. od x 0.020-in. id tubing is a good choice, because it

Table 8.4
Recommended Tubing Spares

1. Tubing cutter
2. File (e.g., 4-in.-long triangular or knife-edge)
3. Razor blade or knife
4. Precut stainless-steel tubing kit containing (minimum):
   4 lengths each of 1/16-in. od x 0.010-in. id x 5-, 10-, and 20-cm long
   4 lengths each of 1/16-in. od x 0.020-in. id x 5-, 10-, and 20-cm long, or
5. Bulk stainless-steel tubing 1/16-in. od
   5 ft of 0.010-in. id
   5 ft of 0.020-in. id
6. Bulk Teflon tubing, 10–20 ft. each of:
   1/8-in. od x 1/16-in. id (for inlet lines)
   1/16-in. od x 0.020-in. id (waste lines)
   1/16-in. od x 0.010-in. id (waste lines)

can be connected to other components, either with stainless-steel fittings or with fittings specially designed for plastic tubing (see Chapter 9). If you use small-bore (e.g., 0.010-in. id) tubing as detector waste-line and restrictor, stock a spare length of this tubing, too.

## 8.5. Problems and Solutions

Tubing problems and solutions are summarized in Table 8.5 at the end of this chapter.

### *Tube Blockage*

Blockage is the primary problem that can occur with tubing. Partial or complete blockage can result from (a) poorly filtered mobile phases, (b) particulates in injected samples, (c) pump- or injector-seal wear, (d) leakage of silica particles from saturator-, guard-, or analytical columns, (e) particles broken off of poorly-deburred tubing, (f) precipitation of mobile phase salts, or (g) any other source of particulate matter in the LC system. Smaller-id tubing is more readily blocked than larger tubing. However, tube

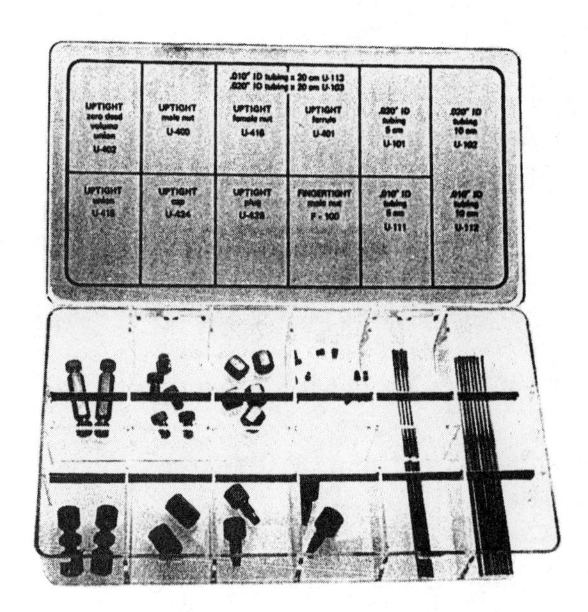

**Fig. 8.4.** Precut stainless-steel tubing kit. Contains 5–10 pieces each of 1/16-in. od tubing: 0.020-in. id x 5-, 10, and 20-cm long; and 0.010-in. id x 5-, 10-, and 20-cm long. Courtesy of Upchurch Scientific, Inc.

blockages are rare in LC systems; blockages more commonly occur at filters or column frits. Various frits (e.g., 2–10-µm porosity) remove particulates that might block the tubing (e.g., 0.010-in. = 250 µm id).

Tube blockage is accompanied by a *rise in pressure.* If the pressure suddenly rises to the pressure limit, blockage is the most likely cause. Partially blocked tubing can give a sudden increase in pressure, or pressure can gradually build up as a minor blockage retains other particulates. Fitting- or seal leakage can also occur when tubing becomes blocked; a fitting that is leak-free at 1500 psi may leak at 6000 psi.

*Isolate* the blocked tube by systematically loosening the fittings in a stepwise manner, starting at the outlet end of the system. Once the blockage is isolated, the blocked tube can be cleared or replaced.

*Clear* a blocked tube by reversing it end-for-end and pumping mobile phase through it. Use eye protection and do not point the outlet of the tubing at exposed skin; the blockage can break loose and leave the tube at a velocity sufficient to penetrate the skin. If the tubing is not readily cleared by reverse-flushing, discard it and replace it with a new piece. Be sure to match the length and id of the old piece so that extra volume is not added to the system.

When the blockage recurs, even in a new piece of tubing, isolate and correct the cause of the blockage before continuing operation. Leakage of silica particles from precolumns or laboratory-packed analytical columns can cause tube blockage. If this is a problem, install an in-line zero-volume filter to trap these particles. Particulates in the mobile phase (bacterial growth, precipitated buffer, etc.) are other possible causes of recurring blockage.

### *Damaged Tube Ends*

Tube ends that are surface-damaged during cutting or removal of a ferrule may not be smooth enough to allow the ferrule to seal properly in a compression fitting. If a fitting continues to leak after it has been tightened, inspect the tube end for damage. To correct the problem, recut the tubing, discard the damaged portion, then reassemble the fitting.

### *Other Problems*

Improper selection of the length and/or id of tubing can give increased bandwidths because of added extra-column volume; see Section 3.4 for a discussion of this problem. Most leaks at fittings are a result of fitting problems, not tube problems; Section 9.4 discusses leaky fittings.

Table 8.5
Tubing-Related Problems and Solutions

| Cause of problem | Symptom | Solution |
|---|---|---|
| Blocked tube | Increased pressure, possible leaks | 1. Isolate blockage<br>2. Reverse-flush tube<br>3. If unsuccessful, replace tube |
| Improper length and/or id | Excessive bandwiths | See Section 3.4 |
| Damaged tube end | Leaky fitting | See Section 9.4 |

## 8.6. References

[1]Dolan, J. W. (1985) *LC, Liq. Chromatogr. HPLC Mag.* **3,** 92.
[2]Stearns, S. Valco Instruments Co., personal communication
[3]Snyder, L. R. and Kirkland, J. J. (1979) *Introduction to Modern Liquid Chromatography,* 2nd ed., Wiley-Interscience, New York.

# Chapter 9

# FITTINGS



## Introduction

Fittings are the "cement" of the LC system, working with the tubing to connect the various modules. The fittings must be inert and leak-free; they also must add no unnecessary volume to the system. Two types of fittings are used in the LC. For pressures below 100 psi, low-pressure plastic fittings are used with Teflon tubing. Compression fittings, generally made of stainless-steel, are used with stainless-steel tubing for pressures up to 6000 psi. Mismatched or improperly-assembled fitting components will result in leaks and/or broadened peaks. A large number of fitting variations is available—some brands are interchangeable; many are not. It is important to be able to readily identify the different fitting types in order to avoid fitting-related problems.

## 9.1. Low-Pressure Fittings

Low-pressure fittings are used to connect Teflon or plastic tubing to the rest of the system; these are found most commonly on the inlet side of the pump and outlet side of the detector. Fittings are available for either 1/8- or 1/16-in. od tubing. Although some manufacturers specify higher pressure limits, it is best to use these fittings below 100 psi.

A wide variety of low-pressure fittings is available; most fittings are interchangeable. Some of the more common fitting designs are described below. These are mentioned by major brandname; similar fittings may be available from other vendors. A summary of low-pressure fittings and their interchangeability is given in Table 9.1.

### *Cheminert*

Cheminert fittings (Fig. 9.1, LDC/Milton Roy) represent a common fitting design. The assembled fitting consists of a flanged piece of plastic tubing backed with a stainless-steel wash-

Table 9.1
Low-Pressure-Fitting Characteristics

| Brand[a] | Type of seal | Pressure limit[b] |
|---|---|---|
| Cheminert | Flared tube-end | 500 psi |
| General Valve | Kel-F insert in tube | n/a |
| Omnifit | Teflon & s.s. gripper | 1000 psi |
| Upchurch | Inverted ferrule | 1000 psi |

[a]All brands shown are fully interchangeable.
[b]Manufacturer's specification with 1/16-in. od Teflon tubing.

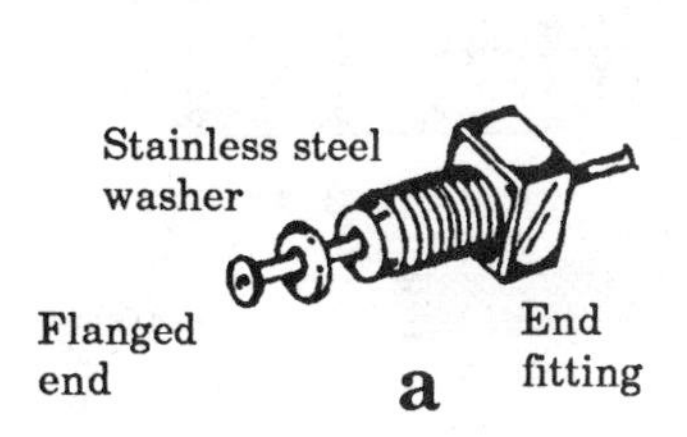


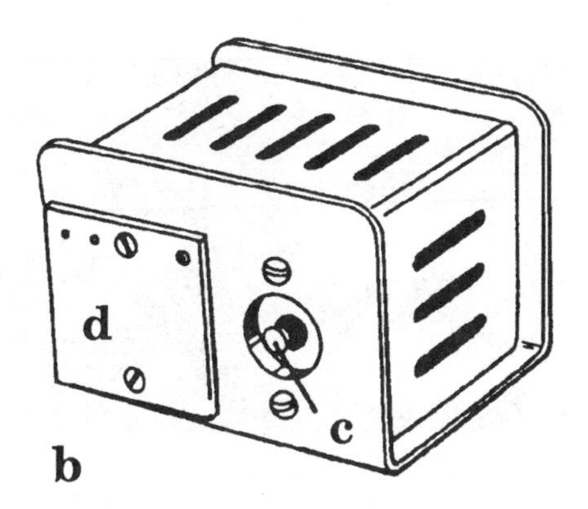


**Fig. 9.1.** Cheminert fitting: (a) tube-end with parts labeled; (b) flaring tool; (c) heated flanging tip; (d) cooling plate. Courtesy of LDC/Milton Roy.

er, held in place with a polypropylene nut (Fig. 9.1a). To make up the fitting, cut the tubing to length and slide the nut and washer over the end. Then push the end of the tube against the heated portion of the flanging tool (Fig. 9.1c) to flare the tube-end. When the tube is flared, push the tube-end against the cooling plate (Fig. 9.1d) to make the flare permanent. A well-shaped end-flare is more difficult to make with 1/8-in. od tubing than 1/16-in. od tubing, so many workers use another fitting type for the larger tubing. Care should be taken to follow the manufacturer's directions when tightening the fittings. Generally 1/8-turn past finger-tight is sufficient.

Overtightening can cause leaks, because the tubing flare is forced off the seat and into the center of the nut. Once the fitting is assembled initially, it can be taken apart and reassembled as often as needed.

### *General Valve*

These fittings (Fig. 9.2) are similar to Cheminert fittings, but instead of flaring the tube end, a Kel-F insert is used to hold the tubing in place and to provide a sealing surface. To assemble the fitting, slide the nut over the tube (Fig. 9.2a) and use the mandrel-end of the insertion tool to stretch the tubing slightly (Fig. 9.2e). Now the Kel-F insert can be forced in with the aid of the other end of the insertion tool (Fig. 9.2d,f).

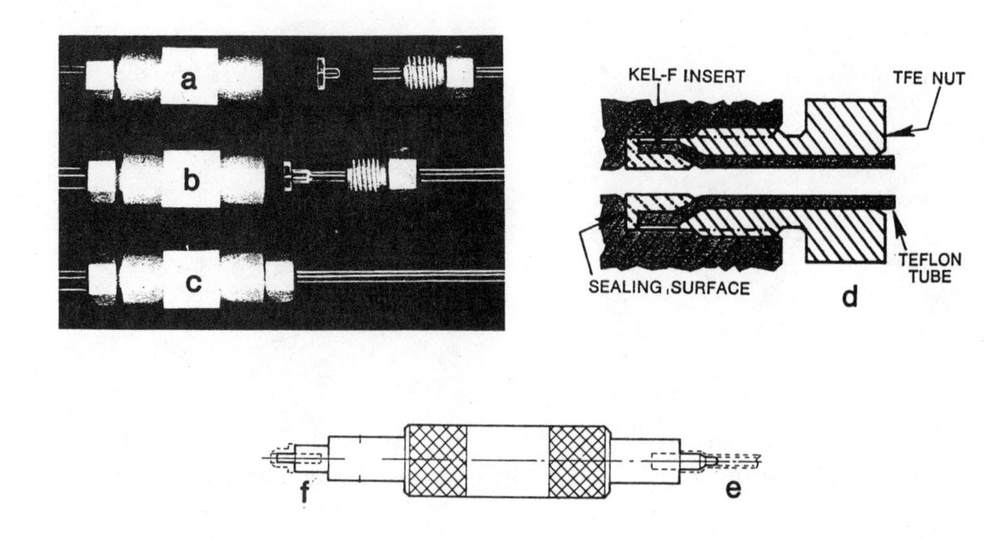


**Fig. 9.2.** General Valve fitting. (a) union, Kel-F insert, and tube end; (b) tube-end with insert in place; (c) assembled union; (d) assembly drawing; (e) insertion tool for stretching tubing; (f) insertion tool for inserting insert. Courtesy of General Valve Corporation.

Teflon tubing often is slippery to handle when preparing a tube-end for a fitting. The manufacturer suggests holding the tubing with a small piece of 600-grit sandpaper to prevent the tubing from slipping through your fingers.

## *Omnifit*

Omnifit fittings (Fig. 9.3) use a Teflon-and-stainless-steel ferrule ("gripper") that is leak-tight to 1000 psi. To assemble the fitting, cut the tubing at a shallow angle with a razor blade and slide the nut over the tubing. Then thread the tube-end through the ferrule and pull it through with a pair of pliers until uncut tubing protrudes. Next, rotate the ferrule several times so that the stainless-steel gripper cuts into the tubing, then cut off the extra tubing flush with the sealing face of the ferrule. Now the nut can be threaded into the body of the fitting to give a leak-free seal. A lock nut is supplied to prevent the nut from loosening due to system vibration—an occasional problem with plastic fittings. This lock nut will also work with other low-pressure fittings.

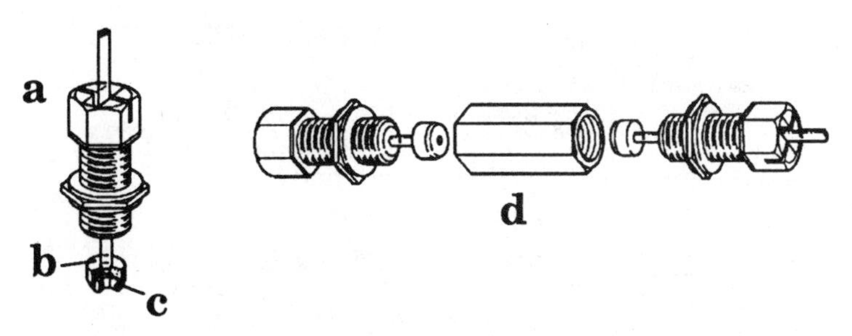


**Fig. 9.3.** Omnifit fitting: (a) tube-end and nut; (b) stainless-steel shell of gripper; (c) PTFE sealing surface; (d) union prepared for assembly. Courtesy of Omnifit Inc.

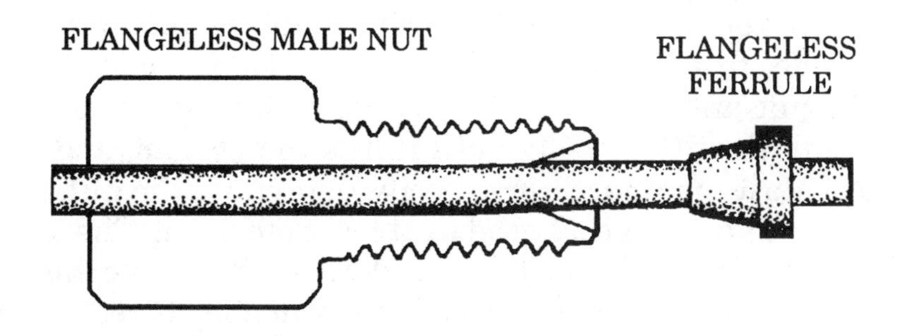


**Fig. 9.4.** Upchurch flangeless fitting. Reprinted from ref. (2) with permission.

### *Upchurch*

The flangeless fitting for plastic tubing shown in Fig. 9.4 is easy to use. Simply slide the nut and then the ferrule onto a square-cut tube end. Whent the nut is tightened into the fitting body, the ferrule slides until it is even with the tube end. Then it grips the tubing as the tapered portion of the nut compresses the ferrule. This is a finger-tightened fitting; no wrench is used.

## 9.2 . High-Pressure Fittings

High-pressure connections between LC modules are made with stainless-steel tubing and compression fittings (finger-tightened fittings are discussed later). Compression fittings can also be used with plastic tubing, but with lower pressure limits. A

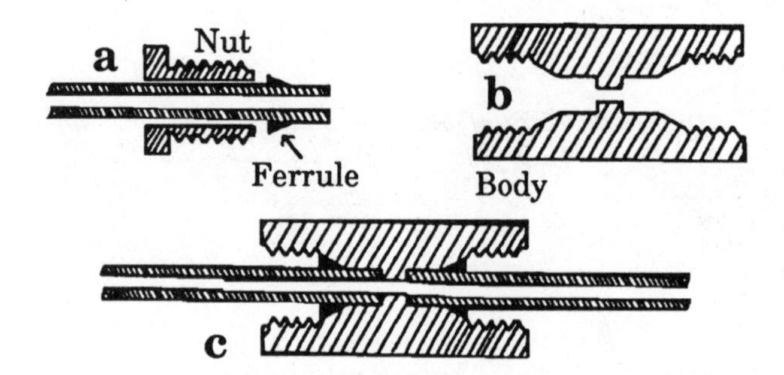


**Fig. 9.5.** Compression fitting parts. Cross-section of (a) tube end; (b) fitting body; (c) assembled union (nuts removed for clarity). Reprinted from (1) with permission.

generalized compression fitting is shown in Fig. 9.5; it includes a nut, ferrule, fitting body, and tube end. To assemble a fitting, slide the nut and ferrule over the tube end, push the end of the tubing into the fitting body until it hits the bottom, and tighten the nut. Most manufacturers specify tightening the nut 3/4-turn past finger-tight for a seal good to at least 6000 psi. The tip of the ferrule seals around the tube, and the tapered portion seals against the taper of the fitting body to form a leak-free assembly. As with low-pressure fittings, overtightening high-pressure fittings can cause leaks by distorting the fitting. Once the fitting has been assembled, it can be taken apart and reassembled as often as necessary. In contrast to the low-pressure fittings, however, different brands of compression fittings generally are *not* interchangeable once they have been assembled; mismatched parts can cause leaks or add dead volume to the system (see discussion of interchangeability below).

### *Internal vs External*

Compression fittings are classified as internal or external depending on the configuration of the nut (see Fig. 9.6). In external fittings (sometimes called external-nut fittings), the nut goes over the outside of the threaded fitting body. External fittings were popular in the early days of LC, because they were readily available from other tubing applications (e.g., gas chromatography). Today, these fittings are used mainly to fasten the column to the column end-fitting. Internal fittings (or internal-nut fittings) are

**Fig. 9.6.** Compression fittings. (a) Internal-union with nuts in place; (b) cutaway internal union; (c) external-union with nuts in place; (d) cutaway external union. Courtesy of Upchurch Scientific, Inc.

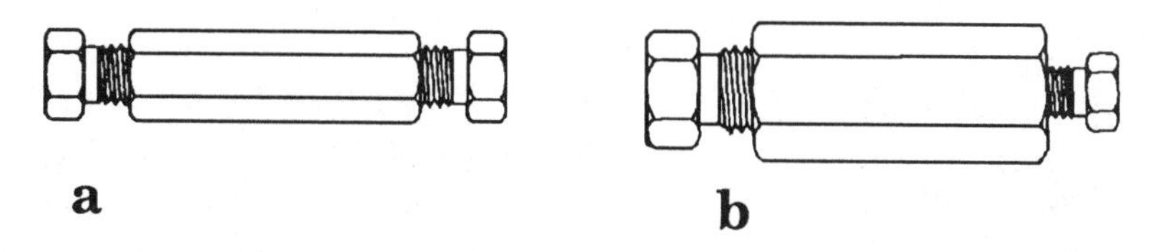


**Fig. 9.7.** Union (a) and adapter or reducer (b). Courtesy of Valco Instruments Inc.

more popular today for 1/16-in. od tubing connections. The nut screws into the threaded fitting body. These fittings are (a) easier to assemble and disassemble, (b) stronger, and (c) more readily examined for damage than external fittings.

## *Unions and Adapters*

For both high- and low-pressure fittings, two types of connectors are common: unions and adapters. Unions (Fig. 9.7a) are used to connect two pieces of tubing of the same outside diameter. Low-pressure unions often butt the tube ends together, whereas high-pressure unions generally have a connecting passage. As seen in Fig. 9.5, the connecting hole in the fitting body of a union should have the same internal diameter as the tubing which it connects. When this hole matches the tubing id, the fitting is called a *Zero-volume* or *Low dead-volume* fitting, because the fitting adds no more volume than a corresponding length of tubing. Smaller-diameter holes generally are no problem, but larger connecting passages may reduce the system performance by adding dead volume.

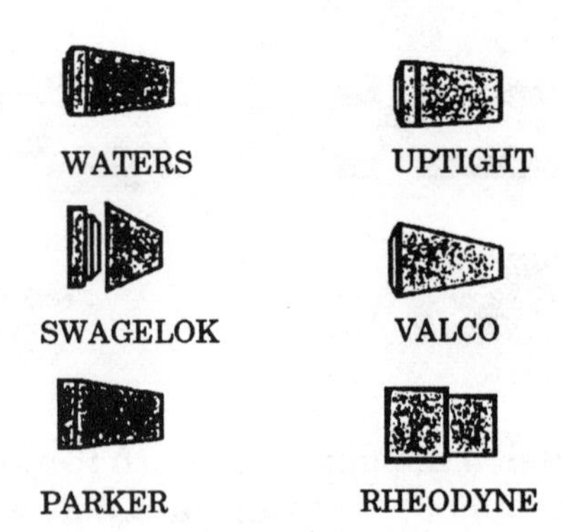


**Fig. 9.8.** Ferrule variations. Reprinted from ref. (2) with permission.

The term *Adapter, Reducer,* or *Reducing union* is used to describe a connector joining two tubing pieces of unequal diameter (Fig. 9.6b). The most common adapter in the LC is the column end-fitting, which connects the column (e.g., 1/4-in. od) to a 1/16-in. od connecting tube. The column end-fitting also contains a frit to keep the packing in the column.

### *Fitting Interchangeability*

Most brands of high-pressure compression-fittings for 1/16-in. od tubing are nominally interchangeable (except for SSI). These fittings consist of a 10–32 threaded nut and body (i.e., size 10 nut, 32 threads per inch), and a tapered ferrule. The most common brands of fittings are pictured in Figs. 9.8–9.10. SSI also makes fittings with similar ferrules, but these fittings are not interchangeable with the others, because the nut and body have 1/4–28 threads (i.e., 1/4-in. nut, 28 threads/in.); an SSI fitting is shown in Fig. 9.14a. The following discussion on interchangeability is limited to fittings for 1/16-in. od tubing, but the same principles hold for other sizes, as well.

Leaks and peak broadening can result if the fitting parts are not matched properly, especially in areas where they contact the sample. The fitting parts nominally are interchangeable before assembly, but may cause problems if the wrong parts are combined after the fitting has been made up the first time.

Look first at the *ferrules* shown in Fig. 9.8. These have a similar overall shape, but differ in detail. For example, Swagelok

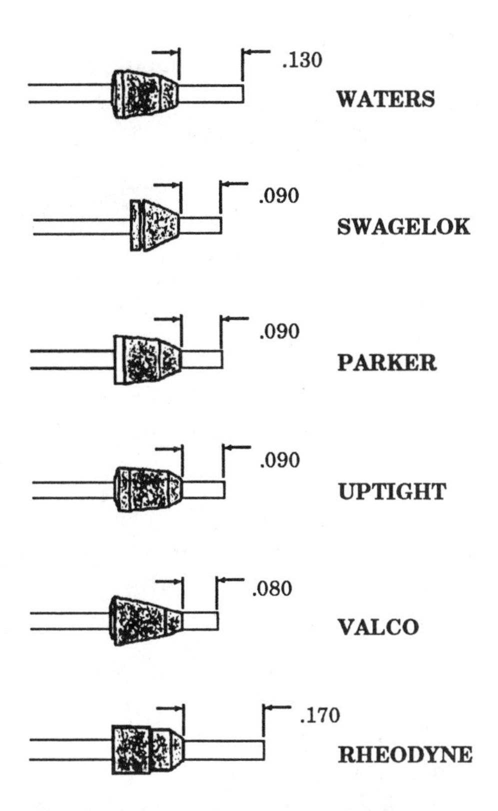


**Fig. 9.9.** Tubing extensions. Measurements in inches; values are typical, but not necessarily manufacturer's specifications; Rheodyne uses different extensions in different fitting types. Note how the tip of each ferrule has been swaged onto the end of the tube; this matches the taper inside the fitting body. Reprinted from ref. (2) with permission.

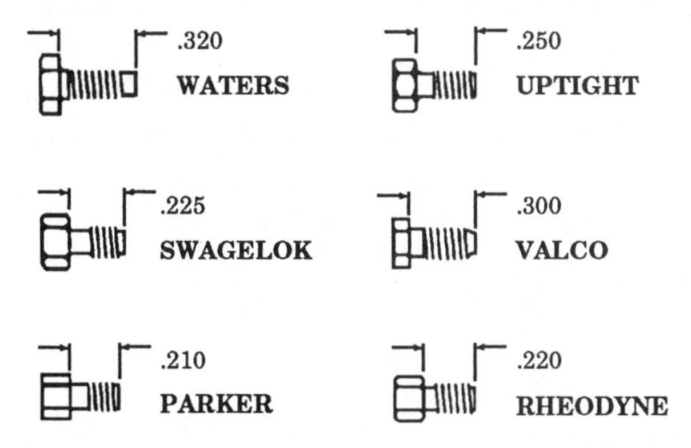


**Fig. 9.10.** Nut variations. Measurements in inches; values are typical, but not necessarily manufacturer's specifications. Reprinted from ref. (2) with permission.

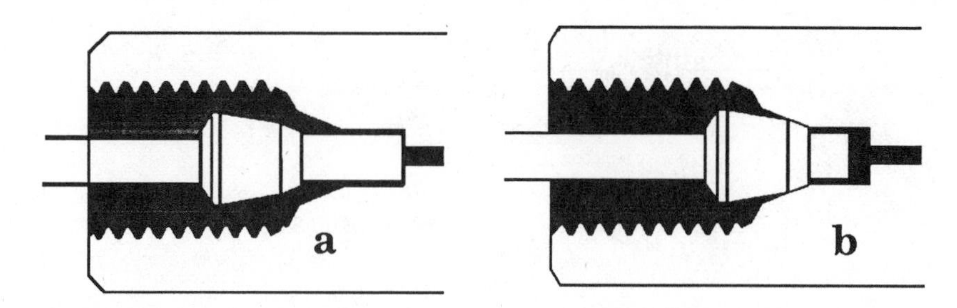


**Fig. 9.11.** Improperly-assembled fittings. (a) tubing extension too long for external port, fitting leaks; (b) extension too short for external port, dead-volume in fitting. Nuts not shown for clarity. Reprinted from ref. (2) with permission.

uses a two-piece ferrule and Rheodyne uses a stepped ferrule, whereas the other brands use single-piece, tapered ferrules. However when the fitting is properly assembled (e.g., Fig. 9.5), the ferrule is somewhat deformed when sealing with the mating tapered-seat. Because the angle of the tapered seat is about the same for all brands, one brand of ferrule usually will provide an adequate seal with any brand of fitting body, but only if it is originally made up in the same brand of fitting body.

The problem with compression-fitting interchangeability is related to the *extension* of tubing past the ferrule, once the ferrule is swaged onto the tubing. The differing length of this extension is caused by differences in the depth of the female port in the fitting body. Different manufacturers use different port depths, and some manufacturers (e.g., Rheodyne) vary the port depth between different products. These differences can be seen graphically in Fig. 9.9. For example, it is obvious that the Waters and Swagelok fittings have different tubing extensions.

Assembling a fitting with the improper extension for the female port can result in one of two problems. *First,* if the extension is too short (Fig. 9.11b), the tubing will not butt properly against the bottom of the port and a dead volume will be added to the system. The fitting will not leak, but the dead volume will contribute to (a) increased peak broadening, and (b) poor system washout. In the *second* case, where a fitting with too long an extension is used (Fig. 9.11a), the ferrule will not seat properly and the fitting will leak.

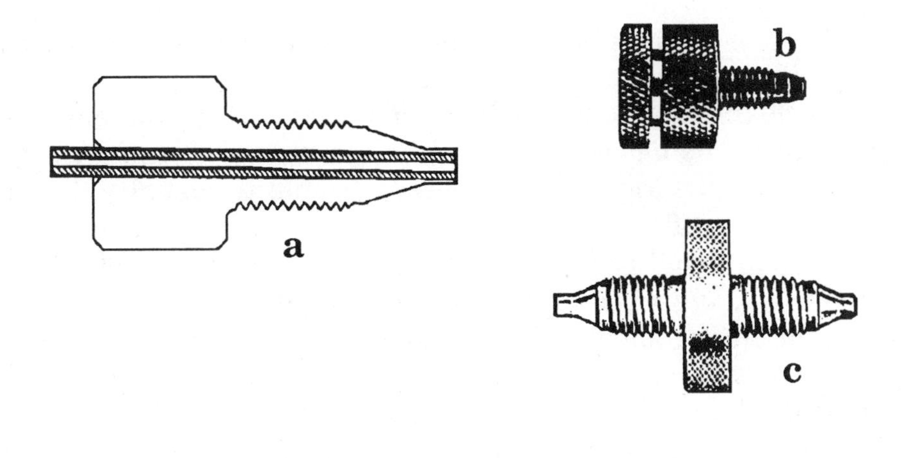


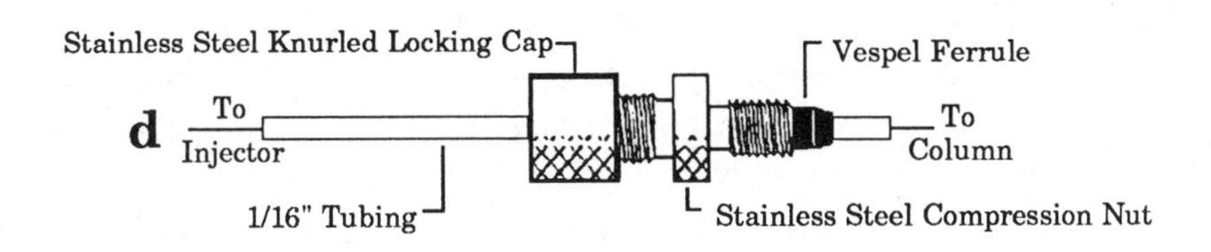


**Fig. 9.12.** Finger-tightened fittings: (a) single-piece Kel-F nut and ferrule; (b) Vespel and stainless-steel fitting, reversible for male or female fittings; (c) single-piece Kel-F column coupler; (d) high-pressure Vespel and stainless-steel fitting. Courtesy of Upchurch Scientific, Inc. (a), Alltech/Applied Science (b,c), and Keystone Scientific (d).

It is also important to choose the correct nut for the compression fitting. All the nuts shown in Fig. 9.10 have the same 10–32 thread, and can be successfully interchanged, with two precautions: be sure that (a) enough threads of the nut grip the fitting body to give a reliable assembly, and (b) the nut does not thread into the fitting so far that the shoulder contacts the fitting body before the ferrule seals. For example, when a nut with a long unthreaded tip (e.g., Waters) is threaded into a fitting body designed for a short nut (e.g., Swagelok), only a few threads of the nut grip the fitting. On the other hand, when a short nut (e.g., Parker) is mated with a deeply threaded fitting body (e.g., Valco), the shoulder of the nut can hit the fitting body before the ferrule has been pressed into place. A more extensive discussion of fitting compatibility can be found in ref. (2).

It is clear from the above discussion that fitting problems can be avoided by using parts from only one manufacturer in each fitting. With experience, you should be able to visually distinguish between different brands of fittings, but labeling each fitting will help avoid problems caused by mismatch. A self-adhesive label (e.g., 3/8 x 1-1/2 in.) wrapped around the tubing makes a convenient fitting label. Standardizing on one manufacturer's fittings also will reduce the probability of mismatch.

One final note of caution: when connecting the various parts of a microbore LC system, use only the fitting parts that were originally assembled as a set. Small differences exist between individual fittings from the same manufacturer. These variations are unimportant in larger columns (e.g., 250 x 4.6 mm); however, these same differences can result in a significant extra-column dead-volume with microbore columns. Always label fittings for microbore connections so that the original parts can be reused together. See ref. (3) for an excellent discussion of fittings use for microbore applications.

### *Finger-Tightened Fittings*

An alternative to stainless-steel compression-fittings is to use fittings with polymeric ferrules. These are available as a single-piece Kel-F fitting (e.g., Upchurch), or a Vespel-and-stainless-steel unit (e.g., Alltech), shown in Fig. 9.12a,b. These fittings have the advantage that the ferrule is flexible and grips the tubing when the fitting is tightened; however, the ferrule does not bite into the tubing and become permanently seated. Thus, the ferrule can be easily moved to allow for a longer or shorter tubing-extension when used with another manufacturer's fitting body. To assemble these fittings, slide the nut-ferrule assembly over the tube end, and seat the tube end firmly in any standard 1/16-in. compression-fitting body. Now, using the knurled nut, finger-tighten the fitting to obtain a fitting that is leak-free to 4000–5000 psi.

An inexpensive, reusable seal similar to the finger-tightened fittings can be made by substituting a nylon or Teflon ferrule (available from Swagelok) for the stainless-steel ferrule in a standard fitting. When this method is used, the ferrule sometimes sticks in the female port after the tube is removed. However, the ferrule can be extracted with a bent paperclip.

Double-ended finger-tightened fittings (e.g., Fig. 9.12c) are available for low-dead-volume coupling of columns to precolumns, or injection valves to columns. The construction, assembly, and sealing action of these couplers are the same as for the finger-tightened fittings just discussed.

One problem with finger-tightened fittings is that they have a lower pressure-limit than stainless-steel fittings (e.g., 4000 vs 6000 psi). Polymeric ferrules are also subject to cold-flow, and can loosen when heated or cooled (PEEK ferrules seem to overcome this problem). A finger-tightened fitting that overcomes these problems, yet still retains the convenience of finger-tightened fittings is shown in Fig. 9.12d (Keystone Scientific). In this case, the seal is made with a Vespel or PEEK ferrule, but a stainless-steel backup ferrule assures that the tube end is properly seated and that the fitting will hold to 10,000 psi.

## 9.3. Problem Prevention

### *Procedures*

Nearly all fitting problems can be avoided by (a) using fitting parts that are designed to work together (e.g., same manufacturer), (b) using fittings that match the application (e.g., zero-volume fittings for sample-carrying lines, and (c) assembling and tightening the fittings according to the manufacturer's directions, taking care to avoid overtightening. Fitting mixup can be avoided by clearly labeling the assembled fittings. A fittings kit, such as the one shown in Fig. 9.13 (or home-made from a partitioned plastic box), helps to organize the spare parts.

When assembling or reassembling fittings, be sure that the sealing surfaces are free of particulates and buffered mobile phases. Particulates can prevent the fitting from sealing properly; in extreme cases, buffers or other salts can cause the fitting to seize.

Finally, use only open-end wrenches when working with fittings. Avoid using adjustable ("crescent") wrenches; these can damage the fittings and are prone to slip.

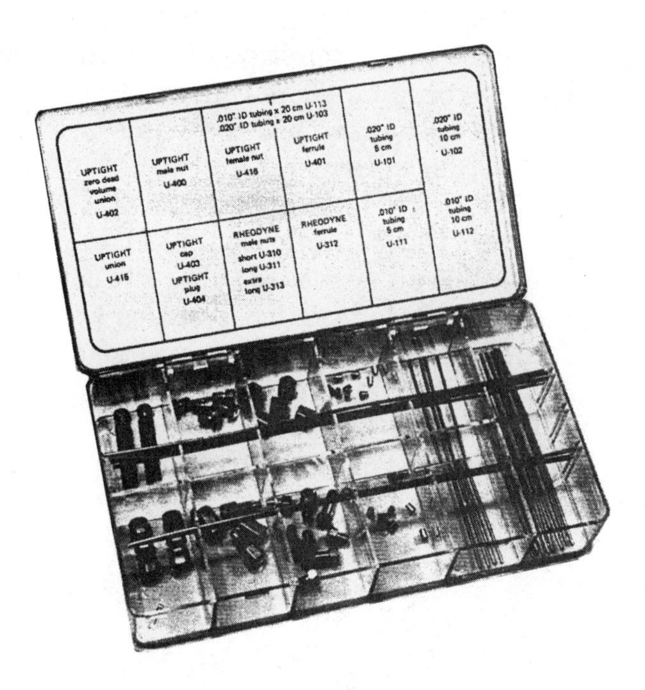


**Fig. 9.13.** Fittings kit. Contains a variety of nuts, ferrules, unions, and precut tubing. Courtesy of Upchurch Scientific, Inc.

## *Spares*

You should stock at least one spare fitting (nut, ferrule, and body) for every fitting-type used in your LC. A list of spares is given in Table 9.2; you will need to customize this list to meet the needs of your particular LC. (The brand names listed in parentheses after each part are the most common brands used for that particular application.) Fittings are stocked most easily in a fittings kit, such as is shown in Fig. 9.13; replenish the parts as the supply dwindles. Kits are available for stainless-steel-, fin-

Table 9.2
Recommended Fittings Spares

1. Column adapters (e.g., Valco-to-DuPont; Rheodyne-to-Waters)

2. Nuts, ferrules, unions
   - -for inlet check-valve (e.g., 1/8-in. Swagelok male nut and ferrule, unless plastic fittings are used)
   - -for outlet check-valve and purge valve (e.g., 1/16-in. SSI male nut and ferrule)
   - -for injector (e.g., Rheodyne, Valco, or Waters male nut and ferrule)
   - -for in-line filters (e.g., Rheodyne, Valco, SSI, or Upchurch male nut and ferrule)
   - -for column (male nut and ferrule to match column brand)
   - -for connecting tubing tubing (1/16-in. nuts, ferrules, unions; standardize on a single brand)

3. Plastic
   - -for low-pressure mixer and proportioning valves (standardize on a single brand)
   - -for inlet check-valve (unless stainless-steel fittings are used)
   - -for high-pressure connecting tubing (optional: plastic ferrules or finger-tightened fittings)

ger-tightened-, and low-pressure fittings. The most-often-used parts are 1/16-in. nuts and ferrules; be sure to stock a sufficient supply (e.g., 10 of each).

If you use columns from a variety of vendors, you will need different adapters for each column brand, to connect the column to the injector and detector. These adapters can be made up from loose parts when they are needed, or they may be purchased preassembled. A variety of these adapters is shown in Fig. 9.14. Alternatively, use a fingertightened fitting (Fig. 9.12); such fittings allow you to interchange columns from different manufacturers while retaining a good connnection to the LC system.

## 9.4. Problems and Solutions

Most fittings-related problems are associated with (a) improperly assembly, (b) improper parts, (c) over- or undertightening, or (d) dirt on the sealing surfaces. Solutions to these problems are discussed in the preceeding sections for each type of fitting and are summarized in Table 9.3.

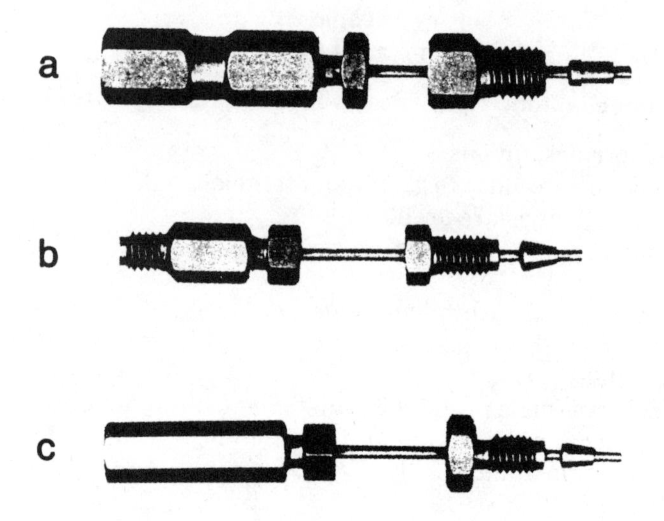


**Fig. 9.14.** Column adapters for connecting (a) Waters internal nut to SSI external end-fitting; (b) Swagelok/Parker external nut to Valco external end-fitting; (c) Swagelok/Parker internal nut to Waters external end-fitting. Courtesy of Upchurch Scientific, Inc.

Table 9.3
Fitting Problems and Solutions

| Cause of problem | Symptom | Solution |
|---|---|---|
| *Low-pressure fittings* | | |
| Loose fittings | Leaks | 1. Tighten<br>2. Use lock-nut<br>3. Replace fitting |
| Damaged tube end | Leaks | 1. Replace fitting or gripper |
| Damaged fitting nut or body | Stripped threads, cracked body | 1. Replace nut and/ or body |
| Overtightened fitting | Distorted fitting, *stripped threads* | 1. Replace nut and/or body |
| *Stainless-steel fittings* | | |
| Loose fitting | Leaks | 1. Tighten 3/4-turn past finger tight |

*(continued)*

Table 9.3
Fitting Problems and Solutions *(continued)*

| Cause of problem | Symptom | Solution |
|---|---|---|
| *Stainless-steel fittings (continued)* | | |
| Overtightened fitting | Leaks, distorted ferrule | 1. Loosen and retighten to specifications<br>2. Replace fitting if still leaks |
| Dirty fitting | Leaks | 1. Disassemble, clean, and retighten<br>2. Replace if still leaks |
| Broken nut | Leaks, physical damage to fitting, extra band broadening | 1. Replacedamaged fitting |
| Poor tubing finish | Leaks | 1. Cut off damaged portion, remake fitting |
| Improperly assembled or mismatched parts | Extra band broadening, leaks | 1. Replace with properly assembled and matched parts |
| *Finger-tightened fittings* | | |
| Loose | Leaks | 1. Tighten |
| Won't seal | Leaks, even when retightened | 1. Replace with stainless-steel fitting |

Table 9.4
Problem Prevention

1. Use only compatible parts
2. Match the fitting to the application
3. Do not overtighten
4. Avoid confusion—label assembled fittings
5. Keep fitting parts organized
6. Use proper tools

## 9.4. References

[1]Snyder, L. R. and Dolan, J. W. (1985) *Getting Started in HPLC, User's Manual,* LC Resources, Lafayette, CA.

[2]Upchurch, P. (1988) *HPLC Fittings,* Upchurch Scientific, Oak Harbor, WA.

[3]Bakalyar, S. R., Olesen, K., Spruce, B., and Bragg, B. G., (1988) *Technical Note 9,* Rheodyne Inc., Cotati, CA.

## Chapter 10

# INJECTORS AND AUTOSAMPLERS

## Introduction

The sample injection-valve introduces measured amounts of sample onto the LC column. By injecting directly into the high-pressure mobile-phase stream, the injection valve avoids the need to stop the mobile-phase flow during injection (as with on-column syringe injection). Injectors may be operated manually or automatically, or may be an integral part of an autosampler. A properly operating valve is critical for precise and accurate sample injection, and thus contributes to the quality of the chromatographic data. Mismatched or damaged injector components can cause broadened peaks, variable sample volumes, leaks, and/or increases in system pressure.

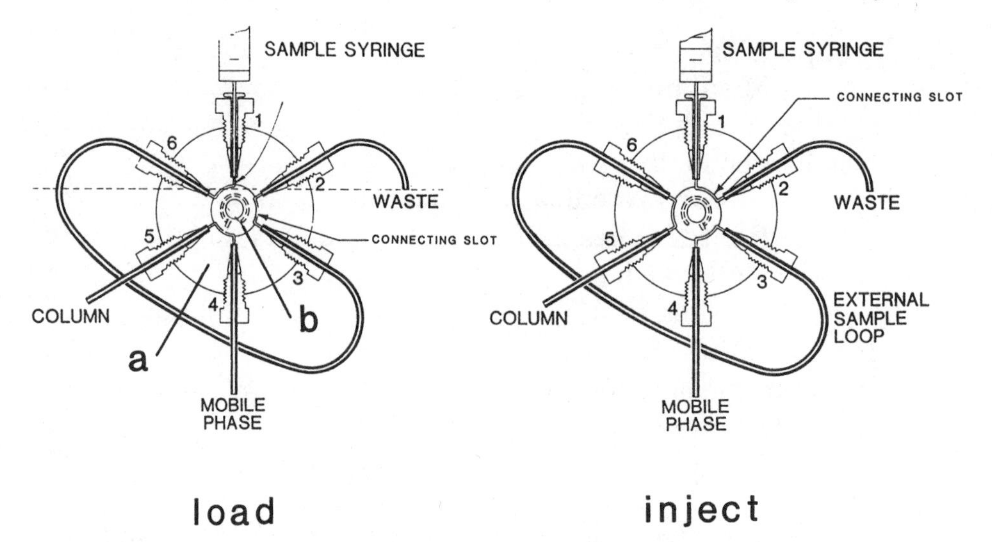


**Fig. 10.1.** Six-port injection valve. (a) Valve body; (b) seal or rotor. Port numbering: (1) injection port, arrow in (a) indicates connecting passage; (2) waste; (3) and (6) connections for sample loop; (4) mobile phase from pump; (5) to column. Courtesy of Valco Instruments Inc.

Autosamplers are automated injection valves that permit unattended injection of samples into the liquid chromatograph. These labor-saving devices have increased laboratory efficiency and allowed for round-the-clock operation of LC systems in labs with large sample loads.

The basic design, operation, maintenance, and troubleshooting of sample injection-valves and autosamplers are discussed in this chapter.

## 10.1. Injection-Valve Design and Operation

The six-port valve shown in Fig. 10.1 is the basis for the design of all sample injection valves. The valve consists of a rotating *Seal* (or *Rotor*) and a fixed *Body*. An *Injection Port* facilitates filling the *Sample Loop* with sample, and the *Waste Port* vents the loop and allows excess sample to be discarded. The pump and column are connected to the two remaining ports.

To fill the loop with sample, move the rotor to the *Load* position (Fig. 10.1a), so that the loop is connected to the injection and waste ports; flow from the pump goes directly to the column. Insert a syringe into the injection port and fill the loop with sample (see discussion below for proper injection techniques). Next, turn the rotor to the *Inject* position (Fig. 10.1b); the loop is flushed onto the column with mobile phase from the pump. The valve now can be returned to the load position in preparation for the next sample injection.

In the position intermediate between the load and inject positions, the rotor seal blocks flow to the column. This blockage causes a sudden pressure increase upstream from the valve and the mobile-phase flow drops to zero. When valve rotation is completed, a pressure (and flow) pulse at the head of the column results. Sudden changes in flow and pressure can damage the column, so the valve should be turned as rapidly as possible. Some injectors are designed with a bypass that shunts the flow around the valve in the intermediate position (see Section 10.3).

### *Injection Ports*

The injection port (also, *sample port, fill port, needle port*) is the interface between the injection syringe and the injection valve, as can be seen in position 1 of Fig. 10.1. An exploded diagram of the injection port is shown in Fig. 10.2. In this case, the port is a finger-tightened fitting composed of a stainless-steel knurled nut, a short length of Teflon tubing, and a polymeric ferrule. The port is screwed into the valve body and the ferrule radially compresses the tubing, so that it seals around the syringe needle. Once the port is adjusted, the needle can be inserted and removed without further adjustment. Because the needle tip is inserted inside the valve body (nearly touching the rotor), this type of connection is called an *Internal fill-port.* Internal needle-ports facilitate partial-loop injections because little or no sample is lost—the needle tip delivers sample directly into the sample loop. Internal ports can be mounted on most injectors, but in order to do so, the injector must be mounted so that the syringe can be inserted conveniently.

The *External fill-port,* on the other hand, can be mounted at a distance from the valve. In this case, the fill port is connected to the valve with a piece of connecting tubing. An example of an external fill-port is seen in Fig. 10.3. External injection-ports are used when it is not possible to insert the syringe directly into the valve body. With external fill-ports, the position of the syringe

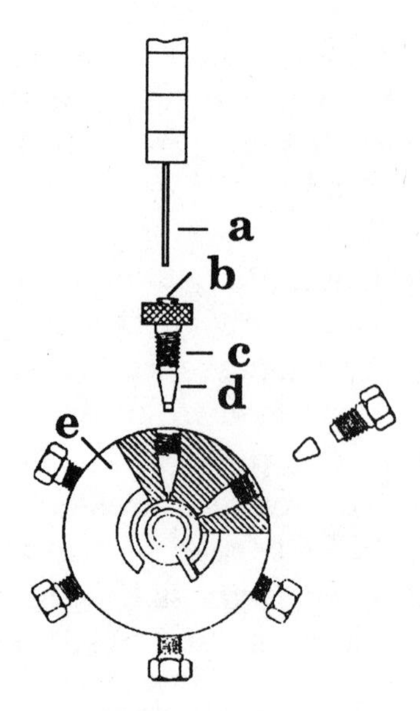


**Fig. 10.2.** Internal injection-port. (a) syringe needle; (b) injection port liner (Teflon tubing); (c) nut; (d) ferrule; (e) valve body. Reprinted from ref. (1) with permission.

needle relative to the valve can be varied (i.e., the connecting tubing can be lengthened or bent). The exploded diagram in Fig. 10.3c shows that a Teflon liner-tube plus a ferrule seal the injection needle in the same manner as the internal fill-port. The stainless-steel connecting-tube joins the external port assembly to the valve body. The sample contained in the connecting tube is not injected, so external fill-ports should be used only for filled-loop injection.

Most sample valves can be used with either internal or external injection ports. The main tradeoffs are between the need for minimum sample loss (internal port) and mounting convenience (external port). The availability of these two options is listed along with other injection-valve characteristics in Table 10.6.

## *Injection Techniques*

Typically, the size of the sample loop dictates the injection volume, although sample volumes less than the loop volume can be injected. With the *Filled-Loop* injection technique (Fig. 10.4), the loop is filled entirely with sample. For maximum precision, a volume of sample at least 2–3 times the loop volume should be

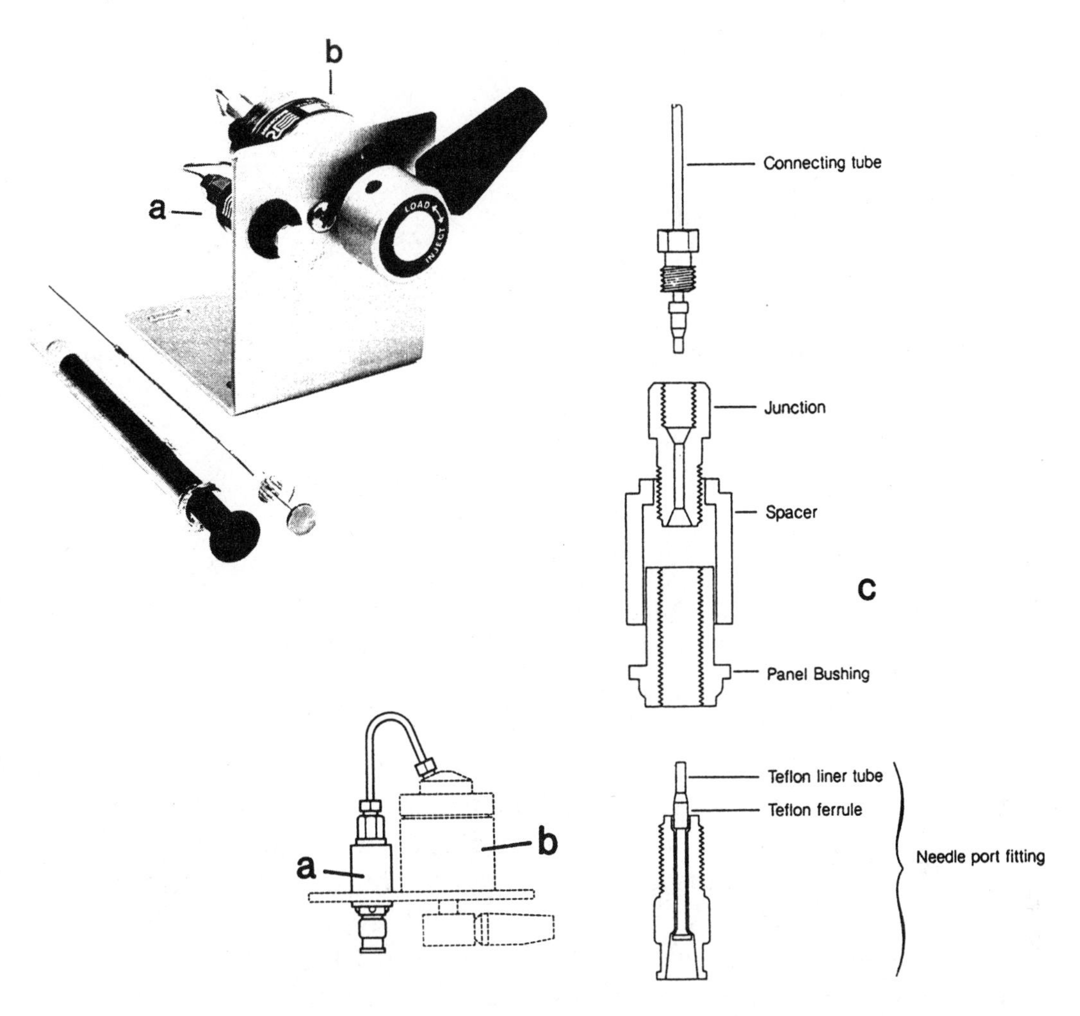


**Fig. 10.3.** External injection-port. (a) external injection-port; (b) injection valve; (c) exploded diagram of (a). Courtesy of Rheodyne Incorporated.

flushed through the loop to assure that the loop is completely full (see "filling characteristics" discussion below). For example, with a 20-µL loop, push 60 µL of sample through the loop with the injection syringe; the excess flows to waste. Now turn the valve (leave the syringe in place) to inject the sample. Remove the syringe after injecting the sample so that the needle is not bent accidentally. Because the loop volume controls the amount of sample injected, the amount of sample in the syringe is not critical. For the best injection precision, however, it is good practice to (over)-

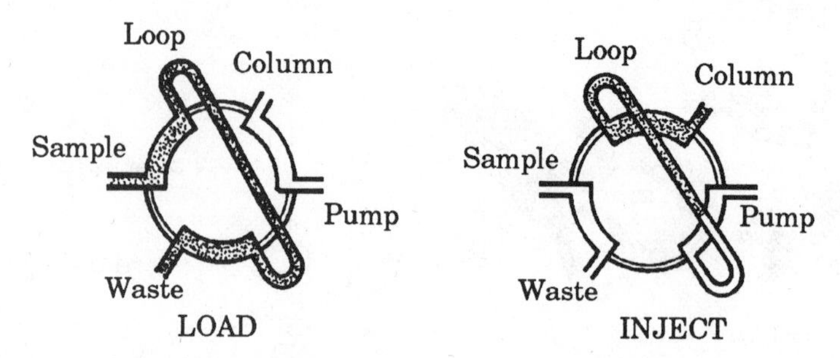


**Fig. 10.4.** Filled-loop injection. Reprinted from ref. (1a) with permission.

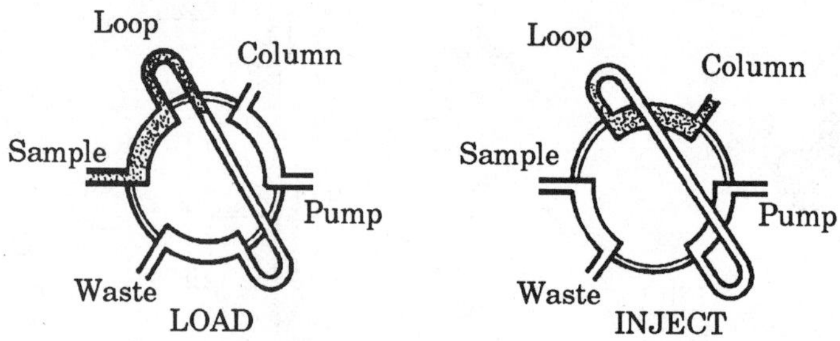


**Fig. 10.5.** Partial-loop injection. Reprinted from ref. (1a) with permission.

fill the loop with the same amount of sample (e.g., ±10%, or 60 ± 6 µL in the present case) each time. To change the sample size with the filled-loop technique, it is necessary to change sample loops.

With the *Partial-Loop* injection technique, a volume of sample less than the loop volume is injected (Fig. 10.5). For example, if a 23-µL injection is desired, push 23 µL of sample into a larger loop (e.g., 50 µL); the rest of the loop remains filled with the previous solvent, usually mobile phase. Now rotate the valve to the inject position, and the sample is placed on the column. Because the injection volume is determined by the syringe, (a) care must be taken when measuring the sample, and (b) sample size should not exceed 50% of the loop volume. The partial-fill method is less precise than the filled-loop technique, but it is much more convenient when the injection volume must be varied, because any volume (up to about 60% of the loop volume) can be injected. The tradeoffs between filled- and partial-loop injection are shown in Table 10.1.

The *Moving-Injection* technique can also be used to precisely inject partial-loop volumes. With this method, the valve is

Table 10.1
Filled- vs Partial-Loop Injection

| Technique | Advantage | Disadvantage |
|---|---|---|
| Partial loop | Flexibility; conserves sample (no sample lost to waste) | Less precision (unless internal standard used) |
| Filled loop | Precision | Less flexibility |

switched momentarily to the inject position and then back to the load position before the loop is fully flushed.[2,3,3a] This technique is especially suited for small (e.g., <2 μL) injections, such as for microbore LC. See refs. (1) and (3) for details on the moving-injection technique.

## *Filling Characteristics*

As a sample is loaded into the loop, it pushes the mobile phase ahead of it out of the loop. In this process, the leading edge of the sample becomes diluted because a fluid traveling through a tube has a parabolic flow profile as a result of *laminar flow.* In laminar flow, the center of the fluid stream travels faster than the part near the tubing walls, as shown in Fig. 10.6. The practical result is that the sample occupies about 2 μL of loop for every 1 μL of sample loaded from the syringe. This means that for filled-loop

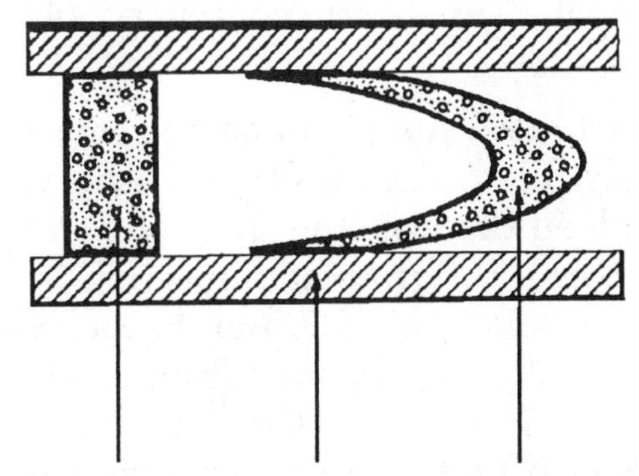


**Fig. 10.6.** Laminar flow profile. Reprinted from ref. (4) with permission.

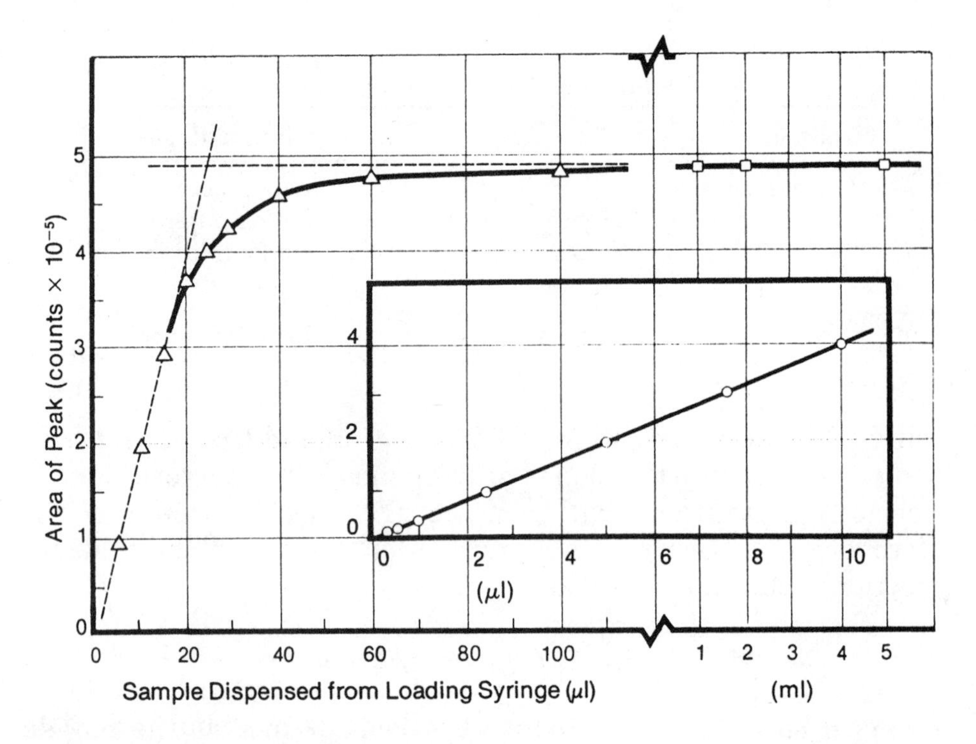


**Fig. 10.7.** Dispensed vs injected volume. Plot of the sample mass injected into the column *vs* the volume of sample dispensed from the loading syringe using model 7125 with 20μl sample loop. Data were obtained using three syringe sizes: 10 μl (0), 100 μl (Δ), and 5 mL ( ❑). The linear regression straight line best fit to the 10 μl syringe data is shown (---). The straight line correlation coefficient is 1.000. Departure from linearity starts around 15 μl, i.e., at about 60% of the actual loop volume. The injector was flushed (INJECT position) with 0.5 mL of solvent after each injection. Reprinted from ref. (4) with permission.

injection, all the diluted sample must be removed before the loop is homogeneously filled. Generally 2–3 loop volumes of sample are sufficient to load the loop and flush out almost all of the previous solvent. When a partial-loop injection is used, the sample size should be kept below about 50% of the loop volume, to prevent the diluted front of the sample band from leaving the loop during injection. Figure 10.7 illustrates the relationship between sample loaded into the loop and sample injected onto the column (4). When less than 10 μL of sample are placed in a 20 μL loop, the curve is linear and when more than 40 μL are used, the injected volume is within 10% of the loop volume (see Fig. 10.7). This is the basis for the recommendation to use (a) less than 50% of the loop volume (partial-loop injection), or (b) more than two times the loop volume (filled-loop injection) for maximum injection precision.

Laminar flow also causes the sample band to be diluted as it is flushed from the loop during injection. It takes 5–10 loop volumes to completely displace the sample from the loop.[1] For this reason, it is wise to leave the injector in the inject position for a sufficient time to be fully flushed. (It should be noted that laminar-flow sample-dilution causes band spreading which is very important in microbore, capillary, and supercritical-fluid chromatography. This can be minimized by using the moving-injection technique. In this case, the injector is not left in the inject position to fully flush the loop.)

An alternative method of partially-filling the loop (the *Leading-Bubble Technique*) allows injection of any fraction of the loop volume with high precision.[5] In this case, the syringe is filled with sample, then drawn back so as to include a small bubble at the syringe tip (e.g., 0.2 μL). When the sample is pushed into the loop, this bubble is injected first and forms a "seal" between the previous loop contents (e.g., mobile phase) and the sample. Because the bubble prevents the sample from mixing with mobile phase, laminar flow conditions do not exist, and samples up to the loop volume can be precisely injected. When the valve is rotated to the inject position, the bubble will go into solution in the mobile phase and not cause any problems. Data to support this technique are given in Table 10.2. If there is a concern about having any amount of air in the system (e.g., with electrochemical detection), the bubble can be pushed out of the loop until it is seen in the waste line (assuming PTFE or polyethylene tubing is used). This wastes less than a microliter of sample, yet removes the bubble from the loop.

## *Enhancing Injection*

For a maximum column-plate-number and minimum loss in sample resolution, the injected sample volume should be less than 20–30% of the volume of the first peak of interest. For practical applications, this means that the sample volume should be kept

Table 10.2
Leading-Bubble Technique for Partially-Filled Loops[a]

| Partial filling of a 40-μl xxternal loop on a 6-port HPLC valve with a 50-μl syringe[b] | |
|---|---|
| Syringe volume, μL | Integrated peak area (x $10^6$) |
| 10 | 0.3340 |
| 15 | 0.5470 |
| 20 | 0.7000 |
| 25 | 0.9031 |
| 30 | 1.0671 |
| 35 | 1.2792 |
| 40 | 1.3571 |

[a]Reprinted from ref. (5) with permission.
[b]Linear regression analysis: μL = −0.127 + (28.443 x peak area); correlation coef. = 0.9962. The valco C6U valve, with 0.016-in. id ports, was connected to a Spectra-Physics 8300 UV detector (254 nm) with 140 ft of 0.010-in. id tubing. The mobile phase was 50% water/50% methanol and was pumped with a Spectra-Physics 8770 pump at a flow rate of 0.95 ml/min; the pressure on the valve was 2000 psig. Data were collected on a Spectra-Physics 4100 computing integrator. The sample was 0.1% toluene in methanol.

as small as possible. With partial-loop injections, the effective injection volume should be minimized by backflushing the loop onto the column. This is important especially when small injections are made from a large-volume loop, as is the case with the Waters U6K injector, which uses a 2-mL sample loop (Fig. 10.13). For the case of a 20-μL injection from a 2-mL loop, the sample would be greatly diluted (because of laminar flow) if it were required to travel through the entire 2-mL loop before reaching the column. This also would result in a considerable delay between the time when the valve is turned and the time when the sample actually reaches the column. However, because the loop of the Waters U6K is backflushed, the 20-μL sample is flushed immediately onto the column—with minimal dilution, and no delay. It is simple to plumb most injection valves so that the loop is backflushed; the valve in Fig. 10.1 is plumbed in this manner. (If the connections at ports 4 and 5 were interchanged, the loop would not be backflushed.) For filled-loop injections, the direction of flow through the loop is not important.

Table 10.3
Tube Volumes for Loop Construction[a]

| id, in.[b] | Volume, μL/cm |
|---|---|
| 0.005 | 0.13 |
| 0.007 | 0.25 |
| 0.010 | 0.51 |
| 0.020 | 2.03 |
| 0.030 | 4.56 |
| 0.040 | 8.11 |
| 0.046 | 10.72 |

[a]To calculate tubing length for a desired volume: (desired volume)/(μL/cm) = length; see text for example.
[b]For metric conversions, see Table 8.1.

## *Loop Design*

*External loops.* Sample loops are constructed typically from 1/16-in. od stainless-steel tubing of various diameters, and are connected to the valve body using standard compression fittings (e.g., Fig. 10.1). Loops that are attached to the outside of the valve in this manner are called *External Loops.* External loops are easily changed, and are available in sizes ranging from 2 μL up to 1 mL (or more). A custom-volume loop can be made from a piece of stainless-steel connecting tubing—just cut the length necessary to give the desired volume. The distance between the two loop ports, and the minimum available id of stainless-steel tubing limits the minimum loop size to 5 μL (Rheodyne) or 2 μL (Valco).

The data in Table 10.3 will help you calculate the length of tubing required for a desired loop volume. For example, to make a 25-μL loop from 0.020-in. id tubing, you need to cut the length to: (25 μL)/(2.03 μL/cm) = 12.3 cm. This does not take into account the contribution of the internal volume of the valve to the injection volume, which can amount to an error of 10% or more for loops less than 10 μL in volume. If you need an accurate loop calibration, use the method described in the following section. Sample-loop ends should be squarely cut. As with other tubing connections, a poor fit between the loop end and valve port will result in added dead volume, which can cause broadened peaks and sample carryover. For convenience and a minimum of loop problems, buy a selection of precut loop sizes.

*Internal loops.* For most applications, external-loop valves are convenient to use. However, some applications, such as microbore, or short, small-particle columns (e.g., <5 cm, 3 μm), require injection volumes smaller than the 2-μL minimum for external-loop valves. For these special applications, a valve using an *Internal Loop* is required.

Internal-loop valves are less prone to blockage, because the loop id is not as small as the 0.005-in. id tubing used for small-volume external loops.

Internal-loop valves use either (a) the volume in the connecting port of the rotor seal, or (b) a loop fastened onto the rotor to contain the sample. The Valco valve (CI4W) uses the volume of the connecting passage in the rotor seal as a sample loop. To change loop volumes, the rotor is removed, rotated to a different volume passage, and replaced; or it is replaced by another rotor with different passage volumes. The Rheodyne valve (7410 or 7413) uses loops of capillary tubing welded onto a loop disc that is mounted inside the valve body. The loop is changed by replacing the loop disc (7410) or rotating the disc so that another loop lines up with the sample fill-port. The SSI injector (3XL) uses three capillary-loops mounted inside the valve. This valve has the unique feature of being able to select any of the three loops by manually rotating the loop disc without valve disassembly. The SSI valve may also be used as an external loop injector. The loop sizes for internal loop valves are listed in Table 10.4.

### *Loop Calibration*

For most procedures, the exact volume of the sample loop is not critical. That is, it doesn't matter if a nominally 20-μL loop is really 19- or 21 μL; the "error" is canceled out in the calculation of the results, because all injections (both calibrators and samples) give the same volume. If you need to know the exact amount of sample (or calibrator) injected, either (a) measure the sample volume with a syringe and use the partial-loop injection method, or (b) use the filled-loop method with a calibrated sample loop. If you need to calibrate a newly-made sample loop, compare the

Table 10.4
Internal-Loop Injection Valves

| Brand | Common loop sizes, μL |
|---|---|
| Valco (CI4W) | 0.06, 0.1, 0.2, 0.5, 1.0 |
| Rheodyne (7410) | 0.5, 1 (std), 2, 5 |
| Rheodyne (7413) | 0.5-1-2<br>0.5-1-5 (std)<br>1-2-5 |
| SSI (3XL) | 0.2-1.0-10 (std)<br>0.2-0.5-1.0<br>2-5-10 |

detector response for a filled-loop injection using the new loop versus the response for an injection using a loop of known volume.

## 10.2. Commercial Examples

The five most common injection-valve brands are discussed below. All of these injectors perform well for routine applications, so our discussion focuses on unique and/or convenient features. Though these standard valves are adequate for most applications, specialty valve components are available from some valve manufacturers. Valve bodies of Hastelloy C, titanium, and tantalum are available for application requirements such as biocompatability or corrosion resistance. Also, different composition rotor seals are available (e.g., Vespel, Valcon H, Roulon) for use when sample and/or mobile-phase compatibility are an issue. Contact the valve manufacturer of your choice for more information about these materials. A summary of injection valve features is given in Table 10.5.

### *Beckman/Altex*

In the Beckman/Altex model 210 injection valve (Fig. 10.8), the ports are configured so that the system connections are on the face of the valve. The loop is attached to the rear of the valve and turns with the rotor when the injector lever is moved. Tubing connections are conveniently accessible, and the large handle makes the valve easy to operate.

Table 10.5
Injection Valve Features

| Brand/ model | Loop sizes[a] | Fill port[b] | Int/Ext loop[b] | Injection style[c] | Other[d] |
|---|---|---|---|---|---|
| Beckman/ Altex 210 | 1–20 µL | I | E | P/F | A, E |
| Rheodyne | | | | | |
| 7125 | 5 µl – 5 mL | I | E | P/F | A, E |
| 7010 | 5 µL – 5 mL | E | E | F | A, E |
| 7410 | 0.5–5 µL[e] | E | I | F | A, E |
| 7413 | 0.5–5 µL[e] | E | I | F | A, E |
| SSI 3XL | 0.2–10 µL[e] | I/E | I/E | P/F | A, E |
| Valco | | | | | |
| C6U, C6W | 10 µL–10 mL | I/E | E | P/F | A, E |
| CI4W | 0.06–1 µL[e] | I/E | I | P/F | A, E |
| Waters U6K | 2 mL | I | E | P | E |

[a] Typical range available, custom sizes available for most. [b] I = internal, E = external. [c] P = partial-loop, F = filled-loop. [d] Options: A = automation, E = event marker. [e] See Table 10.4 for specific loop sizes.

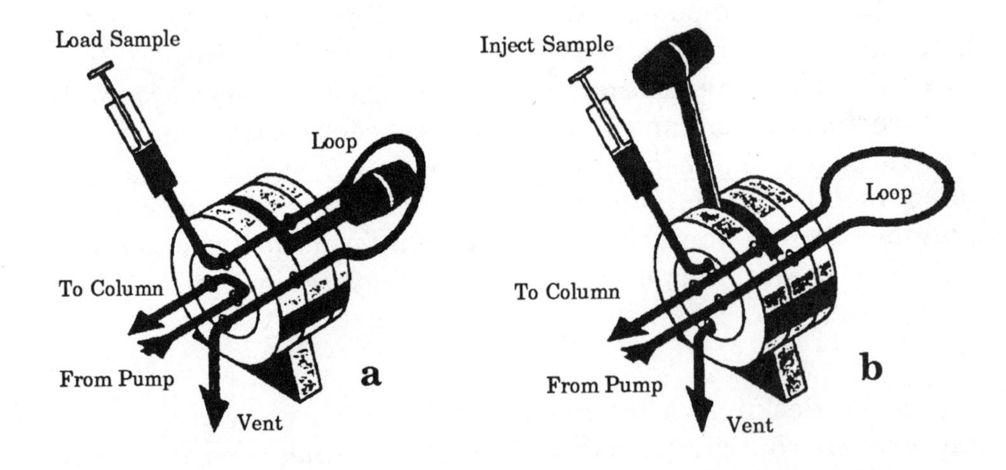


**Fig. 10.8.** Beckman model 210 injection valve. (a) load position; (b) inject position. Courtesy of Beckman/Altex.

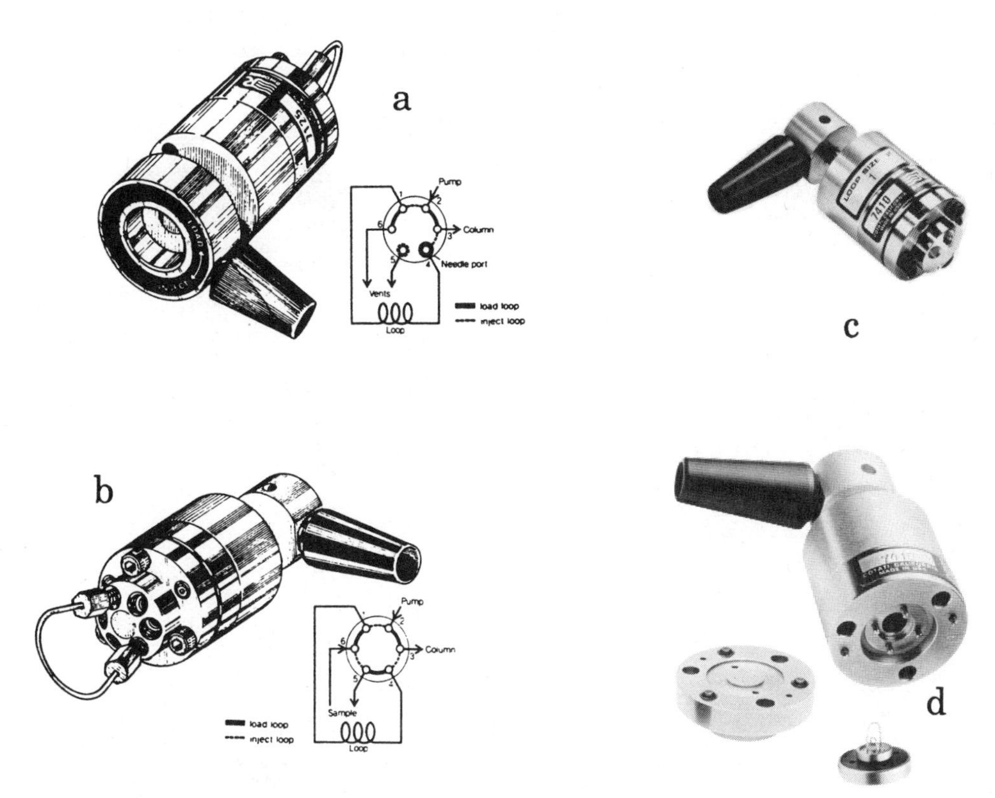


**Fig. 10.9.** Rheodyne injection valves. (a) model 7125; (b) model 7010; (c) model 7410; (d) model 7410 disassembled to show internal loop. Courtesy of Rheodyne Inc.

## *Rheodyne*

Rheodyne offers injection valves with (a) internal or external loops, and (b) internal or external fill-ports; three popular valve configurations are shown in Fig. 10.9. The model 7125 injector uses an internal fill-port mounted in the center of the handle; an external loop and other plumbing connections are made on the rear of the valve. (An exploded diagram of the 7125 is shown in Fig. 10.19.) This configuration is ideal for panel-mounting; the injection port is accessible from the front and all the other connections are out of sight behind the panel. The model 7010 valve is a standard six-port valve (Fig. 10.9b) that requires an external fill-

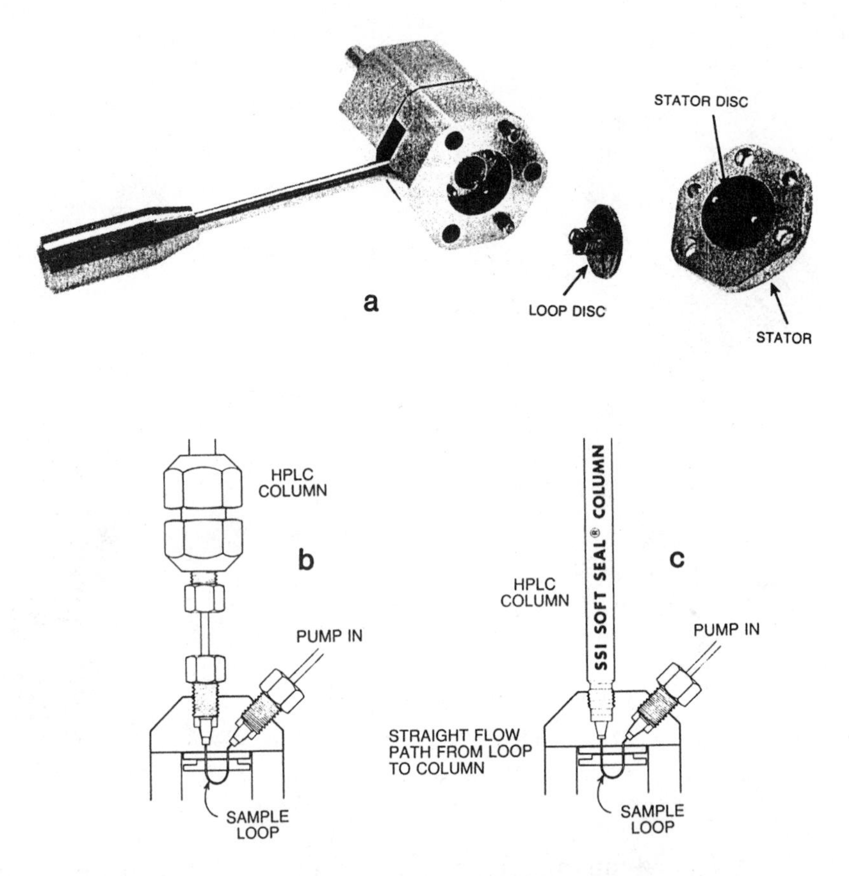


**Fig. 10.10.** SSI model 3XL injection valve. (a) valve disassembled to show loop disc; (b) column connected with tubing; (c) direct column coupling. Courtesy of Scientific Systems, Inc.

port (see Fig. 10.3). The model 7410 valve (Fig. 10.9c) is similar to the 7010, except that it uses an internal loop; the rear plate (stator) must be removed to change sample-loop discs (Fig. 10.9d).

### *SSI*

The SSI model 3XL valve (Fig. 10.10a) is a six-port injector that will operate with either internal- or external loops. In the internal-loop mode, any of the three loops that are mounted on the loop disc can be selected by turning the rotor 60°. This unique feature lets you change loops without disassembling the valve. While the 3XL allows connection of LC columns with transfer

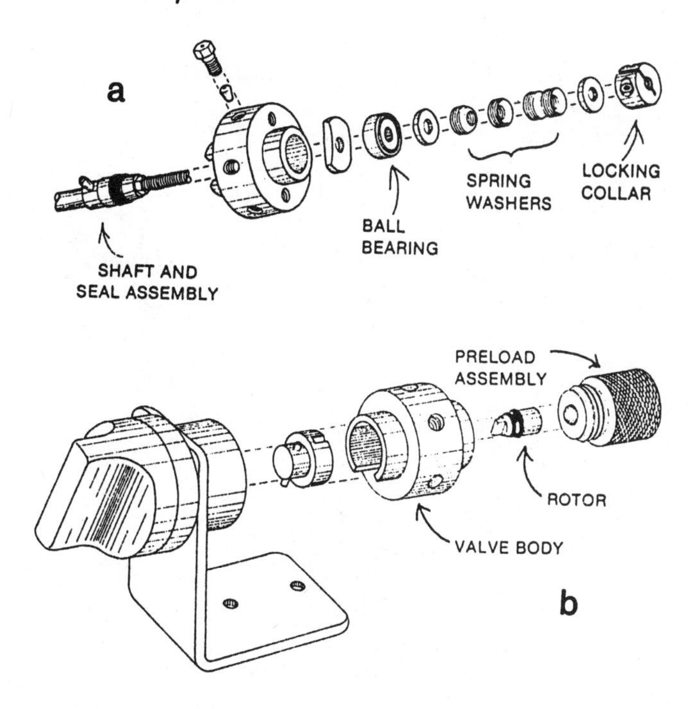


**Fig. 10.11.** Six-port Valco injection valves. Exploded diagrams of (a) model C6U; (b) model C6W. Courtesy of Valco Instruments, Inc.

tubing in the conventional manner (Fig. 10.10b), the valve is also designed for direct connection of SSI's own columns without connecting tubing (Fig. 10.10c).

## *Valco*

Valco offers valves with internal- or external sample-loops; internal- or external fill-ports also can be used. The C6U (Fig. 10.11a) and the C6W (Fig. 10.11b) are six-port external-loop valves, with identical plumbing (see Fig. 10.1). The C6U valve allows adjustment the pressure of the rotor against the valve body by turning the locking collar on the handle shaft. It is convenient and inexpensive to change the disposable rotor-cartridge on the C6W valve. Just remove the pre-tensioned nut (Fig. 10.11b), and replace the old rotor with a new one; no tools are required.

The injection volume of the Valco CI4W internal-loop valve (Fig. 10.12) is determined by the size of the slot in the rotor; no tubing is used. To change the injection volume, the rotor is exchanged for one with a different size loop-slot. The load and

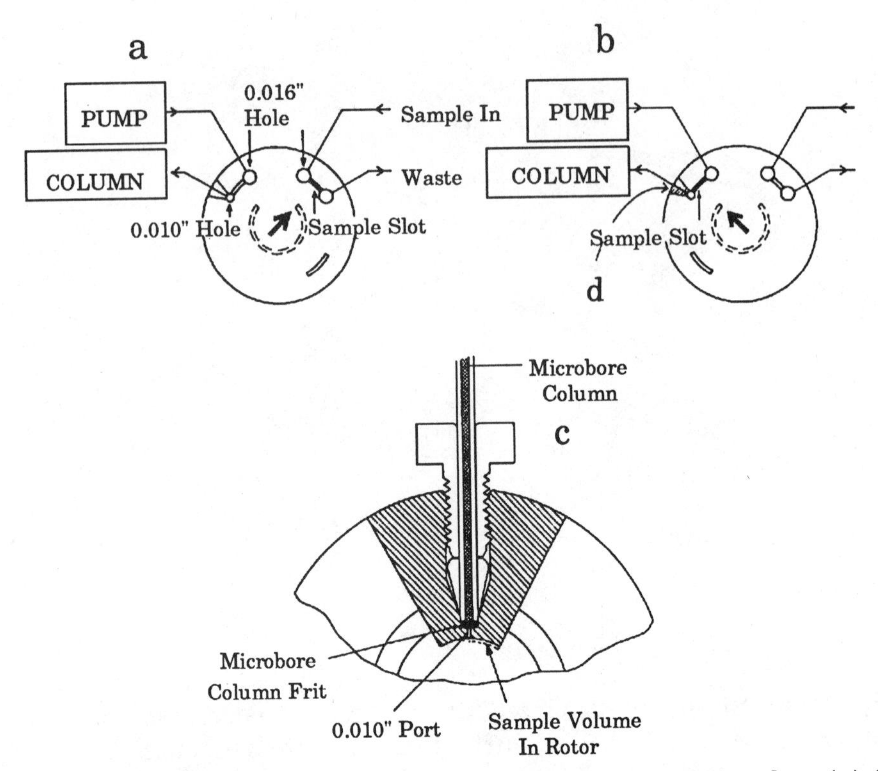


**Fig. 10.12.** Valco model CI4W internal-loop injection valve. (a) flow diagram, load position; (b) inject postion; (c) enlarged view of (d) showing direct connection of microbore column. Courtesy of Valco Instruments, Inc.

inject positions are shown in Figure 10.12a,b. One version of this valve comes with the column port drilled to accept a 1/8-in. compression fitting for direct (zero dead-volume) coupling of microbore columns to the injector (Fig. 10.12c).

### *Waters*

The Waters model U6K injection valve is unique among the valves discussed here because (a) it uses a pressure-bypass circuit to minimize pulses to the column, and (b) it relies only on partial-fill injections. The pressure-bypass feature is discussed in Section 10.3. Figure 10.13 shows a schematic of the flow path during the three stages of valve operation. The U6K uses two valve rotors; the passages at positions 1 and 2 (Fig. 10.13) are mounted on a single rotor; the passages at position 3 operated independently.

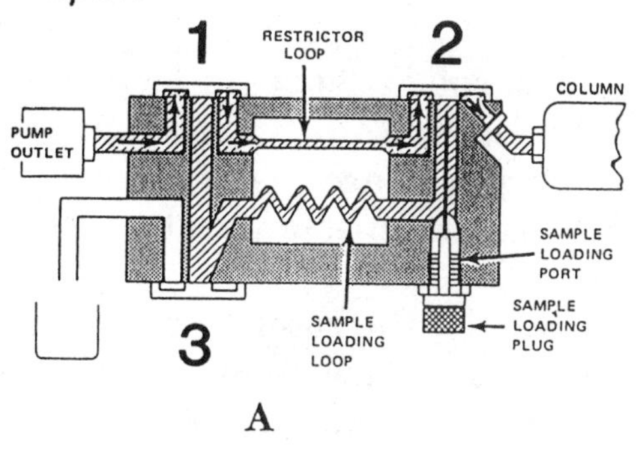


A

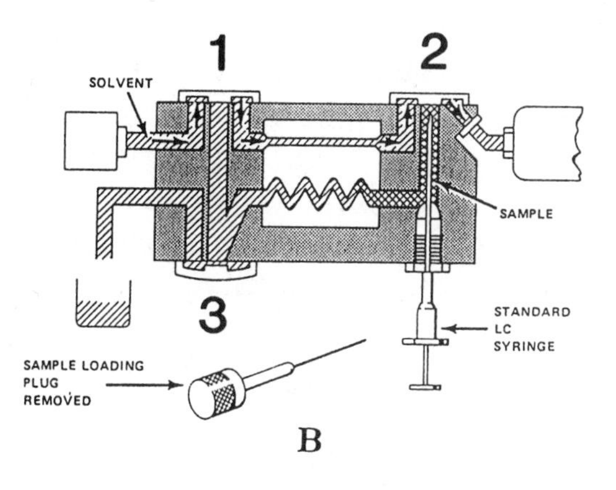


B

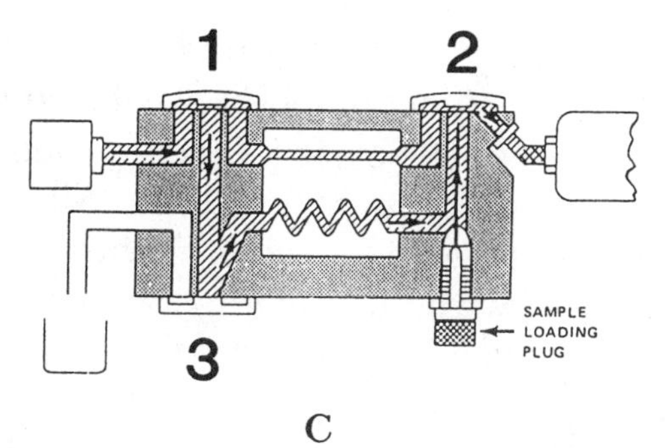


C

**Fig. 10.13.** Waters model U6K injection valve. (a) operate position, pump is diverted to column; (b) load position; (c) inject position. See text for operation. Courtesy of Waters Chromatography Division.

In the Operate mode (Fig. 10.13a), pump flow bypasses the sample loop (via a restrictor loop) and goes directly to the column; the sample loop is blocked by the valve rotor at position 1 and by the sample loading plug.

In the Fill mode (Fig. 10.13b), the pump flow remains unchanged, but the sample loading plug is removed and the valve rotor at position 3 is turned to vent the loop to waste. The sample is now placed in the loop with a syringe (partial-fill method), the sample loading plug is replaced, and the valve at position 3 is turned. This returns the injector to the Operate configuration (Fig. 10.13a)

To Inject the sample, the rotor at postions 1 and 2 is turned and the the sample is backflushed onto the column (Fig. 10.13c). Following injection, the rotor at 1 and 2 is returned to the Operate position for the next injection.

The U6K is a reliable injector, and the partial-fill configuration makes it easy to vary sample size. However, mechanical complexity makes the U6K more expensive, and the loop-filling procedure is more complex than conventional six-port valves. One disadvantage of using a large loop with the U6K (or any injector) is that extra-column band-broadening can occur when viscous samples are injected in the partial-fill mode.

## 10.3. Injector Accessories and Special Options

### *Syringes*

A syringe with a 90° beveled tip is required for LC sample injection (Fig. 10.14a). Injection ports are designed for minimum sample loss, so the tip of the syringe needle should come as close as possible to the seal without touching it. Most valves are constructed so that a squared-off needle tip hits the bottom of the injection port (stator) without touching the valve seal (e.g., Fig. 10.15). However, if a gas chromatography syringe with a sharp, tapered needle (Fig. 10.14b), or a needle with an electrotapered tip (Fig. 10.14c) is used, the tip may touch the seal and scratch it. Some valves are protected from seal damage by using a syringe needle-length matched to the length of the injection port, so that the syringe body hits the needle port, before the tip of the syringe touches the rotor seal. The syringe requirements for several valves are listed in Table 10.6. Syringes with fixed or removable needles can be used, depending on personal preference.

In autosamplers, the needle used to draw sample from the vials generally will be of the style shown in Fig. 10.14b or 10.14d. The beveled point needle (Fig. 10.14b) has a slight hook at the end

Table 10.6
Syringe Requirements for Sample-Injection Valves[a]

| od (in.) | Gage | Length | Tip | Fits |
|---|---|---|---|---|
| 0.028 | 22, 22s | 2 in. | 90° | Rheodyne<br>Altex<br>Valco (port VISF–2)<br>SSI |
| 0.028 | 22, 22s | 0.75 in. | 90° | Valco (port VISF–2) |
| 0.020 | 25s | 1.967 in. | 90° | Waters U6K |

[a]Courtesy of Hamilton Company.

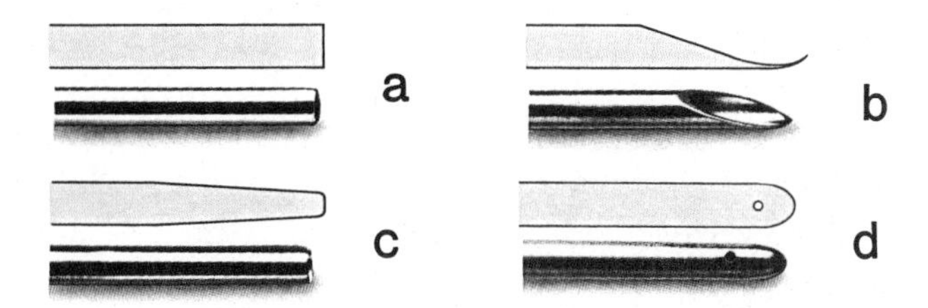


**Fig. 10.14.** Syringe-needle styles: (a) square point, recommended for LC injectors; (b) beveled point, for use with gas chromatography; (c) tapered point; (d) side-port, used with some autosamplers. Courtesy of Hamilton Company.

of the bevel. This hook helps keep pieces of septum from entering the needle when the septum is penetrated. A side-port needle (Fig. 10.14d) is used with some autosamplers. This configuration also prevents pieces of septa from blocking the needle, but it is difficult to clean if blockage does occur. Many autosamplers have specially designed needles, so sample needles are rarely interchangeable between brands. For example, some autosamples use concentric sample and vent needles; others use heavy-duty needles to minimize bending problems.

### *Automatic Actuators*

LC sample-injectors can be controlled remotely using an automatic actuator (e.g., Fig. 10.16). Pneumatic- or electric actuators usually are built into autosamplers, but they can be used alone as well. Pneumatic actuators turn the valve when air pressure

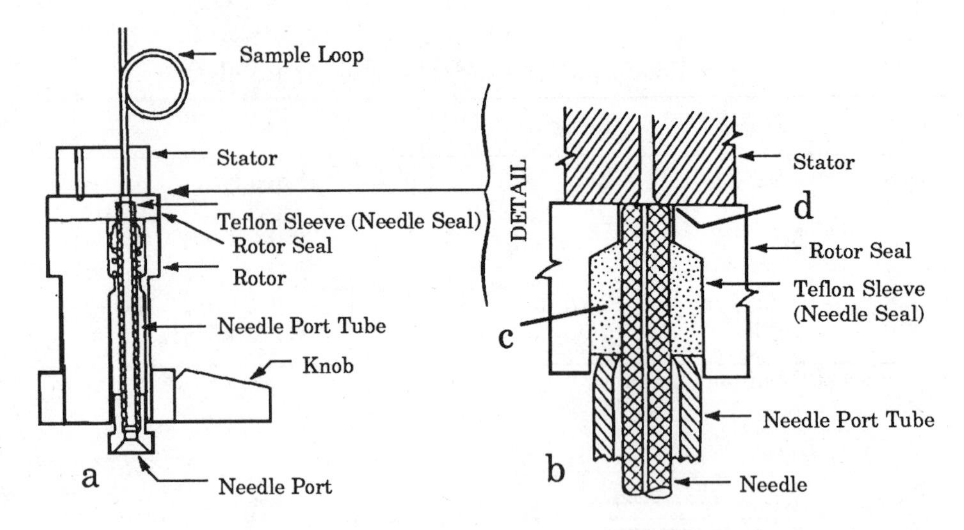


**Fig. 10.15.** Needle port geometry of Rheodyne 7125 injector. (a) cross-section of needle port in valve; (b) detail showing needle seal (c), and showing that the syringe contacts the stator (d), not the rotor seal. Courtesy of Rheodyne Inc.

(40–125 psi) is applied to the actuator. A pair of electrically operated solenoid valves controls the air flow. For faster valve actuation (Section 11.4), helium can be used instead of air. Electric valve-actuators are high-torque motors attached to the valve. Both types of actuators can be controlled by the external-events contact-closures from an LC system controller. The availability of automatic actuators for various valves is shown in Table 10.5.

### *Event Markers*

Event markers sense the valve position and send a pulse or contact-closure when the valve is rotated to the inject position. This signal can be used to mark the point of injection on the chart recorder or to start another event in the LC system (e.g., start gradient formation). Event markers can be mounted on most manual or automatic injectors (e.g., Fig. 10.16b). Table 10.5 indicates which valves can be used with event markers.

### *Pressure Bypass*

As was mentioned in Section 10.3, flow to the column is shut off momentarily when the injector position is intermediate be-

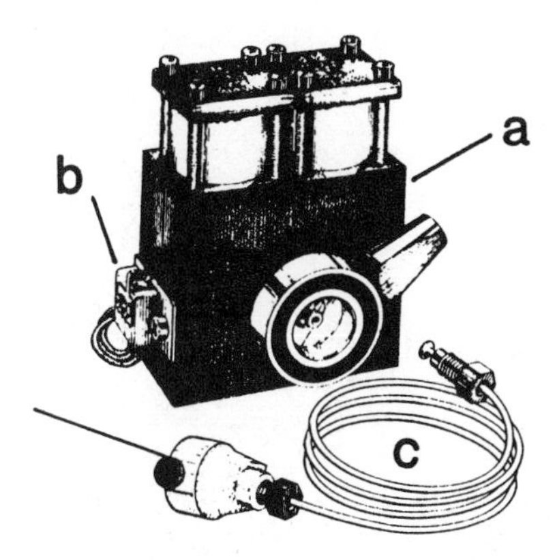


**Fig. 10.16.** Pneumatically operated injection-valve (Rheodyne model 7126). (a) pneumatic actuator; (b) event marker; (c) needle adapter for autosamplers. Courtesy of Rheodyne, Inc.

tween load and inject. This causes a pulse in pressure and flow, which can damage some types of columns. Turning the valve rapidly will minimize this problem. If the pulse needs to be eliminated, a pressure bypass circuit can be added to the valve.

A pressure bypass circuit is an integral part of the Waters U6K valve ("restrictor loop" in Fig. 10.13). In the operate and load positions (Fig. 10.13a,b), flow from the pump bypasses the injector loop and goes directly to the column. In the inject position (Fig. 10.13c), *most* of the flow goes through the loop, but a small portion of the flow still goes through the bypass. Because there always is flow through the bypass, there is (a) no flow stoppage when the valve is rotated, and therefore (b) no pressure buildup that can damage the column.

A pressure bypass can be added to any injector, as shown in Fig. 10.17. The bypass is constructed from two zero-volume tees and a length of 0.007-in. id (or smaller) stainless-steel tubing. For minimium sample dilution during injection, the length of the bypass tubing should be such that less than 5% of the total flow goes through this part of the circuit. Keep the connecting tubing between the tee-fittings and the valve (tubes b and c, Fig. 10.17) as short as possible (e.g., 3 cm). If 0.020-in. id tubing is used on the inlet side of the valve (Fig. 10.17b) and 0.010-in. id tubing on the outlet side (Fig. 10.17c), 3 cm of each will produce about 1/3 the backpressure of 6 cm of 0.007-in. id tubing in the bypass circuit

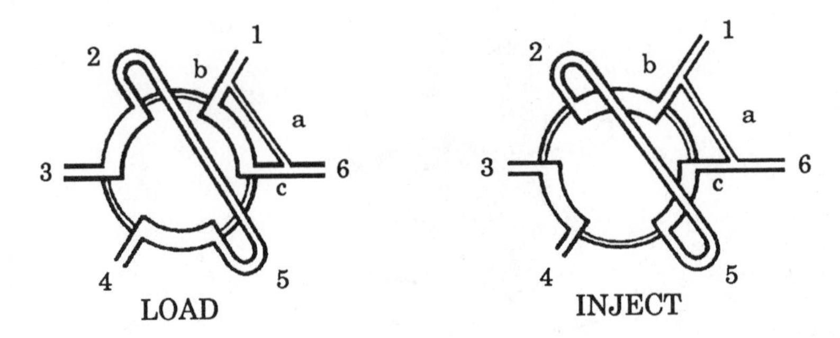


**Fig. 10.17.** Pressure bypass circuit for sample injection valve. Port identification: (1) from pump; (2,5) sample loop; (3) injection port; (4) waste; (6) to column. (a) bypass tube; (b,c) connecting tubing between tee and valve; zero-volume tees at junctions a–b and a–c. See text for operation.

(Fig. 10.17a). To obtain a 1:20 split, this means that about 40 cm of 0.007-in. id tubing are required. Confirm the split ratio by volumetric measurement of the relative flow rates into the outlet tee-fitting from tubes a and c (Fig. 10.17). Just remove the tee and measure the flow rates by collecting the mobile phase in a 10 mL graduated cylinder. At 4 mL/min, tube c should yield 10 mL in about 2.5 min, whereas tube a should give about 1 mL in 5 min. for the present example.

With the bypass installed, the valve works in the same manner as discussed for the Waters U6K above. That is, in the load and inject positions nearly all the flow goes via the conventional flowpath. In the intermediate position, however, the column is protected against flow and pressure pulses because of the bypass.

The major risk of using a pressure bypass is that any blockage of the sample loop (or tubes b or c of Fig. 10.17) can cause most (or all) of the flow to go through the bypass instead of the sample loop; this will reduce the column plate-number and resolution, because the injection is broadened. Blockage of the bypass will not cause an injection problem, but the bypass will no longer protect the column from pulsations.

## *Switching Valves*

Six-port injection valves, as well as valves of other configurations (e.g., 10-port), can be used for switching the flow of mobile phase. Some of these applications include: (a) switching a sample from one column to another, (b) separating the (pure) center of the peak from impurities before and after the peak ("heart-cutting"), (c) selecting a single column from a set of columns, (d) backflushing a column to remove strongly retained contaminants, and (e) selecting a detector from a set of detectors. References (1) and (6) review switching applications.

### *Low-Pressure Valves*

In addition to the high-pressure sample-injection valves discussed in this chapter, there is a large selection of low-pressure valves for LC applications. Many low-pressure valves can be controlled either manually or automatically. Although these valves are limited to about 100 psi, and have poor washout characteristics when compared to high-pressure valves, they are commonly used in two types of LC applications.

First, a low-pressure valve is convenient as a solvent-reservoir selection valve. For example, a single-reservoir isocratic LC can be converted to run several different assays in an unattended manner by adding a low-pressure solvent-selection valve as shown in Fig. 10.18a. After a selected number of assays are run using the mobile phase in reservoir 1, the selection valve can be turned (under timer control) to select the mobile phase from reservoir 2 for another assay. A similar application (Fig. 10.18b) is the automatic flushing of the LC at the end of a series of runs (e.g., to remove buffered mobile phase). A second type of application is a reference-cell flushing circuit for the RI detector, which is described in Chapter 12.

## 10.4. Autosampler Design and Operation

### *Basic Components*

There are five essential parts common to all autosamplers, as shown in Fig. 10.19. These are the injection valve, sample loop, needle, sample vial, and tray. In addition to these five components, commercial autosamplers have a variety of other components and features that make each model unique. In use, the operator places the sample vials in the sample tray and turns on

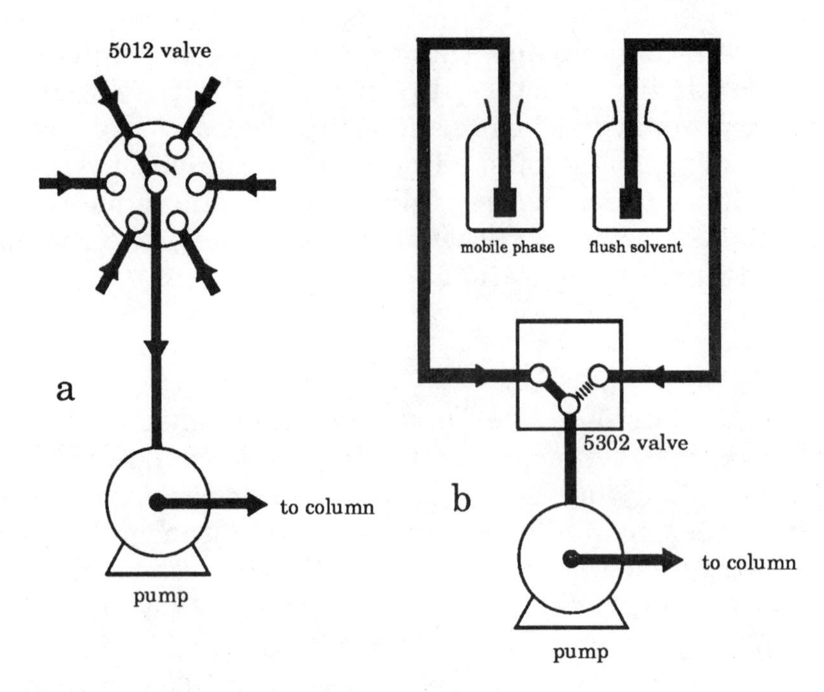


**Fig. 10.18.** Low-pressure valve applications. (a) Six-way selector valve used to select mobile phase from multiple reservoirs; (b) system flushout. Reprinted from ref. (7) with permission.

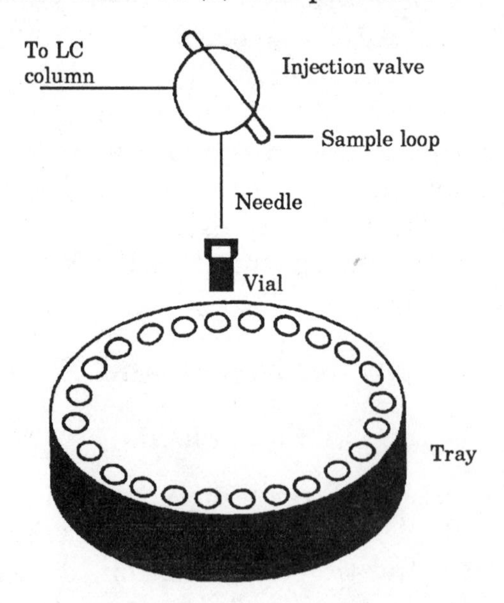


**Fig. 10.19.** Basic components of an autosampler. Reprinted from ref. (8) with permission.

the sampler. To inject a sample, the tray is rotated until the desired vial is under the sample needle, which is lowered (or the vial is raised) until the needle enters the vial. Sample is drawn into a sample loop and the valve is rotated to inject the sample onto the LC column.

The sample *Injection valve* used most commonly in autosamplers is an automated version of the six-port injection valve used for manual injections (e.g., see Section 10.3). The valve is rotated either by air pressure (e.g., Rheodyne, Valco) or by an electric motor (e.g., Valco); an example of an air-operated valve is seen in Fig. 10.16. Thus, the time of injection can be under the control of a timer which signals the valve actuator to rotate the valve to the inject position. The injection valve is plumbed in a similar manner to a manually operated valve (e.g., Fig. 10.1), except that the loop is loaded with an automated syringe or via a displacement mechanism (see discussion of variations below).

The *Sample loops* used with autosamplers are identical to those for manual injectors (see Section 10.1). For maximum precision, the filled-loop injection technique should be used. Autosamplers that can accommodate variable sample sizes use larger loops with backflushing to maintain sample integrity. When partial-loop injection is used, autosamplers can be more precise than manual injection, because the mechanical filling of the loop is more reproducible than a human operator can attain.

The *Sample needle* pierces the septum on the sample vial and provides a path for the sample to be drawn into the sample loop (e.g., Figs. 10.20, 10.21).

A wide variety of *Sample vial* designs are available, but two main types of vials are offered for most brands of autosamplers: standard vials and microvials. Standard vials generally have flat bottoms, hold 1–5 mL of sample, and are used when samples larger than about 50 µL are to be injected. Microvials have tapered bottoms so that injection volumes of as little as 1 µL can be made from sample volumes of 10 µL or more. Vials commonly are made of glass and are discarded after one use, but plastic (e.g., polyethylene) vials are also used. Most vials are sealed with a septum held on by a screw-top or crimped seal. To draw sample into the sample loop, the sample needle is lowered (or the vial is raised) so that it pierces the septum and enters the sample. Sample then is

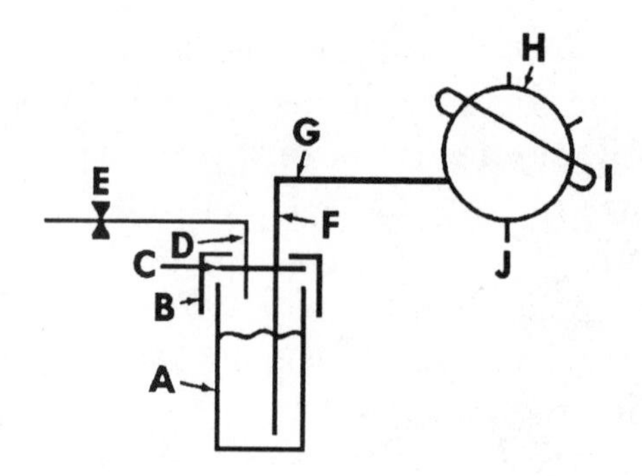


**Fig. 10.20.** Schematic diagram of a displacement autosampler. (A) sample vial; (B) vial cap; (C) septum; (D) pressurizing needle; (E) pressure control valve; (F) sample needle; (G) transfer tubing; (H) injection valve; (I) sample loop; (J) waste line. Reprinted from ref. (9) with permission.

withdrawn through the needle, and the needle is removed in preparation for the next injection.

A *Sample tray* or *Carousel* (Fig. 10.19) is the most common means of holding sample vials until they are used. The tray typically rotates under the needle until the proper sample vial is in position for sampling. In most autosamplers, the sample vials remain in the tray after sampling, but some models drop the vials into a waste bin after the sample is withdrawn.

### *Common Variations*

There are two main categories of autosampler design, based on differences in how the sample is transported from the sample vial to the sample loop. The displacement and syringe designs are shown in Fig. 10.20 and 10.21, respectively. Various models of these two types of samplers usually differ in minor ways (e.g., use similar tray mechanisms or injection valves)

The *Displacement autosampler* relies on air pressure to force the sample from the sample vial, through the needle and transfer tubing, into the sample loop. Because the contents of the sample vial must be pressurized during sampling, displacement autosamplers require that the sample vials be sealed. Two implementations of this design are popular. In one, a pressurizing needle (D in Fig. 10.20) is placed in the sample vial above the surface of

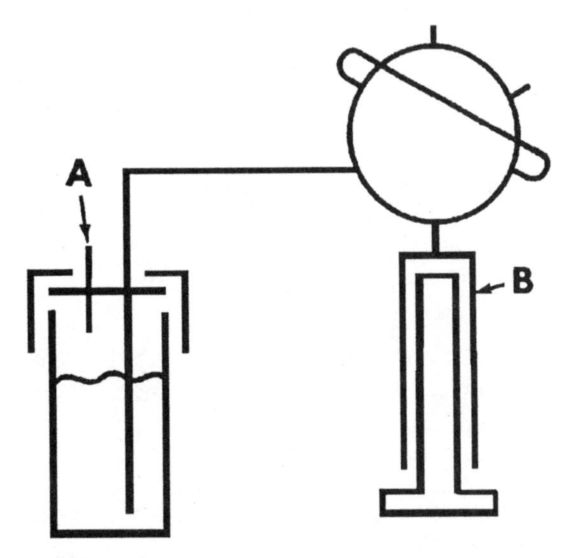


**Fig. 10.21.** Schematic diagram of a syringe autosampler. (A) vent needle (may be concentric to sample needle); (B) sampling syringe; all other parts as in Figure 10.20. Reprinted from ref. (9) with permission.

the sample. This may be a separate needle or a needle concentric to the sample needle. Air pressure is applied to the sample, forcing it through the needle to the loop (the waste port is open to the atmosphere). Clearly, the adjustment of the air-pressure regulator (E in Fig. 10.20) is critical, so that there is enough pressure to displace sufficient sample to adequately flush and fill the loop, yet not so much pressure that the entire sample is forced through the loop to waste. The most common problems with this style of sampler are related to maintaining the proper air pressure. The other style of displacement autosampler uses a one-piece plastic septum and cap that are pressed inside the top of a cylindrical sample vial (Fig. 10.22). To withdraw sample, the needle is lowered until the septum is pierced, then a collar on the needle pushes the cap into the vial (much like a syringe plunger). The increased pressure in the vial displaces sample through the needle to the sample loop. The most common problems with this style of autosampler have to do with the precise positioning of the vial so that the plunger mechanism works properly (rather than crushing the vial).

The *syringe autosampler* uses a motorized syringe to draw sample from the sample vial into the sample loop, as shown in Fig.

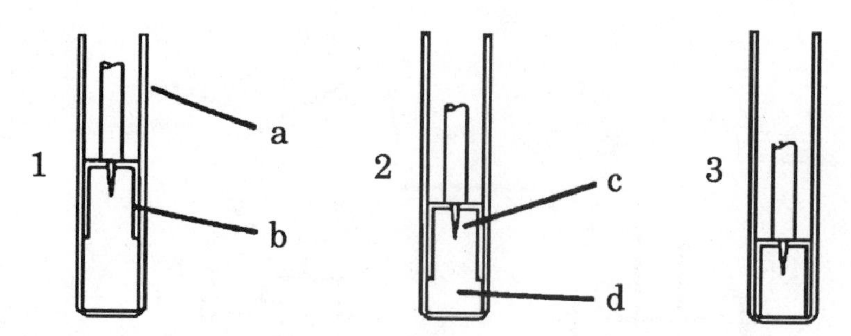


**Fig. 10.22.** Use of a plunger-type vial cap for three consecutive injections, 1–3 (a) glass vial; (b) vial cap/septum; (c) sample needle; (d) sample. Courtesy of Alcott Chromatography.

10.21. The sample needle is lowered (or the vial is raised) to penetrate the vial septum. Simultaneously, a vent needle, either concentric to or separate from the sample needle, penetrates the septum. With the valve in the load position, the syringe plunger (B in Fig. 10.21) is withdrawn, causing sample to flow into the sample loop. The valve is then rotated and the sample is injected. Because the sample is withdrawn by syringe action, it is not necessary to use sealed vials with syringe autosamplers (although it is wise to seal the vials to prevent spillage or evaporation problems). After the injection is complete, the needle is withdrawn from the vial and the syringe is actuated to expel excess sample to waste (either through a diverter valve or the sample needle moves to a waste station). With proper calibration and control, syringe autosamplers can use the partial-loop technique (see Section 10.1) for injection of samples smaller than the loop volume. As was discussed in Chapter 10, for maximum precision the filled-loop injection technique should be used; if partial-loop injection is used, don't use an injection volume greater than 50% of the sample loop.

## 10.5. Problem Prevention

### *Procedures*

There are a number of simple procedures that should be performed in order to increase injector and autosampler reliability and reduce the likelihood of breakdown. It should be clear in the discussion below which procedures apply to injectors and which to autosamplers. As is true for manual injectors, autosamplers work better and last longer if system pressures of ≅ 2500 are not exceeded. Additional procedures may be recommended by the manufacturer—see the operator's manual for details.

*Injection valve.* Autosampler injection-valves are similar or identical to manual injection valves, and should be maintained in a similar manner. Cleanliness and proper adjustment are keys to extended injection-valve life. Follow the manufacturer's instructions when adjusting the tension on the rotor seal—too much pressure on the seal will increase seal wear; too little pressure will make the valve leak-prone. Be sure to flush buffered mobile phases from the valve daily. If you use a precolumn, or suspect another source of particulate contamination upstream from the injector, install an in-line filter just before the valve to trap these particles. Particles can scratch the valve seal and cause cross-port leakage, which means that the valve leaks between adjacent ports. For example, a scratch on the seal between ports 3 and 4 in Fig. 10.1a, would cause the pump to leak into the sample loop in the load position, making accurate sample injection impossible. An injection valve that is properly adjusted and kept clean should last 10,000 cycles (5000 injections) or more before seal replacement is required.

In manual injectors, valve *rotor-seal damage* can also be caused by using an improper type or length of syringe needle. For example, if a gas chromatography syringe (e.g., Fig. 10.14b) is used in a valve with an injection port configured like that shown in Fig. 10.15, the sharp tip of the needle can penetrate too far, damaging the needle and/or valve. Rotor-seal damage by the syringe is less frequent today, because most valves are constructed so that the needle cannot contact the seal. For example, the needle passes through the seal in Fig. 10.15; in Fig. 10.1, the passage between the needle and the rotor seal is 0.006-in. id, so the needle cannot reach the seal.

The Teflon sleeve in the injection port of manual injectors should be adjusted occasionally (e.g., once a month), or whenever any leakage around the needle is noticed. To adjust the needle seal, press firmly on the needle-port tube (e.g., Fig. 10.15) or tighten the injection-port nut (e.g., Fig. 10.2c); this compresses the Teflon sleeve so that it seals snugly around the syringe needle.

*Sample preparation.* Because autosamplers are run unattended, it is essential that samples be *free of particulates* or other material that might block the sample needle, connecting tubing, or injector. Similar problems can occur when samples are injected manually. To see whether a sample should be filtered, hold the sample vial to a light; if any particulate matter, turbidity, or opalescence is seen, filter the sample through an 0.5 μm filter. It generally is not cost-effective to filter every sample, because of the high cost of filters; however every sample should be checked.

Other sample preparation steps for automatic sample injection should be the same as are used for manual injections. Keep in mind that the cleaner the samples are, the less chance there is of an autosampler malfunction because of the sample. (See Section 17.2 for additional information on sample preparation.)

It is possible that *sample matrix effects* and *storage conditions* can cause problems with assay precision. One worker[10] reported the following scenario: Plasma samples were diluted with saline, placed in sample vials, and frozen. Later, the samples were placed on an autosampler tray, thawed and injected. Following the first injection from each vial, subsequent injections showed reduced peak heights, yet manual injections of previously frozen samples gave normal results. It was discovered that the samples froze from the top down, forcing a concentration of the salts and analytes in the bottom of the vial. When the vials were thawed, the density and concentration gradient persisted because the vials were narrow and had not been agitated. The sample needle withdrew samples from the bottom of the vial, so the first injection was at the highest concentration, with subequent injections (from the same vial) at successively lower concentrations. When manual injection was used, sufficient agitation took place to remix the sample. This example should remind us to thoroughly mix samples before they are injected.

In order to minimize band broadening before the sample reaches the column, make the *final sample diluent* no more than half the mobile-phase strength (e.g., if the mobile phase is 50% acetonitrile/water, use 25% acetonitrile/water to make up the sample). Thus, samples in an aqueous matrix may be injected directly, but sample extracted into organic solvents will require dilution. Using a weak injection solvent will allow the sample to be concentrated at the head of the column during injection, reversing the effects of band broadening in the connecting tubing.

*Sample vials.* Sample vials should be clean and free of any materials that might dissolve and thus contaminate the sample. For this reason, most labs use glass vials and discard them after one use. For many samples, plastic (e.g., polyethylene) sample vials are also satisfactory. With some samplers, you can keep the consumables costs down by using (inexpensive) disposable plastic centrifuge tubes.

The vials also should be *sealed,* so that a change in concentration or loss of sample does not occur. Be sure that the composition of the vial septum is such that it does not dissolve or bleed contaminants into the sample solution. Teflon-film septa are the most inert for this purpose, but they do not seal as well as thicker silicone-based septa, and the needle hole does not close well after an injection has been made. Most workers prefer septa that have Teflon film laminated to a silicone disk. When the septum is installed properly (with the Teflon side toward the sample), there is no problem with sample contamination, and the silicone disk provides a seal that may be punctured several times without leaking. Less expensive polymeric septa without the Teflon surface also are available, but these should be tested to be sure they do not contain extractable material that can contaminate the sample. To test this, soak several septa overnight in a vial of sample diluent, then inject some of the diluent. If no peaks show up when a sensitive detector setting is used, you should be able to use these septa with your samples.

Be sure to use vials that are the proper *size and design* for your autosampler. You can be sure of this by buying vials from the autosampler manufacturer. If you buy vials from a generic supply house (e.g., Sun Brokers), be sure to specify the brand and model of autosampler that is to be used; then you should have no problems.

It is wise to individually label all sample vials. Some systems (e.g., Spectra-Physics) use bar-code stickers that are applied to the vials for sample identification. Most other units rely on the tray position for sample identity. Though it is not necessary to label the vials if the tray position is recorded carefully, it is a good precaution to use labels. This facilitates finding a sample that must be rerun, when there is uncertainty about the results or some question is raised concerning sample identity.

*Sample tray.* The sample tray rotates in order to bring the vial to the inject position. The *vial-locating mechanism* typically is a mechanical or optical sensor. *Mechanical switches* require little or no preventive maintence other than to avoid spilling samples or other solutions on them. *Optical sensors* can get coated with materials deposited from the laboratory atmosphere, or from spilled samples. It is wise to clean the optical sensors and the reflective strips on the sample trays regularly (e.g., once or twice a month) to ensure that the optical sensing system works properly. The trays can be cleaned with soap and water; the optical sensor can be cleaned with a cotton swab moistened with alcohol (see the operator's manual for specific recommendations). It is also wise to check the tray alignment once or twice a month to make sure that the tray rotation stops when the vial is centered under the sample needle. The operator's manual will give instructions on how to realign the tray for your model of sampler.

*Sample needle.* The sample needle is the most failure-prone part in the autosampler, so it is imperative that the correct *size and style* of needle is used. Most samplers use either side-port or beveled-tip needles (see Fig. 10.14 and discussion) to minimize problems with blockage by septum fragments. If a straight or angled tip is used, it is possible to punch a piece of material out of the septum and have it become lodged inside the needle. The best way to prevent needle problems is to buy needles from the autosampler manufacturer or a third-party supplier for that brand and model. Be sure to keep several spare needles on hand. Occasionally needles will become bent from misalignment with the sample vial. Once a needle is *bent,* it is best to replace it with a new one rather than to straighten it, because the bend weakens the needle; it probably will soon bend again in the same place. If there is a *vertical needle-adjustment* (or vial-lift adjustment) in your autosampler, it is wise to check the adjustment occasionally. The needle tip should be close to, but not touching the bottom of the vial for most units. When microvials are used with spring-loaded inserts, the needle can be adjusted to touch the bottom of the vial.

*Transfer tubing.* The tubing used to connect the sample

needle and injector (if the needle is not connected directly to the injector) usually is a specialty part from the autosampler manufacturer, so there is little chance of using the wrong size or length of tubing. The tubing connecting the autosampler to the head of the LC column may be several feet long in many setups. This tubing should be kept short and it should be no larger than 0.010-in. id in order to minimize band broadening before the sample reaches the column. Though smaller id tubing can further reduce band broadening, it is more prone to blockage than 0.010-in. id tubing; therefore it should be avoided, unless there is a compelling reason to use it. (See Chapter 8 for more information about the proper use of connecting tubing.)

*Air supply.* When compressed air (or nitrogen) is used with displacement autosamplers to transfer the sample from the sample vial to the sample loop, it is important that the *pressure regulator* be in proper adjustment. The pressure regulator (e.g., E in Fig. 10.20) is mounted in the autosampler or on the supply tank (or in some cases a separate regulator will be mounted on the supply and the autosampler). It is important to use the proper air pressure; too little pressure results in (a) insufficient loop flushing, (b) small sample sizes, or (c) no sample reaching the sample loop, depending on how low the pressure is. Excessive pressure either (a) wastes sample by excessive loop washing, or (b) flushes all the sample through the loop so that the loop is partially or completely filled with air. You may have to readjust the regulator when samples of different viscosity are used. Also, when a two-stage regulator is used (e.g., with a compressed gas cylinder), the ability of the second stage to control the pressure may be inadequate when the pressure to the primary regulator is low (i.e., when the cylinder is almost empty). Therefore, it is wise to replace the supply cylinder before it is completely empty. If you are using helium rather than air to drive a pneumatic valve, you may use this same supply to displace sample. When helium is used for sample displacement, it may be necessary to adjust the pressure so that the proper amount of sample is transferred to the sample loop. Your operator's manual will give specific recommendations for regulator settings to use with displacement autosamplers.

*Washing.* In order to prevent buildup of buffer salts and other residues in the autosampler, you should perform a *system flush* at the end of each day's use. The following procedure can be modified for use with manual injectors. Generally a flush with salt-free sample diluent, water, or mobile phase will be sufficient to clean the system. This is done by performing several injection cycles (or wash cycles) with the wash solution in a sample vial. Be sure to flush the system in the load as well as the inject position. The outside of the sample needle should be wiped with a lint-free wiper (e.g., Kimwipe) moistened with the appropriate solvent to remove any residues from the outside of the needle. Some systems (e.g., Waters WISP) clean the outside of the needle after each injection. Some samples have an end-of-day wash cycle that performs all these steps automatically.

It may or may not be necessary to perform a wash cycle between each sample. Two types of washing are common. First, residual sample needs to be removed from the sample loop and transfer tubing before the next sample is injected. Displacement autosamplers accomplish this by using the first part of the next sample as a wash solution to flush sample residue from the system. Syringe autosamplers force the remaining sample to waste, when the syringe is pushed in following injection. In some cases this is not sufficient to prevent carryover (see carryover discussion in next section), so additional washing is necessary. With syringe autosamplers, you can often program an extra wash cycle between samples. Displacement autosamplers may or may not have the ability for intersample washing. In either case, you can provide a wash cycle between samples by placing one or more vials of wash solution (e.g., sample diluent) between adjacent vials of sample.

*Calibration.* The best way to avoid problems with the autosampler during a series of runs is to make sure that it is working properly before you start running samples. Most labs use a calibration protocol for each method in order to be sure that the assay procedure and LC system are working properly. If the results fail this test, the problem must be isolated to the autosampler or elsewhere in the system.

If you do not have a system-verification protocol, you may want to use the one given in Table 10.7.[11] Two standards (different concentrations) are prepared and run. The first injection is ignored, because it is the one most likely to be affected by startup

Table 10.7
Injection Sequence for Startup Calibrators
Used to Check Autosampler Performance[11]

| | |
|---|---|
| 1. Standard A | 8. Standard B |
| 2. Standard A | 9. Resolution test |
| 3. Standard B | 10. Samples |
| 4. Resolution test | 11. Standard A |
| 5. Samples | 12. Samples |
| 6. Standard A | 13. Standard B |
| 7. Samples | |

problems. To test for precision, the responses for the standards should agree within ±1.5% (after they are corrected for concentration differences). Then one standard is used to determine the contents of the other standard (treated as an unknown). The results of this check should yield a value equal to the known content. The standards should be prepared in a blank sample matrix, so that the normal sample background and other matrix effects will be present. You can use a standard spiked with another test compound to test the resolution of the system. If these standards are run at the beginning of a series of samples, then scattered throughout the samples during the day's run, you can use these to confirm the reproducibility of the assay.

## Injection-Valve Disassembly and Seal Replacement

Valve disassembly should be performed only when cleaning blocked ports or replacing worn parts (some internal-loop valves also must be disassembled to change loops). Problems can arise from improper adjustment or inadvertent contamination by particulates, if the valve is unnecessarily disassembled (and reassembled). The following discussion also applies to the injection valves in autosamplers.

*First,* be sure that you have (a) a new rotor seal, (b) an exploded diagram of the valve, and (c) the manufacturer's rebuilding instructions. Figure 10.11 shows exploded diagrams of two popular Valco valves; Fig. 10.23 is an exploded diagram of a Rheodyne valve. You can see from these figures that the valve assembly can be very simple (e.g., Fig. 10.11b) or more complex (e.g., Figs. 10.11a, 10.23).

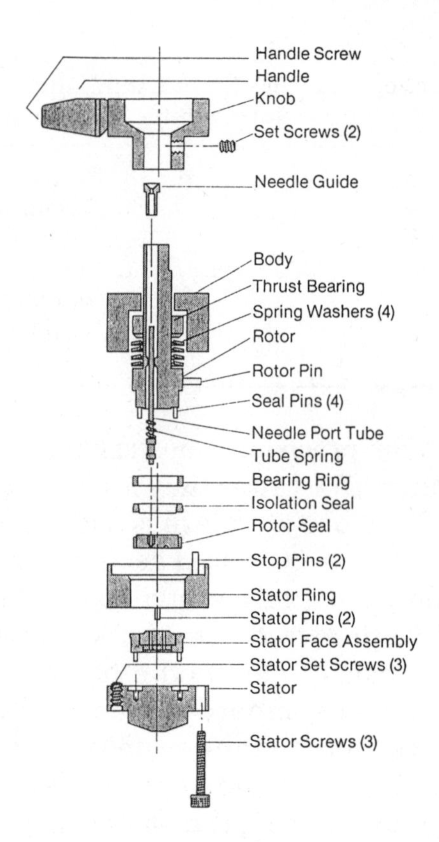


**Fig. 10.23.** Exploded diagram of Rheodyne model 7125 injection valve. Courtesy of Rheodyne Inc.

*Second,* disassemble the valve. Match the parts with the exploded diagram as you take the valve apart; lay them out in order on a paper towel if you are unsure of the reassembly order.

*Third,* clean all parts. Place them in a beaker filled with methanol or a mild laboratory detergent solution (e.g., Micro-cleaner) and sonicate for several minutes. Do not clean in solutions with a pH above 9, because the rotor may contain Vespel, a polyimide that is not stable above pH 10. Rinse the parts in water, then in methanol, and allow them to air dry. Inspect the sealing surfaces (with a hand lens); replace damaged parts, or return the valve to the factory for reconditioning if the sealing surfaces are scratched.

*Fourth,* reassemble the valve, replacing any worn parts (e.g., rotor seal). Follow the manufacturer's instructions, carefully paying attention to (a) orientation of the rotor seal, and (b) adjustment of the seal tension. If the seal is not oriented properly, the holes or grooves will not line up with the valve ports, and the valve will not work. If the seal is not adjusted properly, the valve can (a) leak, (b) be hard to turn, or (c) wear too quickly.

### *Spares*

Suggested spares for injection valves are listed in Table 10.8. A spare rotor seal or a spare injection valve should be stocked for emergency use. Though a valve can last 10,000 cycles or more between seal replacements, it can fail in less than 500 injections if it is contaminated with particulates. In most cases, a damaged valve can be returned to the manufacturer for reconditioning. Keep a variety of sample-loop sizes on hand if you expect to need different injection volumes. Make sure that the spare fittings are the same brand as the valve. A spare injection syringe is essential; syringes can get blocked, bent, or broken without warning.

In order to keep autosampler downtime to a minimum, you should stock the spare parts listed in Table 10.8. Sample vials are the primary consumable item, so be sure to stock plenty of these. In a pinch, you can wash the vials and reuse them, but this is inconvenient. If you use specialty vials, you may need spare microvial springs or other items. When screw-cap vials are used, the caps can be reused, but crimped caps and plunger caps (e.g., Micromeritics) are used only once, so keep an ample supply of caps on hand. A new septum is used each time.

## 10.5. Injection-Valve Problems and Solutions

Table 10.9 summarizes the injector problems that are discussed below and in the preceeding text.

### *Fittings Problems*

Fittings problems at the injection valve show the same symptoms of fittings problems elsewhere in the system: leakage and/or band broadening. The causes of these problems are also the same (e.g., fitting damage, mismatch, or improper assembly). (See Chapter 9 for a detailed discussion of fittings problems.)

Table 10.8
Injector and Autosampler Spares

| |
|---|
| *Injector spares* |
| Rotor seal (or spare valve) |
| Loops (e.g., 10-, 20-, 50-, 100-µL) |
| Nuts and ferrules (e.g., 5 of each, same brand as valve) |
| Syringe |
| *Autosampler spares* |
| Vials, septa, and caps |
| Syringe (syringe autosamplers only) |
| Fuse |
| Injector spares listed above |

## *Blockages*

Blockage of the valve can cause different symptoms, ranging from high backpressure to difficulty in filling the loop. Isolate the blockage in the injector, using a systematic pressure isolation procedure, loosening one fitting at a time and observing the change in pressure. Use Table 10.9 to help isolate the blocked port or loop; note that a blocked or misaligned rotor seal can give the same symptoms as a blocked loop or port.

For example, if the pressure is high in both the load and inject positions, but drops to normal when either end of the sample loop is loosened (inject position), Table 10.9 indicates that a blockage at port 5 (column) is likely.

Blocked ports sometimes can be cleared by backflushing. If this doesn't work, disassemble the valve and sonicate it in a cleaning solution (see above). If the valve is still blocked, return it to the manufacturer for reconditioning. Do not attempt to clean the valve ports with a fine wire; you can permanently damage the valve by scratching the sealing surfaces or by breaking off the cleaning wire in the port.

A blocked loop often can be cleared by backflushing. Replace the loop if this is not successful.

A dirty syringe can masquerade as a valve problem. Before disassembling a valve for cleaning, confirm that the same symptoms appear when a second syringe is used (substitution rule). A

Table 10.9
Valve-Blockage Isolation

| Condition | Position | | Loosen port[a] | | | Blockage location |
|---|---|---|---|---|---|---|
| | Inject | Load | 2 | 3 | 6 | |
| Pressure high | + | – | 0 | –[c] | –[c] | Port 6 |
| | | | 0 | – | + | Port 3 or loop |
| | – | + | 0 | 0 | 0 | Misaligned rotor |
| | + | + | 0 | –[c] | –[c] | Port 5 |
| | | | 0 | +[c] | +[c] | Port 4 |
| Hard to fill loop[b] | + | + | +[c] | 0 | +[d] | Port 1 or syringe |
| | | | +[d] | 0 | –[c] | Port 2 |
| | | | – | 0 | 0 | Waste line |
| | – | + | 0 | 0 | + | Port 6 |
| | | | 0 | + | – | Loop |

[a]Based on Fig. 10.1 configuration; other valve configurations may require some adjustment in table;
[b]i.e., hard to push sample through loop in load position or to waste in the inject position; (+) condition true in this position; (–) condition gone in this position; (0) doesn't apply; [c]test in inject position; [d]test in load position.
Reprinted from ref. (12) with permission.

dirty syringe can be cleaned by (a) flushing thoroughly with methylene chloride or detergent, and/or (b) cleaning the needle with a needle-cleaning wire.

### *Injection-Port Problems*

Leakage around the syringe needle during loop filling, and leakage from the injection port when the syringe is removed, are two problems that appear at the injection port. Leakage during loop filling can be caused by (a) a blockage or resistance downstream (e.g., blocked loop or waste line), (b) improper adjustment of the seal around the syringe (injection port liner), or (c) use of too small a diameter of syringe needle.

Leakage out of the injection port when no syringe is in place can be caused by (a) a siphoning waste line or (b) cross-port leakage. To prevent waste-line siphoning, (a) be sure that the waste line does not dip below the surface of the waste reservoir, (b) place the waste container below the level of the valve, and if necessary

(c) add a restrictor to the waste line (e.g., use 1 m of 0.010-in. id Teflon tubing, or crimp the waste line with a pair of pliers). Cross-port leakage generally will occur in only one position (i.e., either load or inject); rotor-seal replacement is necessary to correct this problem.

### *Sample Carryover*

Sample carryover is indicated when peaks from a previous injection appear following a subsequent blank injection. Sample carryover is caused by poor loop-flushing. This can result from three causes. *First,* if the loop is not left in the inject position long enough, the mobile phase cannot fully wash the sample onto the column, so part of the sample remains in the loop. Be sure to allow the valve to stay in the inject position for at least 10–20 loop-volumes of mobile phase. *Second,* if residual sample is left in the injection port (or syringe), or if sample siphons back from the waste line, contamination of the next injection can result. Rinsing the injection port (and syringe) between each injection may be necessary; flush 5–10 loop-volumes of injection solvent (or mobile phase, or the next sample) through the injection port between injections. If back-siphoning is a problem, add a restrictor, or reposition the line to break the siphon action. *Third,* if the fittings on the loop are not assembled properly, the added dead volume can hold a small amount of sample that can be difficult to flush from the loop. Check the fittings for proper assembly and replace them if necessary.

## 10.6. Autosamper Problems and Solutions

Autosampler problems can be subtle or they can be obvious. When an autosampler problem is suspected, but not certain, the first step is to isolate the problem to the autosampler itself. Do this by (a) making manual injections or (b) substituting a satisfactorily working autosampler. If the performance of the system returns to normal, the problem is with the autosampler. Most of the problems related to sample injection-valves were discussed in Section 10.5 above. The autosampler problems and solutions discussed in this section are summarized in Table 10.11 at the end of this chapter. Autosampler designs vary widely between manufacturers, so specific recommendations for each model are not included here; you should rely on the operator's and service manuals for your unit for further troubleshooting help.

### *Sample-Preparation Problems*

Problems related to sample preparation can result in blocked tubing or sample needles, or in problems in the chromatogram (extra peaks, interferences, etc.). Blockage problems are discussed below (needle and plumbing problems); a complete discussion of sample preparation can be found in Section 17.2.

### *Vial Problems*

When *improper vials* are used, needle damage can result if the vials are shorter than normal. In this case, rather than stopping just off the bottom of the vial, the needle will press into the bottom of the vial and bend. When the vials are deeper than normal, they may be so deep that the needle does not reach the sample, so that no sample or only a partial sample volume is injected. When other dimensions are incorrect, the vial may not fit properly in the sample tray.

If there are *extractable contaminants* in the sample vials, either as a result of improper washing or choosing a plastic vial (when a glass vial should be used), sample contamination can result. This shows up as extra peaks in the chromatogram, even when blank injections are made. The solution is to use only clean glass vials.

Contaminants can also be present when an *improper septum* is used. Silicone and other rubber-like septa sometimes have extractable components, so it is best to use this type of septum with a Teflon disk laminated to the sample side. Be sure to use the septum with the Teflon side toward the sample and avoid splashing the sample solution on the septum when you are loading the sample tray. In some cases, it may be necessary to use Teflon disks alone as septa, but these do not seal as well as the laminated versions.

*Broken vials* can result when the tray or needle alignment is not correct. Autosamplers which use the plunger-cap mechanism (e.g., Alcott) are more susceptible to alignment problems than are other models, but this is less of a problem with the newer versions. Follow the alignment procedures in the operator's manual to correct this problem.

### *Sample-Tray Problems*

Problems related to the sampler tray and its alignment and rotation mechanism usually show up as misaligned vials or the injection of the wrong sample. *Misalignment* of the tray can cause bent sample needles or broken vials. This problem can be solved by mechanical adjustment of the alignment mechanism and/or by cleaning the tray-position sensor (optical or mechanical). A dirty or misaligned sensor also can result in the the sampler "hunting" for a vial, but not finding the proper tray position—the tray will continue to turn or an error message will be displayed. The design of these mechanisms is different with each autosampler, so you are referred to the operator's manual for your unit for specific instructions on tray alignment.

If the *wrong sample* is injected, the first thing to do is to verify that the problem was caused by tray misalignment and not sample mixup. If you labeled the sample vials, it should be easy to check the sequence of injection. If unlabeled vials, or a nonstandard injection sequence was used, it is best to reload the tray with standards in labeled vials. Now run these samples and carefully check to be sure that the proper sample is injected. If the results are OK, it is likely that you mislabeled or misfilled one of the vials. If the results are shifted by the same number of vials (e.g., results for sample 1 report the sample of vial 3), the tray is misaligned. If the results are off in a random manner, consult the operator's manual and/or the manufacturer's service department for help.

### *Needle Problems*

*Blocked needles* generally are the result of particulates in the sample or a piece of septum becoming lodged in the needle. The source of the problem may be the sample, the septum, or the needle type. The symptoms are smaller-than-expected peaks in the chromatogram or no peaks at all. If you are using an internal standard, the ratio of internal standard to known sample level should remain constant. You may be able to clear the blockage by using a syringe-cleaning wire, but most often the needle will require replacement. Prevent this problem by filtering the samples, using a different type of septum, or changing to a different type of needle. These preventive measures are discussed in the preceeding section.

*Bent needles* most commonly occur when the needle hits the vial cap instead of the septum, or when the needle hits the bottom of the sample vial. The first cause can be corrected by adjusting the alignment of the sample tray so that it stops with the vial centered under the needle. The second cause requires adjustment of the needle-lowering (or vial-raising) mechanism so that the needle stops just before it reaches the bottom of the vial. Bent needles rarely can be straightened successfully for long-term use, so it is best to replace bent needles.

*Dull needles* sometimes can cause needle bending or needle blockage because the septum is not punctured cleanly. It is best to replace dull needles.

*Blocked vent needles* will cause a partial vacuum to be formed when sample is withdrawn from the vial. This, in turn, can cause bubbles in the sample loop and lower than expected injection volumes. The vent can become blocked when buffer salts buildup in the needle. Correct this problem by clearing the vent with a needle-cleaning wire or flushing it with solvent or warm water.

### *Injection Valve*

Problems related to the sample injection valve are discussed in Section 10.5 above.

### *Plumbing Problems*

*Blockage* of the connecting tubing in the autosampler is caused either by particulates from the sample or septum fragments, or as a result of using tubing diameters that are too small. If the blockage occurs on the low-pressure side of the injection valve (needle or needle-to-valve transfer tubing), particulate contamination is likely. The symptoms are smaller-than-expected peak heights or a absence of peaks, as discussed above under needle problems. Blockages on the high-pressure side of the autosampler will show elevated system pressures. The location of the blockage can be found by serially loosening the connecting fittings starting at the head of the column and working upstream toward the sampler until a drop in system pressure is observed. The blocked tube is the piece directly downstream from the fitting whose loosening resulted in the drop in pressure. Sometimes the tubing can be cleared by backflushing with the aid of the LC pump, but in most cases, the blocked tube will need to be replaced. To avoid future blockage problems, use a minimum tubing id of 0.010-in. to transfer the sample from the injector to the column. If the blockage is in the injection valve, the information in Table 10.9 can be used to isolate the exact location of the blockage. If blockage is a recurring problem, try filtering all the samples to reduce the chances of getting particulates in the system. Adding a zero-volume in-line filter directly downstream of the injection valve also will help minimize blockage problems.

The use of *large id tubing* in the autosampler can result in excessive extra-column band broadening. The symptoms are broadened peaks in the chromatogram. Tubing through which the sample passes should be no larger than 0.010-in. id in order to minimize extra-column band broadening. (See Sections 8.2 and 8.3 for more information on the proper selection of tubing.)

### *Air Supply*

With *displacement autosamplers,* compressed air is used to force the sample from the vial into the sample loop. If the pressure is too high, excessive sample may be consumed for each injection, and possibly part of the sample will be displaced from the sample loop by air (giving smaller-than-expected peak heights). If the pressure is too low, (a) the sample may never reach the loop (no sample gets injected), or (b) the loop may be inadequately flushed with sample before injection (carryover and/or smaller-than-expected peaks). In both these cases, proper adjustment of the pressure regulator for the air supply should correct the problem.

With *pneumatically actuated valves,* insufficient air pressure will cause the valve to turn too slowly. This will cause a system pressure increase during rotation, sometimes sufficient to shut the pump off because of overpressure. Slow rotation also can result, if the valve rotor is too tight. Increasing the switching pressure (e.g., to 60 psi) will solve this problem in most cases; sometimes it will be necessary to use helium to increase the rotation speed. If the pressure is too high, the valve will turn well, but the increased shock as the valve hits the stop-pin at the end of each cycle may cause increased valve wear; damage to the air switches also can result from excessive pressure. For this reason, pressures above about 100 psi are not recommended with air-actuated valves.

### *Syringe Problems*

A number of symptoms can occur when autosampler syringe is not working properly. If the *wrong size* of syringe is installed, too little or too much sample will be withdrawn from the sample vial. With filled-loop injection, this can show up as (a) a small injection (loop partially filled), (b) sample carryover (insufficient loop flushing), or (c) and injection of air (air drawn into loop). With partial-loop injection, the usual symptom is the wrong amount of sample being injected. In either case, replacement of the wrong syringe with the correct one should correct the problem. If the syringe-drive mechanism is misadjusted, the same symptoms can occur. Adjust the syringe drive (see operator's manual) to fix this fault. If an air leak is present (usually where the syringe connects), air will be drawn in instead of sample and a small or missing sample injection will result. Tighten or replace the fitting, as necessary. A cracked syringe can also leak air.

### *Sample Carryover*

The problem of sample carryover shows up when peaks appear in a blank sample run after a "real" sample has been injected. When carryover occurs, the peaks will be much smaller in the blank injection (typically <5%) than in the original, but the retention times should be the same. Sometimes it is necessary to use a more sensitive detector setting in order to determine the magnitude of carryover. There are three ways to overcome a carryover problem: (a) additional washing, (b) flushing more sample through the sample loop, and (c) adjusting sample order. The problem almost always can be eliminated by adding an extra washing step between samples. This can be programmed into the sample cycle on some autosamplers; other autosamplers require that you insert a wash vial between each sample.

Flushing more sample through the loop (before the sample is captured for injection) is also effective in most cases. For example, if you are drawing 100 µL of sample through a 20-µL loop try flushing the loop with 200 µL of sample. Finally, in some cases you can adjust the order of the samples in the tray so that carryover is not significant. If the carryover is less than 1% (as is the usual case), inject the samples so that the low-concentration samples are first, with higher analyte levels later. In this manner, a small carryover from a low-concentration sample will have an insignificant impact on the accuracy of the higher-concentration sample.

### *Accuracy and Precision Problems*

If results for samples of known content are inaccurate or imprecise, you should isolate the source of the problem. First, manually inject the samples (or use a known good autosampler). If manual injections give improved results, the problem is in the autosampler; if the results are still marginal, there is a problem with sample preparation or elsewhere in the LC system (see Chapter 19 to help isolate the cause).

If the *error is constant* or similar for all samples (eg., +15%), it is likely that there is a calibration problem with the autosampler. If different sample volumes are used for calibrators and unknowns, check to be sure that you have set the sample size correctly and that the correct syringe is in use. In some systems, you must account for the volume of sample contained in the sample needle and tubing between the needle and injector, especially if partial-loop injections are used. Verify that the correct sample loop is mounted on the injector. If you cannot isolate the problem source, it may be necessary to use internal-standard calibration to overcome the problem.

If the *error is random* or erratic, the likely cause is inherent imprecision or an air leak. When partial-loop injections or very small injections (e.g., <5 µL) are made, the sampler will be less precise than where the same unit is operated in the filled-loop mode. If this is the problem, you may be able to improve performance by diluting the samples and injecting larger volumes with the filled-loop method (Section 10.1). If an air leak is present, air instead of sample may be drawn in during sample uptake, which will cause variations in sample size. Carefully tighten all the fittings and replace suspect fittings in order to eliminate air leaks. Improper adjustment of the air pressure in displacement autosamplers can also cause random errors (see previous discussion). Once again, if the problem cannot be eliminated, use internal standardization to compensate for the variations.

### *Error Messages*

Many autosamplers have built-in diagnostics that test the system each time it is started and monitor certain operating parameters during use. When a malfunction is detected, an error message is displayed on the control panel and/or a warning buzzer may sound. When this occurs, consult the operator's manual for specific recommendations.

## 10.6. References

[1]Harvey, M. C. and Stearns, S. D. (1983) "HPLC Sample Injection and Column Switching," in *Liquid Chromatography in Environmental Analysis*, Lawrence, J. D., ed., Humana, Clifton, NJ.

[1a]Snyder, L. R. and Dolan, J. W. (1985) *Getting Started in HLPC, User's Manual,* LC Resources, Lafayette, CA.

[2]Rabel, F. M. (1980) *J. Chromatogr. Sci.* **18,** 394.

[3]Harvey, M. C. and Stearns, S. D. (1983) *J. Chromatogr. Sci.* **21,** 473.

[3a]Harvey, M. C., Stearns, S. D., and Averette, J. (1985) *J. LC, Liq. Chromatogr. HPLC Mag.* **3,** 434.

[4]*Technical Notes 5,* Rheodyne, Dec. 1983

[5]Harvey, M. C. and Stearns, S. D. (1982) *J. Chromatogr. Sci.* **20,** 487.

[6]Majors, R. E. (1984) *LC, Liq. Chromatogr. HPLC Mag.* **2,** 358.

[7]*Technical Notes 3,* Rheodyne, June 1980.

[8]Dolan, J. W. (1984) *LC, Liq. Chromatogr. HPLC Mag.* **2,** 834.

[9]Dolan, J. W. (1987) *LC/GC* **5,** 92.

[10]Bannister, S. J. in reference (8).

[11]Kirschbaum, J., Szyper, M., Perlman, S., Joseph, J., and Adamovics, J., in reference (8).

[12]Dolan, J. W. and Snyder, L. R. (1986) *Troubleshooting HPLC Systems, User's Manual,* LC Resources, Lafayette, CA.

Table 10.10
Summary of Injector-Related Problems and Solutions

| Cause of problem | Symptom | Solution |
|---|---|---|
| *Fittings problems* | | |
| Damage, overtightened, mismatched, loose, | Leaks, band broadening | See Chapter 9 (fittings) |
| *Blocked port* | | Isolate using Table 10.9 |
| Injection port | Hard to push sample into valve in both load and inject position | 1. Clear or repair fill-port<br>2. Clean valve |
| Pump/column port | High pressure in load | 1. Clean or replace valve |
| Loop ports | High pressure in inject, hard to fill in load | 1. Clean or replace valve |
| Waste port | Hard to push sample into valve in both load and inject position | 1. Clean waste line for blockage; clear blockage<br>2. Clean or replace valve |
| Poor loop flushing | Sample carryover | 1. Leave in inject position for 10–20 loop volumes of mobile phase<br>2. Flush inject port with solvent<br>3. Prevent waste-line back-siphoning<br>4. Check/correct loop fitting assembly |
| Blocked pressure bypass | Sample dilution, band broadening, small and irreproducible peaks, pressure surge on valve rotation | 1. Isolate blocked line, clear or replace<br>2. Install in-line filter before valve<br>3. Filter samples |
| Dirty syringe | Hard to fill sample loop | 1. Clean or replace syringe |
| Poor injection technique | Poor reproducibility | 1. Filled-loop: consistently overfill loop with same volume of sample (e.g., 3X) |

*(continued)*

Table 10.10
Summary of Injector-Related Problems and Solutions *(continued)*

| Cause of problem | Symptom | Solution |
|---|---|---|
| *Poor injection technique (continued)* | | 2. Partial-loop: use less than 50% of loop volume; or use leading-bubble technique<br>3. Both methods: leave syringe in injector until turned to inject position; prevent waste-line siphoning |
| Damaged rotor seal | Cross-port leakage, column frit blockage, difficult to load, high system pressure | 1. Disassemle and clean valve; replace seal |
| Wrong loop size | Peaks too small or too large; poor reproducibility (partial-loop injection) | 1. Change to correct loop |

Table 10.11
Autosampler Problems and Solutions

| Cause of problem | Symptom | Solution |
|---|---|---|
| *Sample preparation* | Extra peaks, interferences in chromatogram | 1. Improve sample cleanup or;<br>2. Modify chromatography to move problem peaks |
| | Blocked tubing | 1. Clear blockage<br>2. Filter samples<br>3. Add in-line filters |
| *Vial problems* | | |
| Wrong vial | Bent needles; poor fit in tray; missed injections | 1. Use proper vials |
| Contaminated vials | Extra peak in chromatogram | 1. Better cleaning or don't reuse vials<br>2. Use only glass vials |
| Contaminated septa | Extra peaks in chromatogram | 1. Use septa without extractable contaminants (e.g., use Teflon) |

*(continued)*

Table 10.11
Autosampler Problems and Solutions *(continued)*

| Cause of problem | Symptom | Solution |
|---|---|---|
| Contaminated septa *(continued)* | | 2. Install laminated septa, Teflon side down<br>3. Don't agitate vials when loading, to prevent splashing sample on septum |
| Leaky septa | Changing peak heights | 1. Check for proper seal, adjust cap<br>2. Use laminated septa for better seal |
| Tray misalignment wrong vials | Broken vials | 1. Align tray<br>2. Use only proper vials |
| Sample tray misalignment | Wrong sample injected; bent needles; broken vials; excessive "hunting" for vials | 1. Adjust tray alignment<br>2. Clean and/or adjust tray sensor |
| *Needle problems* | | |
| Blocked needle | Smaller than expected peak size or no peaks | 1. Clear with syringe cleaning wire; or<br>2. Replace needle<br>3. Filter samples; use better sample cleanup<br>4. Change septa and or needle type |
| Bent needles | Bent needles | 1. Replace needle<br>2. Align tray<br>3. Adjust needle |
| Dull needle | Torn septa, blocked needle | 1. Replace needle |
| Blocked vent needle | Lower than expected peaks, reduced precision | 1. Clear needle with wire or washing<br>2. Don't splash sample or buffer on vent |
| Injection valve | Various symptoms | 1. See Table 10.10 |

*(continued)*

Table 10.11
Autosampler Problems and Solutions *(continued)*

| Cause of problem | Symptom | Solution |
|---|---|---|
| Blocked connecting tubing | On low-pressure side of valve, smaller peaks or no peaks; in high-pressure side, elevated system pressure | 1. Clear tube by backflushing 2. Replace tube 3. Filter samples, use other septa and/or use zero-volume inline filter |
| Too large of tubing id | Excessive band broadening | 1. Use short runs of 0.010-in. id tubing |
| Air pressure for sample displacement too low | Excessive sample injected; carryover small peaks | 1. Adjust pressure regulator |
| Injector air pressure too high | Damage to stop pin; air leakage | 1. Adjust pressure regulator regulator |
| Injector air pressure too low | Valve turns slowly; system pressure surge during turning; possible overpressure shutoff on injection | 1. Adjust pressure regulator 2. Use helium to drive valve 3. Install valve pressure bypass (see Sect. 10.3) |
| *Syringe problems* | | |
| Wrong size | Wrong size sample injected; carry-over; air injected | 1. Replace with proper syringe |
| Misadjusted | Wrong size sample injected; carry-over; air injected | 1. Adjust syringe drive |
| Air leak | Air injected; small sample size injected; decreased precision | 1. Tighten fittings 2. Replace cracked syringe |
| *Sample carryover* | Bands from previous sample present in chromatogram | 1. Improve washing between samples (add wash cycle or wash vials) 2. Flush loop more thoroughly before loading 3. Adjust sample order |

*(continued)*

Table 10.11
Autosampler Problems and Solutions *(continued)*

| Cause of problem | Symptom | Solution |
|---|---|---|
| *Calibration problems* | Constant error (bias) in results | 1. Check/adjust sample size or syringe<br>2. Check/replace a sample loop<br>3. Use internal standardization |
| | Increased random error | 1. Check for air leaks; correct<br>2. dilute samples and inject larger volume (for < 5 μL injections)<br>3. Use internal standardization |
| *Diagnostics warnings* | Error mesages or warning buzzers | 1. Consult operator's manual |

## Chapter 11

# COLUMNS

## Introduction

The column plays a central role in obtaining separations by LC, and many separation problems originate with the column. Some of these problems can be avoided or minimized by starting out with the right column. Other problems can be delayed by appropriate preventative maintenance. Finally, some problems occur despite our best efforts—these problems present a wide range of different symptoms that must be diagnosed and then cured.

The origin of a column problem may be mechanical, chromatographic, or chemical in nature, and the effective diagnosis of such problems often requires a thorough understanding of the LC separation process. It is therefore important to review Sections 3.1–3.3, which deal with this aspect of the column. It is also useful to know something about the column itself—and about different types of columns. For this reason the present chapter begins with a detailed review of this area.

In many companies, the cost of individual columns ($300 or more) is perceived as a major part of the expense of running an LC laboratory. There may be considerable apprehension over avoiding damage to the column, plus a desire to use a column for as long as possible. Certainly it is prudent to attempt to get as much use as possible out of each column. However, other costs are involved in attempting to extend column life indefinitely. For this reason, in this chapter we analyze the costs and benefits of column repair and of the continued use of columns that have degraded in performance. As with other areas of troubleshooting, we want to minimize the cost/benefit ratio.

## 11.1. Column Types and Column Hardware

Columns are available in a large number of configurations, and are sold by about a hundred different suppliers. It is therefore useful to start by reviewing these column differences and their significance to the separation:

1. *Columns for Different LC Methods.* Many different kinds of columns are available for each of the commonly used LC methods (Sect. 3.5). Some of these are reviewed in references (1–3). The choice of a given LC method usually is determined by the type of sample; e.g., reference (1).
2. *Columns with Different Stationary Phases.* Reversed-phase and polar-bonded-phase packings have a variety of organic coatings to choose from; e.g., alkyl phases (C8, C18), cyano, phenyl. The choice of phase is sometimes dictated by the need for maximizing sample resolution, but certain phases (e.g., C8, C18) apply to a much broader range of practical applications.
3. *The Chemical Composition of the Packing Matrix.* Most packings in use today are made from porous silica particles. However, certain columns are made from nonsilica matrices; e.g., polystyrene, polyethylene oxide.
4. *Columns from Different Suppliers.* Packings of the same nominal description (e.g., C8-silica, 10-nm pores, 5-μm particles) can give different separations, depending on the manufacturing process. Similar differences in the column sometimes are seen for different lots from the same manufacturer. These differences are caused, in part, by differences in the starting silica (surface area, minor impuri-

ties, special treatment), the bonding chemistry (mono-, chloro- vs trichlorosilane reaction reagents), and proprietary techniques (pretreatment and bonding processes, packing techniques, etc.). Because of such differences, only columns from the same lot (and manufacturer) can be assumed to be strictly identical.

5. *Matrix Characteristics.* Individual packings can differ, depending on the starting matrix: particle size, pore size, surface area, irregular vs spherical shape, etc. In practical terms, this can affect column plate-number, selectivity, sample capacity, column backpressure, and other operational parameters.
6. *Column Configuration.* The column itself can vary in its dimensions (length and internal diameter), and in the column hardware (normal columns, cartridge columns, radial-compression columns, etc.).

### *Columns for Different LC Methods*

There are five widely-used LC methods, as summarized in Section 3.5. A given type of column usually is designed for one of these LC methods; e.g., a C18 column for reversed-phase separation. In addition, specialty columns for separating certain kinds of compounds are also available; chiral columns for separating mixtures of optical isomers (enantiomers), protein columns for separating protein samples, etc. The choice of LC method—which determines the columns that can be used with that method—usually is not determined by the pros and cons of different columns. Rather the LC method is selected on the basis of ease of separation, compatibility with the sample, etc. Good reviews of the various column types available are given in refs. (1–3); some representative column types are summarized in Table 11.1, along with additional information.

### *Columns with Different Stationary Phases*

The kind of stationary phase (see Table 11.1) used for a given LC separation usually is not selected to minimize column problems—but rather to optimize sample resolution [e.g., as in ref.

Table 11.1
Column Packings for Different LC Methods[a]

| LC method | Column packing | Comments |
|---|---|---|
| Reversed-phase and ion-pair LC | C18, octadecyl | Widely available, more rugged packing; accounts for about half of all packings |
| | C8, octyl | Similar to C18 packings |
| | C1, trimethylsilyl, TMS | More susceptible to loss of bonded phase, shorter column life |
| | Phenyl, diphenyl | Specialty column |
| | Perfluorodecyl | Specialty column, selective for flouro-substituted samples |
| | Cyano | Can be used in either reversed- or normal-phase LC |
| | Polysytrene[b] | Stable for mobile phases with 1 < pH < 13; claimed to give longer column life than silica-based packings |
| | Amino | Used mainly for carbohydrate samples, less stable |
| Normal phase | Silica[c] | More rugged, less expensive; less convenient except for preparative LC |
| | Cyano | Increasingly used for normal-phase LC; rugged packing |
| | Diol | Similar to cyano column |
| | Amino | Less stable |
| Size-exclusion | Silica[c] | More stable, less convenient (sample absorption, Section 14.7) |
| | Diol-silica | Less stable, most convenient for aqueous SEC (gel filtration) |
| | Polystyrene[b] | Less stable, most convenient for organic SEC (gel permeation) |
| Ion exchange | Polystyrene[b] | Less efficient (lower N-values), more stable, more reproducible |
| | Bond-phase ion-exchangers | Less stable and reproducible |

[a]Silica-based bonded phases unless noted.
[b]Nonsilica phase.
[c]Nonbonded silica.

(4)]. Within a group of related columns, however, some columns may be more desirable with respect to stability, reproducibility, etc. For example, in reversed-phase separations, column stability (lifetime) is often greater for C18 columns vs short-alkyl-chain packings such as trimethylsilyl (TMS) (Table 11.1). Likewise, amino-phase columns tend to be less stable, and normally would not be used except for some special application.

### *Matrix Composition*

Today most packings for LC columns are made from porous silica particles. These particles usually are spherical and can be made in a wide range of particle sizes, pore diameters, and surface areas. Porous polymeric (usually polystyrene) particles also are available for reversed-phase work. These were first introduced by the Hamilton Co. in the early 1980s, and are today offered by a number of suppliers. Early polymeric columns were less efficient than their silica-based counterparts, but today the plate numbers of both kinds of columns appear to be similar. Polymeric columns can tolerate a wide range of mobile phase pH values ($1 < pH < 13$), whereas silica-based particles are unstable outside the pH range $2.5 < pH < 7$). It is said that polymeric columns are more rugged, but this has not been documented. Secondary-retention effects (Section 14.7) may be less significant on the polymeric columns, but contradictory reports on this exist.[4a] Polymeric particles have also been used instead of silica for LC size-exclusion and ion-exchange packings (Toya Soda, Pharmacia), especially for separating protein samples. Here it is claimed that secondary-retention effects are less important vs corresponding silica-based packings, but again without much hard evidence.

In principle, polymeric packings might be expected to overcome several limitations of present silica-based packing materials. However, this is offset by the much greater experience we have with the latter packings, and the broader availability of these silica-based materials. Further work will be required to establish any overall superiority of polymer-based packings.

### *Columns from Different Suppliers*

Practical workers are aware of significant differences in columns of the same type obtained from different suppliers. Presumably these reflect differences in the overall manufacturing process: (a) synthesis of the starting silica particles, (b) reaction of these particles to form bonded-phase packings, and (c) packing this material into columns. As a result, columns from different sources exhibit variations in separation characteristics, in lot-to-lot reproducibility and in column life. Personal experience usually is the basis on which individual laboratories arrive at preferred column suppliers.

### *Matrix Characteristics*

Packing materials show quite different separation performance, depending on the physical properties of the starting silica particles. Smaller-diameter particles (i.e., 3 vs 5–10 μm) generally are favored for faster separation (with similar resolution), but smaller-diameter particles are also more trouble-prone (column blockage, extra-column effects, etc.).

Most packings for separating samples within the 100–2000 dalton range have pore diameters of 6–10 nm and surface areas of 200–400 $m^2/g$. Larger pore-diameters are favored for the separation of macromolecules (mol wt > 10,000 daltons). Today, 5-μm particles with pore-diameters of about 8–10 nm are generally preferred for most small-molecule separations.

### *Column Dimensions*

Commonly used column sizes include lengths of 3–30 cm and internal diameters of 0.4–0.8 cm, with 0.46cm representing the most common diameter. Wider-diameter columns are used for preparative LC. The choice of column dimensions has little to do with troubleshooting, except that smaller-volume columns are more likely to experience extra-column band broadening (Sections 3.4 and 14.4).

### *Column Hardware*

Most LC columns use stainless-steel tubing, with compression fittings and frits at either end of the column to contain the column packing. Analytical columns are typically 1/4-in. od x 0.46 mm id, and come in 10, 15, 25, and 30-cm lengths. Shorter columns are available from some manufacturers. Column endfittings typically are reducing unions (1/4- to 1/16-in.). Fittings are more fully described in Section 9.2.

Recently, less-expensive column hardware has become available in the form of cartridge columns. Cartridge columns may use the same column tubing as standard columns, but the packing is kept in the tubing by pressed-in frits. In use, the cartridge is placed in a column holder that provides a pressure-tight seal to a 1/16-in. id compression fitting for tubing connections. The performance of these columns is similar to conventional columns. The cost reduction occurs because the same set of endfittings can be used for each column rather than a new set of endfittings being required for each column.

A variation on the cartridge column concept is the radial-compression column from Waters Chromatography Division. These columns are available only in two sizes (10 x 0.8 cm or 10 x 0.5 cm id), and loosely packed in a polyethylene sleeve. In use, the cartridge is placed in a column holder that hydraulically compresses the column packing to form a stable bed.

In addition to the column configurations discussed above, specialty columns are available in intermediate lengths, with glass linings, etc. for special applications.

## 11.2. Column Specifications: Evaluating The Column

In Section 11.1 we discussed the factors involved in choosing a general type of column. The selection of column length, diameter, and particle size usually is determined by the requirements of the separation—as reviewed in Chapter 3 and covered in detail in ref. (5). Once a certain kind of column has been selected, there usually are several possible suppliers for that column. Columns from different sources often will be found to vary in the quality of their columns. In this section we will discuss the various specifications that tell us how well a column is likely to perform in practice. We will also examine other considerations in selecting a column supplier.

The main performance specifications of the column include:

1. Plate number, $N$
2. Band asymmetry factor, $A_s$ (generally measured as shown in Fig. 11.2)

3. Pressure drop, $P$
4. Retention reproducibility as measured by a mixture of polar and nonpolar test compounds
5. Bonded-phase surface concentration (if applicable)

Usually some or all of these column specifications are reported in a data sheet that comes with each new column. An example is shown in Fig. 11.1 for an 8-cm, 3-µm-particle C18 column from Du Pont.

### *Plate Number*

The column plate-number, $N$, probably is the most important feature of a column, because it is directly related to how well the column can separate samples in general. The measurement of column plate-number was discussed in Section 3.3. The most common procedure is to measure the width of the band at half-height ($W_{0.5}$) and the retention time $t_R$, from which

$$N = 5.54 \ (t_R/W_{0.5})^2 \tag{11.1}$$

There are a number of other ways that column plate number can be measured [e.g., review of ref. (6)], but all of these are essentially equivalent for the case of nontailing bands (see Band Asymmetry below). Columns that give tailing bands are of limited value, regardless of their plate number.

The column plate-number $N$ varies markedly with column length, particle size, flow rate, and other separation conditions. Therefore, it is not possible to specify a single $N$-value that a particular kind of column should have. However, if the test separation (for measuring $N$) has been optimized to give as large a value of $N$ as possible (for that separation and column), a good column should have a value of $N$ close to $N_{max}$:

$$N_{max} = 4000 \ L/d_p \tag{11.2}$$

Here $L$ is the column length is cm, and $d_p$ is particle diameter in microns.* Thus, a 25-cm column of 5-µm particles should have a plate number of about 20,000 when run under ideal conditions.

*Eq. (11.2) is equivalent to a reduced plate-height $h$ of the column equal to 2.5 particle-diameters (Section 3.3).

**GOLDEN SERIES**
**ZORBAX® ODS QC**
**CHROMATOGRAM**

**OPERATION CONDITIONS**
Instrument: Du Pont HPLC
Column: Golden Series Zorbax® ODS, 6.2 mm i.d. x 8 cm
Mobile Phase: 75% Methanol/25% Water
Flow Rate: 1.2 cm³/min
Temperature: 23°C
Sample Volume: 5 mm³
Detector: UV Abs. (254 nm), 0.2 AUFS

**PEAK IDENTITY**
1. 5 µg/cm³ Uracil ($t_0$)
2. 200 µg/cm³ Phenol
3. 25 µg/cm³ 4-Chloronitrobenzene
4. 850 µg/cm³ Toluene
   Dissolved in Mobile Phase

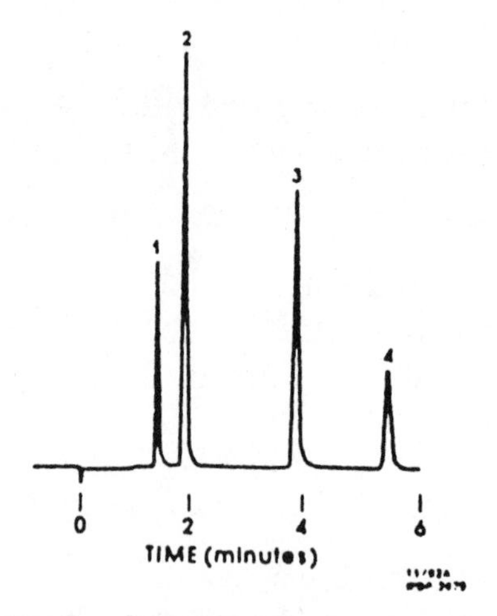


COLUMN PERFORMANCE REPORT

COLUMN: 8CM X 6.2MM 3MICRON ZORBAX ODS
PART NO.: 880959902
SERIAL NO.: KQ1288

TEST CONDITIONS:

MOBILE PHASE = 75/25 MEOH/H2O
PRESSURE = 90 BAR
FLOW = 1.24 ML/MIN
LINEAR VELOCITY = 0.13 CM/SEC
TEMPERATURE = 24.0 C

PERFORMANCE RESULTS FOR TOLUENE ( K' = 3.54 )

| PARAMETER | SPECIFICATION | COLUMN:KQ: |
|---|---|---|
| THEORETICAL PLATES | 8000 MIN | 11290 |
| SKEW* | 1.20 MAX | 0.77 |
| SELECTIVITY** | 1.50- 1.80 | 1.70 |

* PEAK SYMMETRY FACTOR BASED ON TAU/SIGMA RATIO
** CALCULATED AS K' PEAK 4 / K' PEAK 3

**Fig. 11.1.** Column specifications for the Du Pont Golden Series ODS Column. Courtesy of Du Pont.

Another way of assessing column performance (in terms of $N$) is to use the Drylab I program[5] from LC Resources to evaluate the column-parameter $A$. The $A$ value of a column measures how well the column has been packed, with good columns giving $A$ values in the range of 0.5–1.0. The $A$ value can be determined from any test

Table 11.2
Maximum Plate Numbers (Ideal Conditions) for Different Column Lengths and Particle Diameters [Eq. (11.2)]

| Particle size, µm | $N_{max}$ for different column lengths | | | | | |
|---|---|---|---|---|---|---|
| | 3 cm | 5 cm | 8 cm | 15 cm | 25 cm | 30 cm |
| 3 | 4000 | 6700 | 10,700 | 20,000 | 33,300 | 40,000 |
| 5 | 2400 | 4000 | 6,400 | 12,000 | 20,000 | 24,000 |
| 10 | 1200 | 2000 | 3,200 | 6,000 | 10,000 | 12,000 |

chromatogram, not just one that is run under ideal conditions. The column of Fig. 11.1 should have a maximum $N$ value of 10,700, vs the experimental value of 11,300. Therefore this is a good column so far as its plate number. Its $A$ value was also determined from the DryLab program: $A = 0.55$, which likewise indicates a good column. Table 11.2 summarizes values of $N$ that should be expected (under ideal run-conditions) for various combinations of column length and particle diameter.

The value of $N$ for a column generally is smaller than the values of Table 11.2, when real samples are being separated [see discussion of ref. (7)]. This is normal, inasmuch as $N$ depends strongly on separation conditions, and ideal conditions for maximum $N$ often are not optimum conditions for a particular LC application. The value of $N$ for a column also decreases with use. After the injection of 500–1000 samples, it is not unusual for $N$ to decrease by 50% or more—compared to the $N$ value for the new column. Ways of minimizing this effect are discussed in Sections 11.2 and 11.3.

### *Band Asymmetry*

Band-tailing can be measured by the asymmetry factor $A_s$, defined in Fig. 11.2. Some manufacturers report peak skew instead of $A_s$ (see Fig. 11.1). The relationship between these two quantities is:

| Skew | $A_s$ | Skew | $A_s$ |
|---|---|---|---|
| 0.0 | 1.0 | 1.00 | 1.45 |
| 0.50 | 1.1 | 1.25 | 1.65 |
| 0.75 | 1.25 | 1.50 | 2.2 |

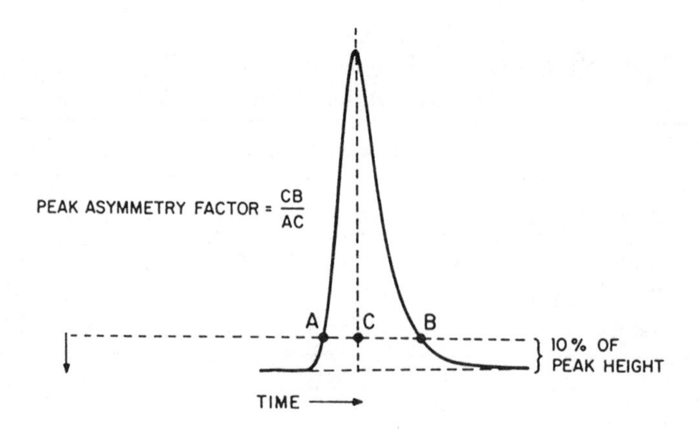


**Fig. 11.2.** Measurement of band asymmetry $A_s$. Reprinted from ref. (1) with permission.

New columns should have $A_s$ values between 0.9 and 1.2. The closer $A_s$ is to 1.00, the better. Tailing bands reflect poorer separation, as well as numerous other problems (Chapter 14). The column of Fig. 11.1 is seen to have a skew of 0.77, meaning an $A_s$ value of about 1.3. This is larger than desirable, but small-particle or short columns often exhibit larger $A_s$ values because of extra-column effects (Section 14.4).

Tailing bands, as in Fig. 11.2, can arise either from a poorly packed column (or one that has developed voids), or from chemical effects that reflect interactions between the sample molecule and the stationary phase (see Section 14.7). Chemical tailing generally does not depend so much on the column as upon separation conditions and the kind of sample being separated. We can adjust separation conditions in most cases to minimize tailing caused by chemical effects (Section 14.7). Band asymmetry due to the column can be measured by selecting test compounds that do not interact with silanols or other functional groups in the stationary phase. Various hydrocarbons, such as toluene or naphthalene, commonly are used for this purpose (e.g., toluene in Fig. 11.1).

When significant tailing of a band exists, then the plate number measured by Eq. (11.1) will be artificially high.[6] In this situation, other ways of measuring $N$ may be more reliable. For example, the Dorsey-Foley equation

$$N = 41.7\ (t_R/W_{0.1})^2/(A_s + 1.25) \qquad (11.3)$$

has been recommended for this case.[6] Here, $W_{0.1}$ is the bandwidth at 10%-height. The importance of band asymmetry on the column plate number is shown below, where the Dorsey-Foley plate numbers are shown for bands with various degrees of tailing.

| $A_s$ = | 1.0 | 1.1 | 1.2 | 1.3 | 1.4 | 1.5 | 1.7 | 2.0 |
|---|---|---|---|---|---|---|---|---|
| $N$ = | 20000 | 17400 | 15200 | 13300 | 11800 | 10500 | 8400 | 6200 |
| | 15000 | 13000 | 11400 | 10000 | 8800 | 7900 | 6300 | 4600 |
| | 10000 | 8700 | 7600 | 6700 | 5900 | 5200 | 4200 | 3100 |
| | 5000 | 4300 | 3800 | 3300 | 2900 | 2600 | 2100 | 1500 |

### *Pressure Drop*

Well-packed columns of spherical particles usually have similar permeabilities. That is, the column pressure will be comparable when separation conditions are the same. We can predict the pressure drop for a good column from the following relationship:

$$P = 7500\, L^2\, \eta / d_p^{\,2}\, t_0 \qquad (11.4)$$

Here $P$ is in psi, column length $L$ is in cm, mobile phase viscosity $\eta$ is in cpoise, particle diameter $d_p$ is in µm, and column dead-time $t_0$ is in seconds. To obtain values of $\eta$, see Table 11.3, ref. (7), or Appendix I of ref. (1).

New columns should give pressure drops within 30–50% of the value predicted by Eq. (11.4). For example, the column of Fig. 11.1 is predicted to have a pressure drop of 1160 [Eq. (11.4)], vs the experimental value of 90 bar = 1320 psi. This is adequately close to the predicted value in this case. With continued use, the column pressure drop generally will rise—as discussed in following sections.

### *Retention Reproducibility*

Column manufacturers generally evalute the retention reproducibility of their product by running a test mixture of several compounds—including both polar and nonpolar molecules. Figure 11.1 illustrates this in one particular case. Here the compounds uracil, phenol, 4-chloronitrobenzene, and toluene are run as test compounds. This particular sample is seen to comprise a

Table 11.3
Viscosity at 25°C of Organic/Water
Mobile Phases Used in Reversed-Phase LC[7]

| Organic solvent, %v | η, in cpoise | | |
|---|---|---|---|
| | Methanol | Acetonitrile | Tetrahydrofuran |
| 0 | 0.89 | 0.89 | 0.89 |
| 20 | 1.40 | 0.98 | 1.22 |
| 40 | 1.62 | 0.89 | 1.38 |
| 60 | 1.54 | 0.72 | 1.21 |
| 80 | 1.12 | 0.52 | 0.85 |
| 100 | 0.56 | 0.35 | 0.46 |

nonpolar compound (toluene), two polar solutes (uracil and 4-chloronitrobenzene), and an acidic compound (phenol). Different batches of column-packing should show similar retention times for the test compounds when run under standard conditions.

Additional test compounds have been suggested to further evaluate column characteristics. For example, silanols are associated with a variety of column problems (Sections 11.4, 14.7), and some column packings are known to be more acidic (stronger silanol interactions) than are others. Various basic compounds (e.g., benzylamine, diethylaniline) have been used to evaluate the relative acidity of different packings. The compound DMDPC (5,14-dimethyl-7,12-diphenyl-1,4,8,11-tetraazacyclotetradecane) has been suggested as an especially effective probe for measuring silanol effects,[8] as discussed further in Section 11.4. Some workers have suggested that trace metals are also responsible for tailing bands and retention variability.

The test-compound acetylacetone has been used to determine the presence of trace metals in LC packings.[9]

### *Bonded-Phase Concentration*

Good bonded-phase columns will have a high concentration of silyl groups attached to the particle surface. Typically about 2.8–3.3 μmol bonded-phase/m$^2$ of surface should be present. Lower concentrations suggest incomplete bonding, leading to poorer retention reproducibility and less stable columns. It is often diffi-

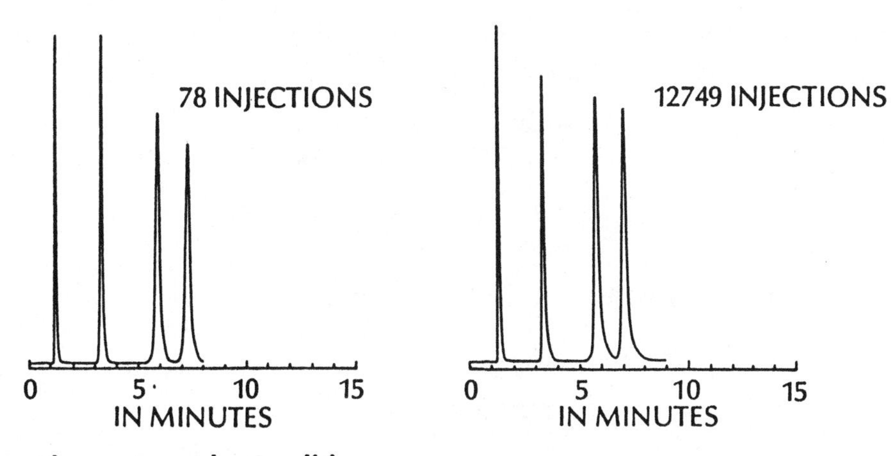


**Chromatography Conditions:**

| | |
|---|---|
| **Column:** | Reliance 5 Phenyl (80 × 4.0 mm) |
| **Guard Column:** | Reliance Guard Column Phenyl |
| **Mobile Phase:** | 30% $H_2O$; 70% MeOH |
| **Flow Rate:** | 0.75 ml/min |

**Fig. 11.3.** Over 12,000 sample injections on a Zorbax Reliance 5 cartridge column.

cult to know what the surface coverage is for a particular column, because most manufacturers report %-carbon, not concentration of bonded phase when describing a column packing.

## 11.3. Preventive Maintenance

The lifetime of an LC column can range from a few samples to several thousand injections. There are many factors that influence column life and create other column problems. Some of these are not under the control of the operator—for example, when the samples to be separated are very dirty, and it is not practical to clean them up completely. However, in most cases we can follow procedures that minimize column problems and make possible the use of the column for at least a thousand injections. Figure 11.3 shows that very long column lifetimes are possible—this particular column survived almost 13,000 injections! This is not typical, but usually we should not be satisfied with only a few hundred samples per column. Preventive maintenance and good LC practice include several recommendations:

1. Use in-line filters and guard columns
2. Avoid pressure shocks to the column
3. Follow column manufacturers' recommendations with respect to separation conditions (temperature, pH, mobile phases); use pre-columns where appropriate
4. Clean up dirty samples
5. Flush the column regularly with a strong solvent

### *In-Line Filters and Guard Columns*

In-line filters are designed to be placed immediately after the sample injector—and before the LC column. Typically, these consist of a half-micron frit in a low-dead-volume holder, and they are available for both regular and microbore columns. Their purpose is to trap particulate matter that normally would block the column inlet. The frit is relatively inexpensive, and can be discarded after every few hundred injections, or when indicated by pressure increase or band deterioration (see Section 11.4). They are available from Rheodyne, Upchurch, and other suppliers.

Particulate matter can arise from the samples, the mobile phase, or from wear on the polymeric seals within the pump and/or sample injector. Particulates in the sample can be minimized by filtering the sample before injection, but this can be expensive if a special LC-filter is used for each sample. Likewise, the mobile phase can be filtered to free it of particulates (Section 6.3). However, it is impossible to avoid particulates from system-wear, except with an in-line filter. For these reasons, in-line filters are essential for small-particle columns (3–5 μm), and are advisable in most cases.

Guard columns are short columns that are placed between the in-line filter and the column inlet. Thus the normal sequence in an LC system is:

sample-injector : in-line filter : guard-column : main column.

Guard columns are intended to collect chemical "garbage" (from the sample) that would stick to the inlet of the main column and eventually degrade its performance. Thus, like the in-line filter, guard columns are sacrificial in nature. That is, they must be replaced every 50–100 injections when used with particularly dirty samples. Guard columns have a secondary purpose, namely to serve as a backup to the in-line filter and to catch any particulates that leak through the filter.

The reason for using guard columns with dirty samples is that guard columns are inexpensive: typically \$25–50, vs 5- to 10-times more for the main column. In some cases, guard columns are packed by the user, in which case their out-of-pocket cost is even less (but the labor required to pack the column may be significant). It is important that the guard column not degrade the performance of the main column. That is, the guard column must not lower the plate number of the total column-system, relative to $N$ for the main column alone. To achieve this, two approaches are generally used in designing the guard column: (1) keep the total volume of the guard column small, and/or (2) pack the guard column with the same efficiency packing used in the main column (3- or 5-μm particles). When a guard column is properly designed, its presence should make little or no difference in the separation. This is illustrated in Fig. 11.4 for two separations of the same sample. In Fig. 11.4a, a single 8-cm cartridge column is used. In Fig. 11.4b, a guard column has been added. It is hard to see any difference in the two separations. Guard columns are combined with in-line frits in some units (Upchurch) for greater convenience. Du Pont and other companies sell column-protection kits that combine the guard column, in-line frit holder and packing for the guard column (18-μm C18 particles) into one handy package.

Several guard-column designs are commercially available. These can be classified as (a) separate, (b) integral (combined into one unit with the main column[10]), or (c) user-packed. Both the separate and integral guard columns are available in a cartridge configuration in which column blanks are inserted into a column holder. These are particularly convenient, and the separation of Fig. 11.4 involves a guard column of this type. User-packed columns are available as quite small column-blanks (e.g., 3 x 0.2 cm) that can be dry-packed with 15–20-μm particles, which is particularly convenient. Pellicular packings also are used in the user-packed configuration, but their capacity for removing chemical contaminants is more limited, and they provide less efficient

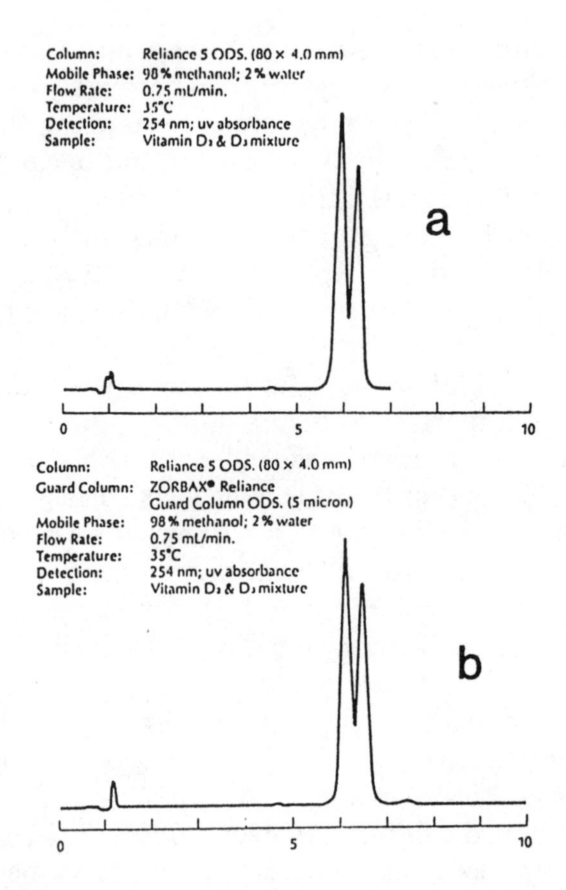


**Fig. 11.4.** Separation of vitamins $D_2$ and $D_3$ on Reliance 5 ODS cartridge column (a); and cartridge column plus guard column (b). 8 x 0.40-cm columns of 5-µm particles, 98%v methanol/water, 35°C. 0.75 mL/min. Courtesy of Du Pont.

protection against particulates. For further information on guard columns, see ref. (10a).

### *Avoid Pressure Shock*

LC columns are designed to tolerate high pressures, but most columns can be damaged by sudden changes in pressure. These pressure shocks can arise from:

1. Slow rotation of the sample-injection valve
2. Fast startup of the pump
3. Column-switching operations

When a sample is injected with the usual six-port sample-valve, the flow from pump to column is momentarily interrupted during the rotation of the valve. This results in an increase in pressure on the pump side of the valve, and a loss in pressure on the column side. When flow is resumed (after completion of valve rotation), there is a sudden surge in flow through the valve, plus a fluctuation in the pressure reading (usually measured between pump and column). The magnitude of the resulting pressure shock to the column is directly proportional to the time it takes the valve to rotate.

These pressure changes during injection often are not very severe for manual injection. However, some autosamplers have valves that turn more slowly. This is then seen as a sharp change in the pressure reading when a sample is injected. A change in pressure of more than 20% during sample injection is an indication of possible damage to many columns. There are various remedies to this situation: (a) use helium for air-actuated valves; (b) use a different valve design for faster rotation; (c) use a bypass around the sample injector. These solutions are discussed in more detail in Section 10.3.

Some pump designs allow a rapid pressure buildup when first switched on with a high flow rate. This can be monitored by observing the system pressure during startup. Where a rapid pressure change is observed for this situation, good practice is to adjust the flow rate from zero to the final value in steps. For example, if the final flow rate is 3 mL/min, go through 1, 2, and then 3 mL/min in steps. Little time needs be spent between the successive adjustments (e.g., 10–15 s). Some pumps (e.g., Beckman Instruments) are capable of including a rapid flow-rate ramp on startup to avoid this problem.

Column-switching is a technique that uses two columns in series, with a switching-valve in between [Chapter 16 of ref. (1)]. During the switching operation, sharp pressure changes are possible. This will be especially true when a column is switched in and out of flow from the pump, in which case the pressure at the inlet rapidly fluctuates between zero and some large number. The solution to this problem (which can kill columns very quickly) is to design the system so that such pressure changes do not occur. This can be achieved either by the use of dummy columns, extra pumps, or special valve configurations.[12] In any case, the first step is to analyze the particular column-switching system-design to be used for such pressure effects. Appropriate corrective action then can be considered.

## *Separation Conditions*

Most LC columns allow a wide range of experimental conditions to be used, in turn permitting the achievement of most separations within a reasonable time. However, each column also has certain limits in the choice of separation conditions, as specified in the manufacturer's literature for that column. Therefore, it is important to be familiar with what you can or can't do when using a particular column. There also are general guidelines that we will review next.

*pH.* The most important "do" when using bonded-phase silica-based packings is: use mobile phases whose pH is between 2.5 and 7. Bonded-phase packings are rapidly degraded at pH extremes, with loss of bonded phase and (eventually) dissolution of the silica. The initial effect is a change in sample retention, with nonbasic compounds showing decreased retention, and bases showing (usually) increased retention. This may be accompanied by broadening of sample bands, especially for basic compounds.

If high or low-pH mobile phases must be used, their adverse effect on the column can be reduced by using a *Precolumn* (or *saturator* column). The precolumn is packed either with the same packing as used in the column, or (more commonly) with ordinary silica. It is connected between the pump and the sample injector, and is intended to saturate the mobile phase with silica—thus minimizing further dissolution of the column packing by mobile phase. The precolumn need not have a high plate number, and can be loosely packed with inexpensive, laboratory-grade silica. The precolumn should be checked occasionally, to ensure that the silica has not dissolved completely. Precolumns can have adverse effects, if the mobile phase is changed frequently (for different LC procedures). Thus, the large amount of silica in the precolumn may be slow to equilibrate with the new mobile phase, leading to retention variability (see also Section 15.4). Also, an 0.5-µm in-line filter should be used directly after the precolumn to prevent any particulate matter from causing problems downstream. Precolumns are discussed further in ref. (10a).

*Temperature.* The temperature of the column should be held within certain limits. Silica-based packings and anion-exchange columns are subject to increased chemical attack by the mobile phase at temperatures above 60°C. The combination of a higher temperature plus extreme pH (2.5 > pH > 7) leads to even faster deterioration of the column. Operating some small-particle columns at higher temperatures can lead to bed collapse and loss in plate number. Temperatures of above 40°C for 3-µm particles, and 70°C for 5-µm particles have caused as much as a 50% decrease in *N*. However, this was not observed in other 3-µm columns (e.g., Du Pont 3-µm columns).[11]

*Mobile-Phase Problems.* A particular column may be intolerant of certain solvents for the mobile phase. This is true particularly of small-particle polystyrenes used for non-aqueous size-exclusion chromatography (gel permeation). In extreme cases, once the column has been equilibrated with a certain solvent (e.g., tetrahydrofuran) it is unwise to change that solvent. For these packings it is imperative to be thoroughly familiar with the manufacturer's literature.

Another mobile-phase problem is blockage of the column by microbial growth, caused by storing columns filled with aqueous mobile phases. Microbial growth is inhibited by organic solvent, so the column normally should be stored in either pure organic solvent or mixtures of at least 50% organic plus water. An exception is columns for gel filtration, which should be stored in aqueous buffer solutions. In this case, 0.01% azide should be added to inhibit microbial growth.

### *Dirty Samples*

Many samples consist of a pure solvent, plus a few components that are completely eluted from the column within the time allotted for a single run (5–10 min). Such samples rarely present any danger to the column. Other samples may contain (a) particulate matter, (b) components that will precipitate inside the column (e.g., proteins), (c) compounds that are retained very strongly on the stationary phase, (d) solvents that will attack the column packing, and so on. In every case it is important to know the composition of the sample, and the likely effect of the sample on column life. When the sample is likely to damage the column, various preventative remedies must be sought.

Any sample that shows a cloudiness or opalescence when held up to the light should be filtered before injection. In-line filters and guard columns can remove particulates before they reach the main column, but they are not intended to replace sample pretreatment. That is, an excessive amount of particulates in the sample will overload the in-line filter and guard column, and either block them quickly or result in particulates passing on to the main column.

If it is suspected that the sample may precipitate when it contacts the mobile phase (and thereby block the column), this should be tested by mixing sample and mobile phase together and looking for any evidence of precipitation (haziness, opalescence, etc.). If a sudden increase in pressure is observed after each sample injection (followed by a gradual decrease in pressure), this is strong evidence that either particulates or precipitation is a problem with your samples. In cases of sample precipitation, it usually is advisable to modify the separation conditions (change the mobile phase or sample solvent), or to pretreat the sample (remove insoluble components—see Chapter 17) so as to eliminate the problem. In some cases, the use of a smaller sample size may be adequate.

Sample components that are sorbed very strongly onto the column packing can present other kinds of problems: (a) loss in plate number, (b) change in sample retention with time, and (c) subsequent elution over a long period of time, accompanied by an erratic baseline. The best approach is to remove these strongly retained sample components before the sample is injected; i.e., use some form of sample pretreatment (Chapter 17). Use of a guard column can greatly extend column life and minimize these problems. Finally, regular cleaning of the column as described in the following section is always helpful.

Occasionally the sample is dissolved in a solvent that could damage the column. For example, samples from a production area might be dissolved in 6*M* sodium hydroxide. Injection of 50–100 μL of such a sample onto a silica-based column could quickly dissolve enough silica to ruin the column. In these cases, the sample solution should either be neutralized, or the original solvent removed, so as to minimize column damage.

### *Regular Column Flush*

It is good practice at the end of each day to flush the column with a strong solvent; e.g., methanol or acetonitrile in the case of reversed-phase LC. This serves to remove any strongly retained compounds that may have built up on the column during the day, and provides a clean column for the start of the next day's work. When particularly dirty samples are being run, this clean up procedure can be extended to include stronger solvents (see Table 15.4). When flushing the column with organic solvents, be sure to remove any salts or buffers first by pumping water or non-buffered mobile phase, in order to prevent buffer precipitation (Section 5.3).

Table 11.4 summarizes the main preventative maintenance procedures for protecting the column.

## 11.4. Problems and Solutions

The inital selection of a good column (Section 11.2) plus effective preventative maintenance (Section 11.3) is the best approach to avoiding column problems. However, this does not guarantee that the column always will perform adequately, and any column will degrade in time to the point where it cannot be used. The symptoms of column malfunction are: (a) loss in plate number, (b) deterioration of band shape, (c) increase in pressure, and/or (d) change in retention. These can occur in various combinations.

### *Plate-Number Loss*

A gradual loss in column plate-number is an inevitable result of continued use of the column. A rough rule is that a 50% loss in $N$ is to be expected after 2000 samples are injected. If the loss in $N$ is much faster than this, then the various proposals in Section 11.3 should be reviewed for possible application to your situation. When the loss in $N$ is sudden, e.g., overnight or within a day, a void may have developed (see below), or a bad sample may have been injected. If the loss in $N$ occurs when using the column on a different LC system, extra-column effects may be involved (Sec-

Table 11.4.
Summary of Preventive Maintenance Procedures for the Column

1. Use in-line filters for all columns
2. Use guard columns for dirty samples
3. Use helium to speed the rotation of air-actuated injection valves to minimize pressure pulses when sample is injected
4. Change flow rate in 1 mL/min steps
5. For silica-based columns, keep mobile-phase pH between 2.5 and 7; use a precolumn outside these pH limits
6. Keep column temperature < 60°C (exept for gel permeation with polymer samples)
7. Check samples for particulates and compatibility with the mobile phase; take appropriate action when indicated
8. Pretreat dirty samples (Chapter 17)
9. Flush the column daily with strong solvent; use stronger solvent protocol (Table 15.4) for dirty samples, as needed
10. Store columns in 50–100% organic/water mixtures or add azide (gel filtration columns)

tion 3.4). A low plate-number for a new LC procedure may be inherent in the sample and conditions used (Section 15.1); in this case there may be nothing wrong with the column at all.

### *Poor Peak Shape*

The appearance of badly tailing bands, split peaks, or other non-Gaussian bands can arise from numerous causes, as reviewed in Chapter 14. Poor peak shape almost always is combined with a drastic decrease in plate number. Badly distorted bands without change in retention time are usually caused by blocked frits (Section 14.1) or a column void. Repairing column voids is discussed in Section 11.5.

### *Pressure Increase*

Just as a column shows a gradual decrease in plate number with continued use, the column pressure also will slowly rise. If pressure becomes excessive, two possibilities should be considered: (a) sample deposited (precipitated) inside the column, or (b) blocked column-inlet frit.

If it is suspected that deposited sample has blocked the column, try flushing the column with an appropriate solvent (one that is known to dissolve the sample). Disconnect the column from the detector, and reverse its position in the system (original outlet connected to line from sample-injector). Flush the column with 30–50 mL of solvent and collect the effluent in a container placed at the column outlet. Do not run this effluent through the detector, since it can result in a dirty flowcell. Check the pressure during this operation, and continue flushing until the pressure stabilizes.

If backflushing was ineffective in reducing the backpressure, or if sample deposition is considered unlikely, then the inlet frit can be replaced with a new one. This is a simple procedure, but care must be taken not to disturb the column packing during the operation. Remove the column from the system, then detach the inlet fitting from the column while holding the column vertical (inlet end up). Carefully remove the inlet frit and replace it with a new frit (it is not recommended to clean the old frit). At the same time, inspect the exposed packing at the inlet for any imperfections; the surface of the packing should be perfectly smooth and level with the end of the column. Replace the inlet fitting and reinstall the column in the system. *Be Very Careful Not To Disturb The Top Of The Column Packing During This Operation.*

For a thorough discussion LC column-pressure problems, see ref. (13).

### *Change in Retention Time*

Two kinds of retention variability exist: (a) change in retention times ($k'$-values) for the same column from sample to sample, and (b) change in retention times from column to column. Changes in retention of the first kind (same column) are discussed in Section 15.4. A change in retention from column to column usually reflects differences in the column packings, as a result of lot-to-lot variations in the manufacturing procedure. This is a major problem today in many laboratories, which we will discuss in some detail.

### *Column-to-Column Retention Variations*

These can range from small shifts in retention for all bands (without change in band spacing or resolution) to major rearrangement of the peaks within the chromatogram. Although this area is not well understood, the following discussion points out some of the areas which are important in retention variations. Minor changes in retention often can be adjusted for by changing the flow rate or solvent strength (more or less water in the mobile phase for reversed-phase LC). These are not of major concern. When the bands within the chromatogram change place with each other, or when two bands fuse together, the situation is more serious.

Major changes in retention caused by the column usually are the result of silanol interactions or other secondary-retention effects (see discussion of Section 14.7). These can arise for any silica-based packing [similar effects have been reported for polystyrene-based packings as well, ref. (4a)]. The problem is that different kinds of silanols exist on the silica surface, and different silica particles can have varying concentrations of each silanol type. In the case of bonded-phase packings, some (but not all) of these silanols are removed by silanization. The residual silanols can vary in type and in relative concentration. Acidic or basic sample molecules interact with these silanols to varying degrees, determined by (a) the sample, (b) the particular silica used for the column packing, and (c) the experimental conditions.

The effect of silanols on sample retention thus varies from one lot of silica to the next, and can vary further from one lot of bonded phase to the next (even for the same silica). The resulting change in retention from one lot of packing to the next generally will be most serious for cationic or basic samples. Several approaches to solving this problem are summarized in Table 11.5.

The key in each case involves (a) anticipating that such problems will arise (e.g., when separating basic or cationic samples), and (b) taking appropriate action *before* the problem occurs.

Because the problem of retention variability occurs for different lots of column packing, one solution is to obtain several columns from the same lot and to use only these columns for a particular LC procedure. Usually columns are identified by lot number,

Table 11.5
Preventing Column-to-Column Retention Variability

1. Use only columns from the same manufacturing lot (not always practical)
2. Design the separation conditions to minimize retention variablity (should be a standard part of method development)
3. Select LC methods or columns that are less sensitive to secondary-retention effects
4. Map retention vs separation conditions to allow you to correct for column variability if necessary (by changing conditions) when a new column is installed

and an order for several columns usually will be filled from the same lot. The column supplier generally can provide any (reasonable) number of columns from the same lot, if you specifically request this at the time your order the columns. Some column suppliers will guarantee that columns ordered for a particular LC assay will provide the same separation as did original columns.

The second approach to minimizing column-to-column variability is to select separation conditions that minimize silanol interactions. This is discussed in detail in Section 14.7 (see text for Fig. 14.12). The use of mobile-phase additives such as triethylamine plus higher buffer concentrations generally suppresses the effects of silanols, which in turn leads to better retention reproducibility among columns from different lots. This can be seen clearly in the comparisons of Fig. 14.12 (repeated in part as Fig. 11.5), where the use of appropriate mobile-phase conditions gives identical separations on two entirely different C18 columns (Fig. 11.5a,b), but a mobile phase without additives gives quite different separations (Fig. 11.5c,d).

Once an LC method has been developed, separations should be run on at least two columns from different lots to see if retention variability is likely to be a problem. If identical separations are obtained on columns from different lots, this is a good sign that retention reproducibility will be maintained from column to column.

The third approach is to consider LC methods and columns that appear to be less affected by silanol interactions. Thus, sev-

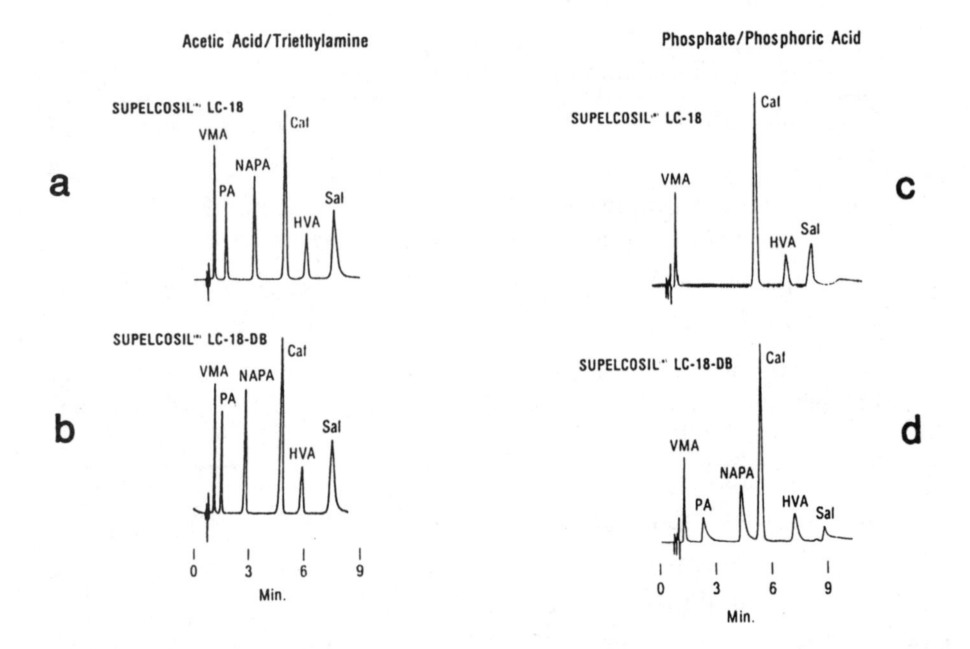


**Fig. 11.5.** Separation of mixture of acids (HVA, VMA, SAL), neutrals (CAF), and bases (PA, NAPA) on two different C18 columns, with two different mobile phases. (a) Supelcosil LC-18 and (b) LC-18DB, using mobile phase with triethylamine and acetate; (c) Supelcosil LC-18; and (d) LC-18 DB without triethylamine or acetate. For other conditions, see Fig. 14.12.

eral studies have shown that ion-pair separations are more reproducible (for the same sample) than reversed-phase separations.

Columns from different manufacturers also show general differences in the kinds of silanols present. Table 11.6 summarizes the relative acidity of various reversed-phase packings in use today. Columns such as Zorbax Rx and Vydac are relatively nonacidic and are most useful for separations of bases and cations. More acidic columns such as Micropak and ordinary Zorbax are less useful for such samples.

When problems in retention variability are anticipated (for whatever reason), a good approach to minimizing problems in routine use of the LC procedure is to map the effects of various separation conditions on the resolution of adjacent bands—especially those bands that have the poorest resolution and are most likely to be affected adversely by retention changes. This proce-

Table 11.6
Relative Acidity of Different Reversed-Phase Columns[a]

| | Column |
|---|---|
| Less acidic | Zorbax Rx |
| ↓ | Vydac |
| | Nucleosil |
| | Supelcosil DB |
| | μBondapak |
| | Novapak |
| | Prtisil |
| | RSil |
| | Polygosil |
| | Shedrisorb |
| | Lichrosorb |
| | Chrompack |
| | Rainin |
| | IBM |
| | Hypersil |
| | Perkin-Elmer |
| | Supelcosil |
| | Zorbax |
| More acidic | Micropak |

[a]Reprinted from ref. (8) with permission

dure is illustrated for the routine analysis of PTH amino acids shown in Fig. 11.6. Because of the complexity of these samples (as many as 20 components present), any change in retention is likely to have a catastrophic effect on the separation. The very polar nature of these compounds, which include both acids and bases, make such retention shifts likely.

By studying the effects of organic solvent composition (relative proportions of acetonitrile vs THF), pH, and buffer concentration on critical parts of the chromatogram, retention can be systematically varied as needed to restore resolution. This is illustrated in the examples of Fig. 11.7

## *Other Changes in Column Performance*

Various combinations of plate-number loss, pressure increase, peak shape change, and retention variation can occur. Their diagnosis is summarized in Table 11.7.

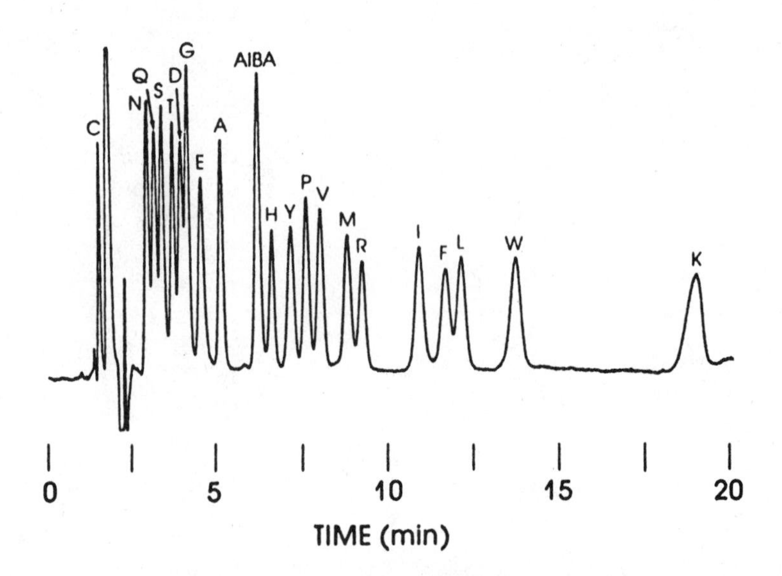


**Fig. 11.6.** Separation of PTH-amino acids. Zorbax PTH column (25 x 0.46-cm); flow = 1 mL/min; temperature = 35°C; detection = 254 nm, 0.005 AUFS. Mobile phase = 66/18/16 buffer/acetonitrile/tetrahydrofuran (buffer = 6 mM phosphoric acid, pH 3.15 with NaOH). Sample 10 μL containing 30 pmol of each amino acid. Reprinted from ref. (17) with permission.

## 11.5. Column Life: When Should A Column Be Discarded

Obviously, a column should be replaced when a major problem develops that proves to be uncorrectable. Many workers attempt to avoid replacing the column if at all possible, because the column represents a significant expense. However, it is often not economically justifiable to attempt to repair a failing column. The time and effort expended may be of greater value than the cost of a new column. This will become even more true as more workers shift to the new cartridge columns—which have the potential of cutting column costs by a factor of three or more.

Table 11.7
Problems Caused by the Column (see also Tables 14.4 and 15.5)

| Problem | Symptom | Cure |
|---|---|---|
| Blocked frit | Pressure increase and/or poor peak shape plus loss in $N$ | Reverse column and/ or replace frit |
| Column void | Poor peak shape, loss in $N$ | Discard column |
| Loss of bonded phase | Change in retention, poor peak shape, loss in $N$ | Discard column |
| Column blocked by sample | High pressure | Reverse-flush column with solvent that dissolves sample |
| Strongly sorbed sample | Loss in $N$, decreased retention | Reverse-flush column with strong solvent (Table 15.4) |

Column repairs can be classified as follows:

1. Column reversal
2. Replacing inlet frits
3. Flushing the column to remove chemical garbage or precipitated sample
4. Remove a small amount of contaminated packing and replace it with fresh
5. Filling voids
6. Replacing bonded phase
7. Repacking the column

These are ordered in roughly decreasing promise; that is, column reversal usually is justified, while repacking the column is seldom worthwhile. A detailed discussion of remedies #1–4 is given in ref. (15).

## *Column Reversal*

Column reversal (Section 14.1) is intended to deal with the problem of blocked column frits. Deposition of particulate matter on the column frit can result in pressure increase and/or severe peak distortion. Reversing the column and passing mobile phase through it for 15 min or so is a simple remedy that either solves the problem (about one time out of three) or doesn't. When this works, there is little likelihood of further problems, requiring

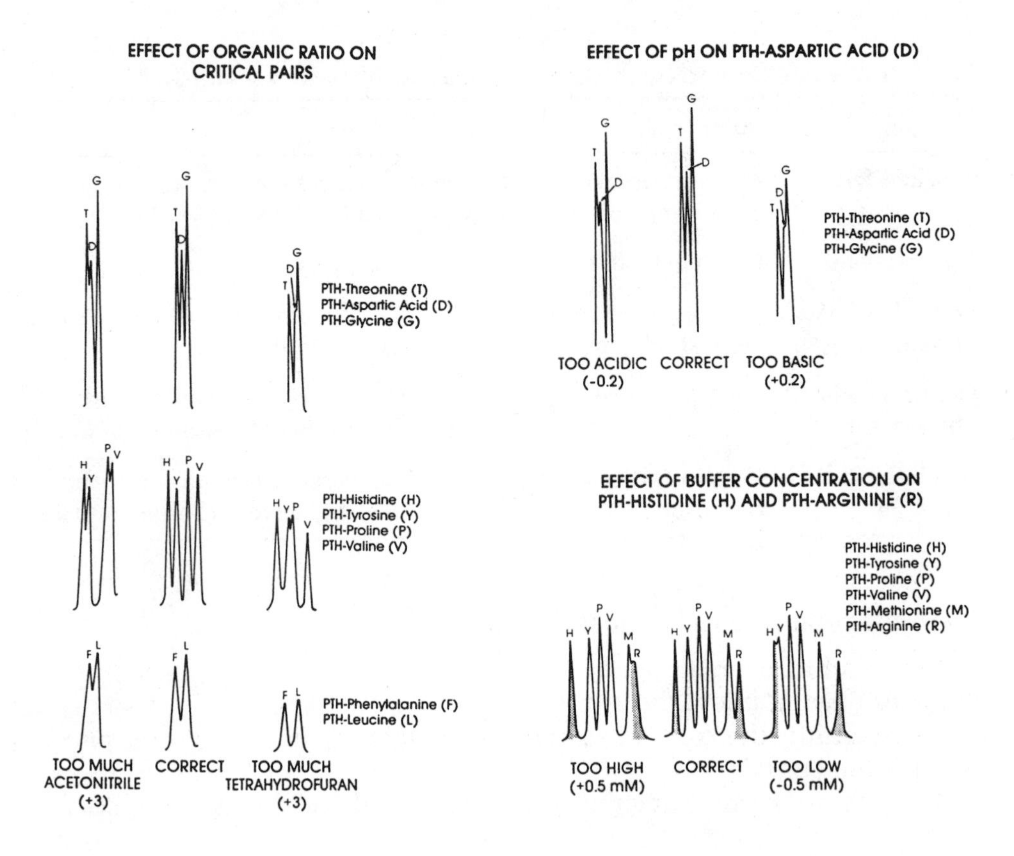


**Fig. 11.7.** Correcting changes in column-to-column retention variation. PTH-amino acid sample of Fig. 11.6. Reprinted from ref. (17) with permission.

continued testing of the column to make sure it works. Column reversal will be cost-effective in most LC laboratories. Not all columns are stable to reversal; if you have a question about this, check with the column manufacturer first.

## *Replacing Column-Inlet Frits*

This is the next step (Section 14.1), when column-reversal fails to restore good band shape and/or lower column pressure. It requires about the same amount of effort as column-reversal, and has about the same success rate. It, too, is a cost-effective procedure. Many workers prefer to replace the frit first, and backflush as a second step. This allows them to check for voids (see below) before the column is reversed.

### *Column Flushing*

This procedure is recommended when the following situation exists: (a) plate number decreases, (b) retention for all peaks decreases, and (c) dirty samples are being run. In fact, this situation calls for column flushing on a regular basis (preventative maintenance). Column flushing is effective at preventing long-term buildup of contaminants. See Table 15.4 for a recommended column-flushing protocol.

### *Replace Part of Packing*

If you are fairly certain that the inlet end of the column is contaminated with chemical garbage, you may be able to improve column performance by replacing a portion of the column packing. Remove the endfitting and frit from the inlet end of the column. With a small spatula, scoop out a small amount of the contaminated packing (generally < 5 mm in depth). Now refill the void that you have created, using the procedure below.

### *Filling Voids*

When the inlet frit is removed and a void is apparent in the column packing, either the column must be discarded or the void must be repaired. Filling the void with new packing is a simple procedure. The new packing is slurried with mobile phase to make a thick paste, then carefully used to fill the void (with some excess left over). A razor blade or spatula then is used to plane the top of the packing so as to create a smooth surface level with the column end. A new inlet frit is added, the fitting is reinstalled, and the column is connected into the LC system. In the case of large voids, this procedure may have to be repeated one or more times; i.e., the voids tend to settle upon flow of mobile phase through the column. The use of high flow rates can accelerate this process, and speed up the final column repair.

Many workers repair column voids in this manner. However, it is often found that the original plate number of the column is not restored fully. Furthermore, a repaired column often lasts for only a short time before the plate number is again unacceptably low. During this period, runs may have to be repeated, and the column may require frequent checking. Furthermore, columns that develop voids are often near the end of their useful life for other reasons. Bonded-phase may be partly gone from the silica packing, chemical garbage may have contaminated the packing, etc. The bottom line is that often more time is spent in the repair process than can be justified by the column cost. For this reason, we feel that repairing column voids is generally not worthwhile.

One possible exception to the above conclusion can be noted for the case of certain very expensive columns, especially those used for size-exlusion chromatography (SEC). SEC columns show less retention variability with age, and are often used for less demanding separations. In this case, it may well be worthwhile to top off the column with packing, when a void develops. Finally, some workers[16] have found that topping-off a voided column can be successfully performed if the repaired column is operated in the reverse direction (repaired end at the new column outlet). A stable column bed with the original performance characteristics was observed with this practice.

### *Replacing Bonded Phase*

Procedures have been described for restoring the bonded-phase coating to columns which are partly stripped of bonded phase.[17] As in the case of column voids, the problem of stripped bonded phase generally is associated with older columns that are ready to fail for other reasons. If extensive replacement of bonded-phase is involved, it is unlikely that the rebonding will restore the same retention properties to the final column. Furthermore, one can expect that runs will have to be repeated in some cases, column performance must be checked frequently after rebonding, etc. Again, the cost effectiveness of this approach is open to question. One group[18] was able to successfully rebond the stationary phase for reversed-phase columns used for peptide separations. The mobile phase used in the method caused a rapid loss in column performance, which was recovered fully with an overnight silanization procedure.

### *Repacking Columns*

Because few laboratories today pack their own columns, even fewer laboratories repack columns. However, there are commercial firms that will repack used columns, usually for about half the cost of a new column. The repacked columns often are claimed to be better than the originals (which is unlikely). There also is a serious question about the equivalence of the original and repacked column, because the packing material must be supplemented with fresh material for the repacking operation.

In summary, we believe that some column repairs are worth doing, and some are not. Bear in mind the cost of the actual repair, the possible consequence if the fix does not hold up, the need for extra runs to check column performance after the repair, and finally the uncertainty attached to results from a repaired column during the interim period following the repair. These are potentially major expenses that must be balanced against any savings in column costs.

## 11.6. References

[1]Snyder, L. R. and Kirkland, J. J. (1979) *Introduction to Modern Liquid Chromatography,* Wiley-Interscience, New York, Ch. 5.

[2]Majors, R. E. *LC, Liq. Chromatogr. HPLC Mag., Column Watch,* a monthly feature article

[3]Majors, R. E., Barth, H. G., and Lochmuller, C. H. (1984) *Anal. Chem.* **56,** 300R (biennial review of column liquid chromatography).

[4]Antle, P. E. and Snyder, L. R. (1984) *LC, Liq. Chromatogr. HPLC Mag.* **2,** 840.

[4a]Stuurman, H. W ., Kohler, J., Jansson, S. -O., and Litzen, A. (1987) *Chromatographia* **23,** 341.

[5]Snyder, L. R. and Anlle, P. E. (1985) *LC, Liq. Chromatogr. HPLC Mag.,* **3,** 98.

[6]Bidlingmeyer, B. A. and Warren Jr., F. V. (1984) *Anal. Chem.* **56,** 1583A.

[7]Snyder, L. R. and Antle, P. E. (1985) *LC, Liq. Chromatogr. HPLC Mag.* **3,** 98.

[8]Leach, D. C., et al. (1988) *LC/GC,* **6,** 494.

[9]Verzele, M. and Dewaele, C. (1984) *Chromatographia* **18,** 84.

[10]Majors, R. E. (1985) *LC, Liq. Chromatogr. HPLC Mag.* **3,** 10.

[10a]Majors, R. E. (1984) *LC, Liq. Chromatogr. HPLC Mag.* **2,** 16.

[10b]van der Wal, Sj. (1985) *LC, Liq. Chromatogr. HPLC Mag.* **3,** 488.

[11]Stout, R. W., DeStefano, J. J., and Snyder, L. R. (1983) *J. Chromatogr.* **261,** 189.

[12]Patrick, D. W. and Kracht, D. W. (1985) *J. Chromatogr.* **318,** 269.

[13]Abbott, S. R. (1985) *LC, Liq. Chromatogr. HPLC Mag.* **3,** 568.

[14]Sjut, V. and Palmer, M. V. (1983) *J. Chromatogr.* **270,** 309.
[15]Wehr, C. T. (1983) *LC, Liq. Chromatogr. HPLC Mag.* **1,** 270.
[16]Vendrell, J. and Aviles, F. X. (1986) *J. Chromatogr.* **356,** 420.
[17]Du Pont Zorbax PTH Column Users Guide (1985) Du Pont, Wilmington, DE.
[18]Mant, C. T., Parker, J. M. R., and Hodges, R. S. (1986) *LC/GC* **4,** 1004.

# CHAPTER 12

# DETECTORS

## Introduction

The detector is the "eye" of the LC system—it measures changes in the concentration (or mass) of sample leaving the column. Photometric detectors measure changes in the optical absorbance, fluorescence, or refractive index of the column effluent. Electrochemical and conductivity detectors measure changes in the electrical characteristics of the solution. Other detectors are available for special applications. Unless the proper detector is used, the sample information obtained and the quality of the data may make the analysis results unusable.

## 12.1. Detector Characteristics

The detector is a transducer that converts a change in sample concentration (in the column effluent) into an electrical signal that is recorded by a data processor or strip-chart recorder. Interpretation of this record gives quantitative and qualitative data about the sample. The most common LC detectors are UV-photometers; these measure changes in the UV-absorbance of the column effluent.

The detector parameters of main interest are sensitivity, selectivity, precision, and accuracy. *Sensitivity* is defined as the detector signal-output per unit concentration of sample compound; more sensitive detectors give larger signals (bigger peaks). With optical detectors, longer flowcell pathlengths increase sensitivity (though not necessarily minimum detectability because of added noise). Since the baseline noise determines detectability, it is important to compare the *Signal-to-noise* ratio when comparing detector sensitivity; the detector with the larger signal-to-noise ratio (for a given analyte concentration) is more sensitive, even if it doesn't produce larger peaks. Detectors that respond to unique molecular features (e.g., fluorescence or electrochemical reactivity) tend to be more sensitive than bulk property detectors (e.g., refractive index).

*Selectivity* is the relative ability of a detector to detect one compound and not another; more selective detectors give larger signals for one compound than for the same mass of another compound. For example, a fluorescence detector is more selective than a UV photometer because it responds only to compounds that fluoresce.

*Precision* is the repeatability of a measurement; a precise detector will give the same repeatable answer (or very close) for a given sample concentration. A malfunctioning detector can give less precise results. For example, when a deuterium lamp is nearly worn out, filament flicker often occurs, increasing the baseline noise. The noisy baseline, in turn, makes it more difficult to determine the true value for the peak height. Thus, repeated injections of the same sample will give poorer precision than when a properly functioning lamp is used.

*Accuracy* is the measure of how close a measurement is to the true value; most detectors rely on standardization to provide accurate results. That is, most detectors do not give absolute quantitative results, they just give sample peaks that are measured relative to a known quantity of a standard. If the detector is not calibrated properly, the absorbance maximum of a sample can be in error. Some detectors, such as diode-array detectors or scanning UV detectors, however, rely on internal calibrations for wavelength accuracy.

All properly operating detectors should give precise and accurate results. Changes in the precision or accuracy of an assay can be caused by a large number of variables in the LC assay, including detector problems (Chapter 16).

Discussion here is limited to UV absorbance, fluorescence, refractive index, and electrochemical detectors. Specialty detectors, such as infrared, mass spectrometric, and mass detectors, are not discussed; operator and service manuals are alternative

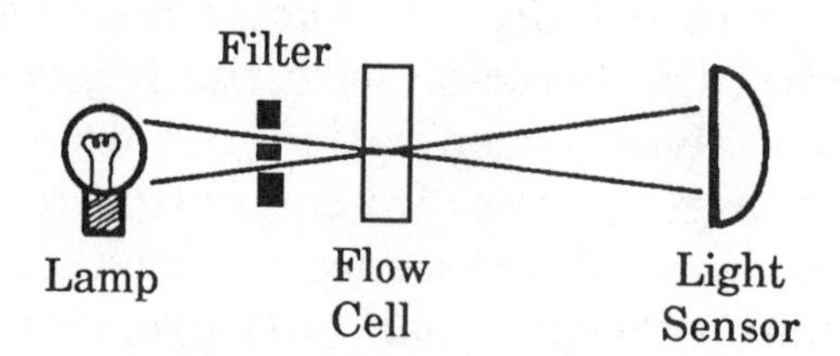


**Fig. 12.1.** Generalized UV detector. Reprinted from ref. (1a) with permission.

information sources for troubleshooting these detectors. An extensive discussion of the design and operation of commercially available detectors can be found in ref. (1)

## *UV Absorbance Detectors*

UV photometers are the most widely used LC detectors.[2] These detectors measure changes in the absorbance of light in the 190–350 nm region as the column effluent passes through the detector flowcell. The same detector design can be used in the visible region (e.g., 350–700 nm), although sensitivity usually is lower because most LC samples absorb weakly or not at all in the visible region. Some detectors are designed to provide absorbance measurements in both the UV and the visible regions; these are sometimes called UV/VIS detectors. The present discussion concentrates on detectors that operate in the UV region (190–350 nm), but the same operation and troubleshooting practices apply to detectors operating in the visible region as well.

All UV absorbance detectors are based on the design shown in Fig. 12.1. The basic parts are a *Lamp, Flowcell,* and light *Sensor.* Additionally, a *Filter* or *Grating* monochromator can be used to select a specific wavelength of light. If a filter is used, the detector is called a *Fixed-Wavelength* detector, because a single wavelength (or narrow band) is selected by the filter. When a grating is used, a variety of wavelengths is available, so the detector is called a *Variable-Wavelength* or *Spectrophotometric* detector.

The lamp can be a tungsten light bulb that emits radiation in the visible region, but more commonly a UV-emitting lamp is used for increased sensitivity. Lamps can be classified in two catego-

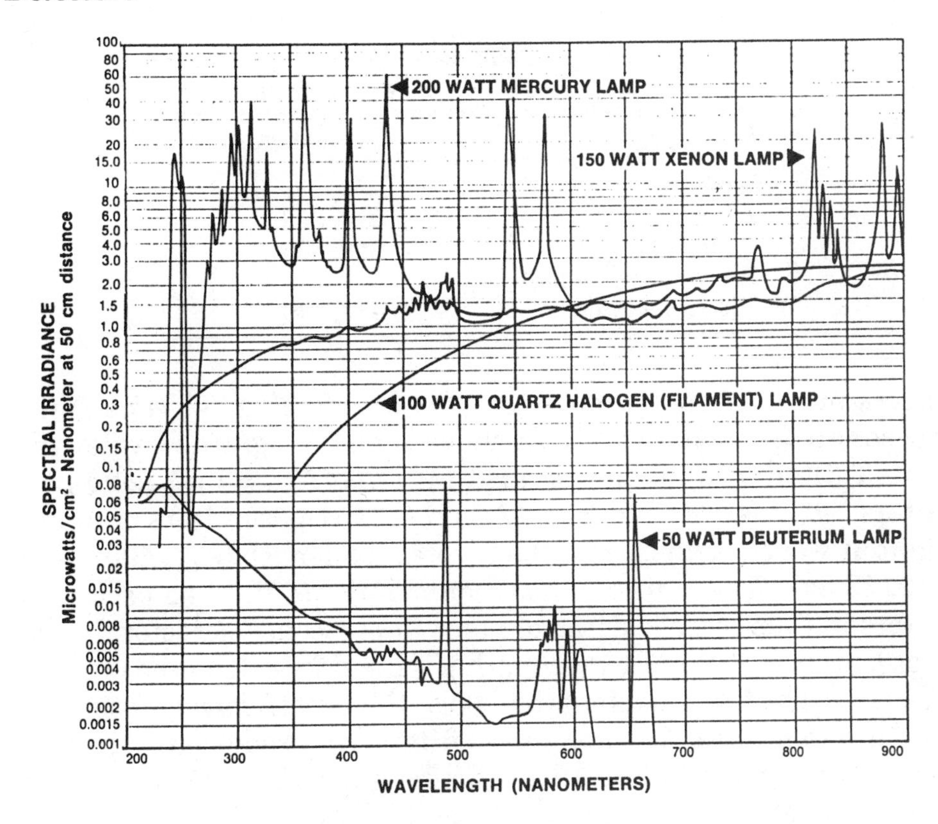


**Fig. 12.2.** Typical output spectra for several detector lamps. Courtesy of Oriel Corp.

ries: line and continuum lamps. The line sources are combined with filters in fixed wavelength detectors. Low-pressure mercury lamps are the most common line source, and are used in fixed-wavelength detectors for detection at 254 or 280 nm. Additionally, xenon, zinc, or cadmium lamps can be used for other wavelengths. Continuum lamps are used with monochromators to select the desired wavelength. The most common continuum source is the deuterium lamp, which is used for variable-wavelength detectors in the 190–360 nm region. Xenon lamps are also used as a continuum source. A comparison of emission spectra for several of these lamps is shown in Fig. 12.2. The common UV lamps and wavelengths used for fixed- wavelength UV detectors are listed in Table 12.1.

A filter or grating is used to select a specific wavelength or range of wavelengths for absorbance measurements. Two types of

Table 12.1
Common UV-Detection Wavelengths

| Lamp | Wavelengths, nm |
|---|---|
| Mercury | 254, 280, 365 |
| Zinc | 214, 308 |
| Cadmium | 229, 326 |
| Black fluorescence | 360 |
| Blue fluorescence | 410, 440 |

filters are used in UV detectors. A *Cutoff filter* (or *edge filter*) passes all wavelengths of light above or below a given wavelength. Cutoff filters are further divided into *Short-pass* and *Long-pass* filters, as illustrated in Fig. 12.3a,b. Short-pass filters transmit light at wavelengths *below* the cutoff wavelength (e.g., 400 nm in Fig. 12.3a). These filters most commonly are used as excitation filters in fluorescence detectors (see fluorescence-detector discussion), and not in fixed-wavelength UV detectors. Long-pass filters transmit light at wavelengths *above* the cutoff wavelength (e.g., 350 nm in Fig. 12.3b). Long-pass filters are used used as second-order filters (see Section 12.2, wavelength problems) and in fluorescence detectors as emission filters.

A *Bandpass* filter passes light in a narrow range or band defined by the filter (e.g., ±5 nm). For example, the 546 nm bandpass filter that was used for the spectrum in Fig. 12.3c transmits light only in the 540–555 nm range. Bandpass filters are used in fixed-wavelength UV detectors to select the detection wavelength and in fluorescence detectors as emission filters.

A grating can be used either before or after the cell in a variable wavelength detector as shown in Fig. 12.4 and 12.5. When placed between the lamp and the detector flowcell (e.g., Fig. 12.4), the grating directs a single wavelength (or very narrow range of wavelengths) of light through the cell. In the most popular configuration, a wavelength is selected manually via a dial or software control. Scanning detectors allow the grating to be rotated during a run (e.g., 20 nm/s) so that the spectrum of a peak can be obtained. This spectrum must be corrected for changes in sample concentration during the scan if much useful information is to be obtained from the spectrum. The scan rate depends on the detector design, and in some detectors is user-selectable. In the detec-

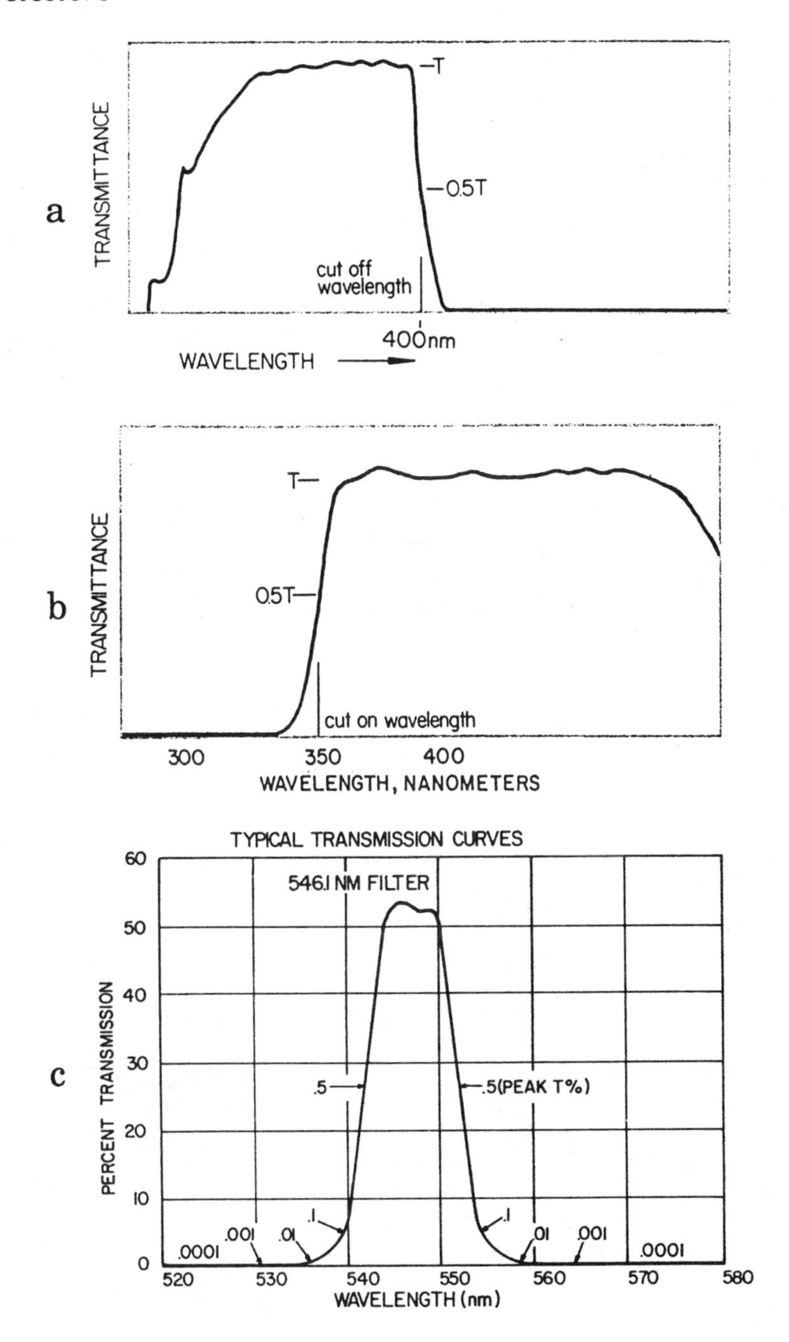


**Fig. 12.3.** Typical spectra from interference filters. (a) transmission of a typical short-pass cutoff filter; (b) transmission of a long-pass cutoff filter; (c) transmission of a bandpass filter. Courtesy of Oriel Corp.

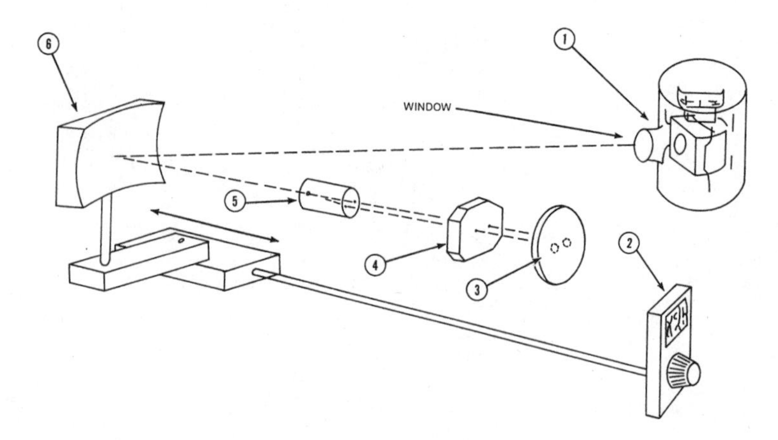


**Fig. 12.4.** Variable-wavelength UV detector configuration with an optical grating mounted between the lamp and the flowcell. (1) deuterium lamp; (2) wavelength selection control; (3) photodiode; (4) flowcell; (5) fiber-optic beam splitter; (6) holographic grating. Courtesy of LDC/ Milton Roy.

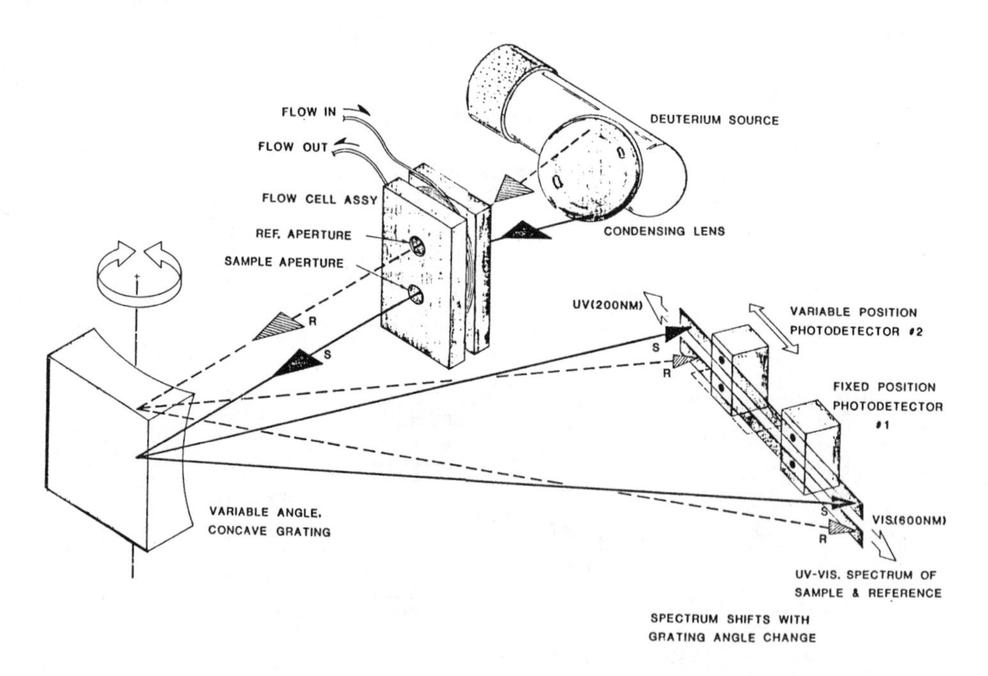


**Fig. 12.5.** Variable-wavelength UV detector configuration with an optical grating mounted between the flowcell and the photodiode. This detector (Micromeritics model 788) can simultaneously monitor two wavelengths. Courtesy of Alcott Chromatography.

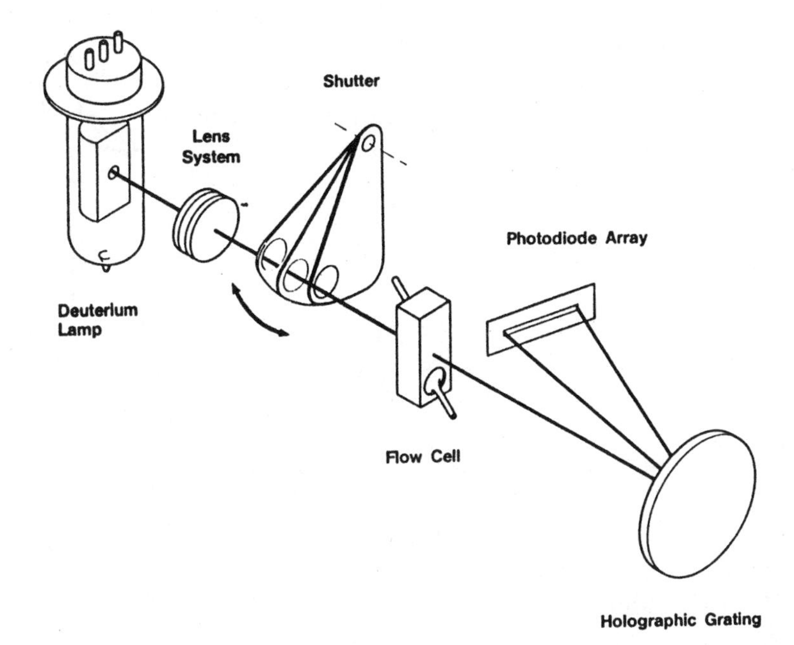


**Fig. 12.6.** Schematic diagram of a photodiode-array UV detector. Courtesy of Hewlett Packard.

tor configuration shown in Fig. 12.5, the grating is mounted after the flowcell, and two discrete photodiodes are used to simultaneously detect two wavelengths. This allows the use of wavelength ratioing to determine peak purity without requiring a diode-array detector.

When a grating is placed between the detector cell and the light sensor (e.g., Fig. 12.5), three possible configurations result: (a) a single wavelength (narrow band) can be selected (as in Fig. 12.5), (b) the grating can be rotated to scan a range of wavelengths, producing results similar to the pre-cell configuration discussed above, or (c) multiple detector elements can be configured in an array so that each element is exposed to a different portion of the spectrum (Fig. 12.6). The *Diode Array* detector uses a photodiode array (e.g., 254 elements) to detect many wavelengths simultaneously. This simultaneous detection gives spectra that do not need to be corrected for changing sample concentration. Because of restrictions in the size of individual photodiode elements, early diode array detectors were less sensitive than other UV detectors, but the newest models have competitive sensitivity.

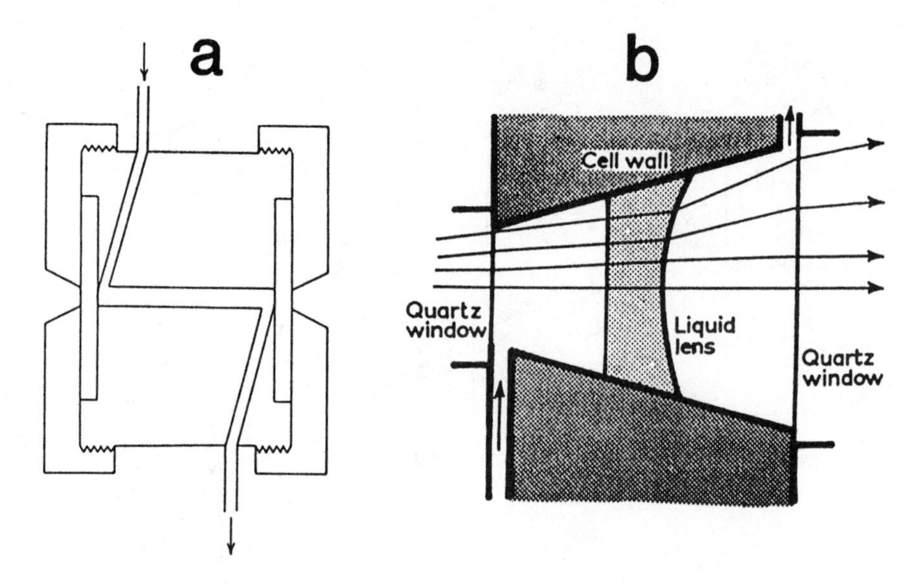


**Fig. 12.7.** UV detector flowcell designs. (a) "Z-path"; (b) tapered cell. Reprinted from ref. (3) with permission.

The detector flowcell is a miniaturized flow-through cuvet through which the sample stream passes. The flowcell is mounted in the optical path as seen in the examples of Figs. 12.4–6. Two detector cell designs are used commonly for UV detectors; one uses a cylindrical cell, and the other uses a tapered cell. The cylindrical, or "Z-path" flowcell (Fig. 12.7a) consists of a block of stainless-steel or Teflon with two holes drilled through it, one hole for the sample cell and one for the reference cell. Quartz windows are sandwiched on either side of the cell body and sealed with a Teflon gasket. The sample passes through the cell in a Z-shaped pattern for good flow characteristics. The most common cell configuration is 1 mm id x 10 mm long, giving a volume of about 8 µL. Smaller-volume cells give less band-broadening, but also have lower signal-to-noise levels.

A second flowcell configuration is the tapered- or stepped flowcell (e.g., Fig. 12.7b). The tapered design improves the signal-to-noise ratio by (a) reducing internal reflections, and (b) minimizing refraction effects.

The sample stream from the column is directed through one passage of the flowcell. The second of the two detector-cell passages is used as a reference cell. This reference cell can be used empty (air reference), or filled with solvent when highly-absorbent mobile phases are used. Because an air reference is most commonly used, some detectors are designed without a reference cell.

A *Heat Exchanger* just before the flowcell reduces detector noise by stabilizing the temperature of the column effluent entering the detector. For example, a piece of stainless-steel capillary tubing (e.g., 0.005 in. id x 0.5 m long) can be wrapped around the detector cell as a heat exchanger. As the sample stream passes through the heat exchanger, its temperature equilibrates with the flowcell temperature, thus stabilizing the temperature within the cell. Although heat exchangers can reduce temperature fluctuations, they can also add undesirable dead volume to the system and degrade peak shape.

Photometric detectors use UV-enhanced photodiodes as the sensor elements. One photodiode is used for the sample cell and one for the reference cell in fixed- and variable-wavelength UV detectors. A photodiode array with 35–1028 elements is used in diode array detectors.

### *Fluorescence Detectors*

Fluorescence detectors measure changes in the fluorescence of the column effluent when it is exposed to selected wavelengths of light. Fluorescence detectors are based either on the straight-path design (similar to UV photometers), or on the more common right-angle design shown in Fig. 12.8. Lamps are the same as those used in UV detectors. A photomultiplier tube is commonly used as the photodetector (vs photodiodes in UV detectors). A short-pass cutoff filter or wide-band bandpass filter is used between the lamp and the cell to limit the maximum excitation wavelength. A bandpass filter or a long-pass cutoff filter is used as an emission filter between the cell and the photodetector. The lower limit of the wavelengths passed by the emission filter should be greater than the upper limit of the excitation filter, or stray light from the lamp may reach the detector.

Fluorescence detectors that use filters to select the excitation and emission wavelenths are called *Filter Fluorometers. Spectrofluorometers* are fluorescence detectors that use a diffraction grat-

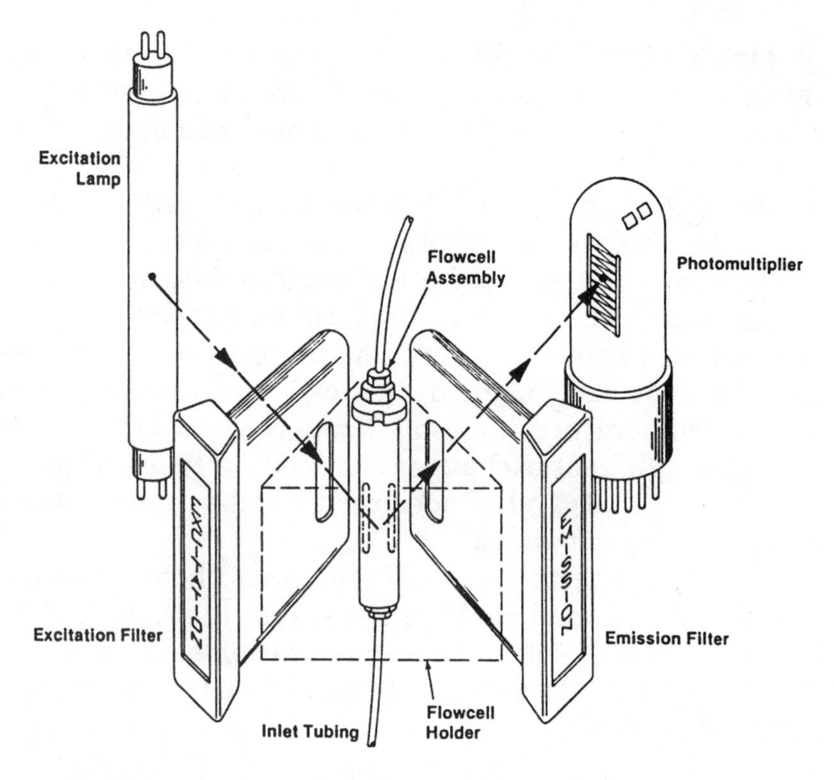


**Fig. 12.8.** Optical layout for right-angle filter fluorescence detector. Courtesy of Applied Biosystems (formerly Kratos).

ing to select the excitation and/or emission wavelengths. If the grating(s) is used in the scanning mode, the detector is a *Scanning Spectrofluorometer.*

With the straight-path design, a standard UV cell can be used, but the specifications of the filters must be selected carefully to prevent stray light from reaching the photodetector. Without the proper blocking filter, UV light will pass through the cell to the photodetector, giving a high background fluorescence reading (because of excessive background light). Thus, the filter must prevent any excitation wavelengths from reaching the detector, but pass all the emission wavelengths, or a poor signal-to-noise ratio will result.

Right-angle fluorometers often use a cylindrical cell made from a piece of quartz tubing (Fig. 12.9). This design is less efficient than the straight-path cell because light scattering and light-piping problems result in less of the light reaching the photodetec-

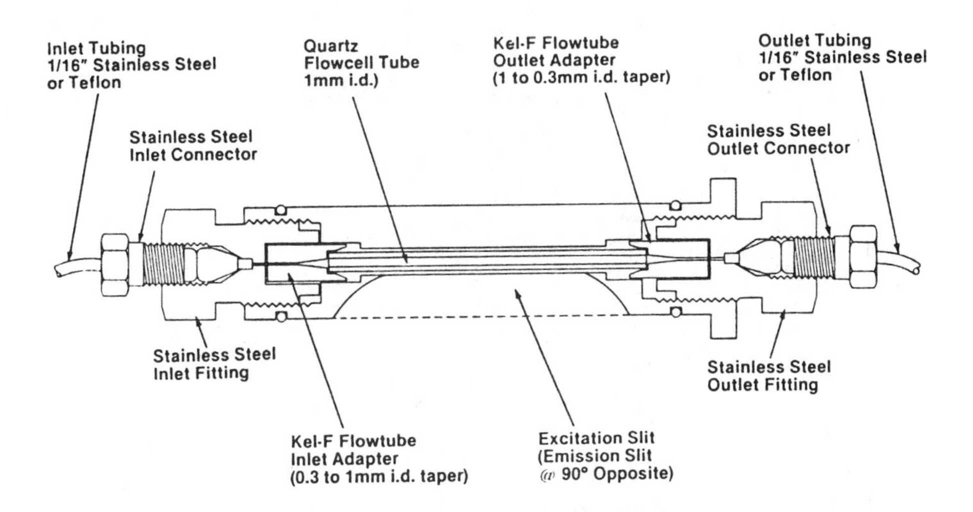


**Fig. 12.9.** Fluorescence detector flowcell using a cylindrical quartz tube for the optical cell. This cell is used for right-angle fluorescence detectors, such as shown in Fig. 12.8. Courtesy of Applied Biosystems (formerly Kratos).

tor. The right-angle design, however, is less susceptible to interference from stray light from the lamp, because the photodetector is not directly in line with the lamp.

Figure 12.10 shows a third design of fluorescence detector cell, one which incorporates a reflective cell and a concave mirror. The reflective back surface on the cell prevents the excitation wavelengths from passing through the cell to reach the photodetector. A concave mirror collects scattered (fluorescing) light emitted from the sample cell and focuses it on the photodetector for increased collection efficiency.

## *Refractive Index Detectors*

Refractive index (RI) detectors measure changes in the refractive index of the column effluent. The RI detector is a bulk-property detector which responds to all compounds (i.e., it is universal, and *not* selective). This characteristic makes the RI detector useful for detecting unknown compounds and compounds that are not amenable to other types of detection. The RI detector is less sensitive; RI detectors therefore are used when more sensitive detectors will not respond to the compounds of interest. RI

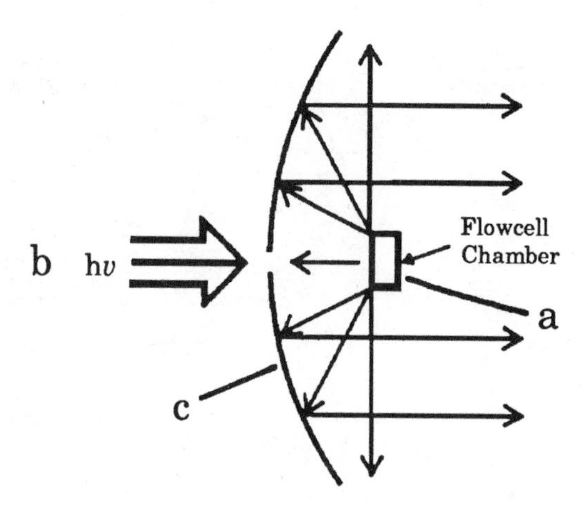


**Fig. 12.10.** Fluorescence detector cell using a reflective back surface (a) to prevent radiation from the lamp (b) from reaching the photomultiplier (PMT). Concave mirror (c) aids in increasing the efficiency of collecting the emission radiation. Reprinted from ref. (4) with permission.

detection is used most commonly when detection sensitivity is less important (e.g., preparative LC and size-exclusion chromatography).

Three RI detector designs are common: (a) *Fresnel,* (b) *deflection,* and (c) *interferometric.* The Fresnel refractometer (Fig. 12.11) is based on Fresnel's law of reflection, which states that the amount of light reflected at a liquid–glass interface depends on the refractive index of the solution and the angle of incidence of the light.[3] In practice, light from a tungsten source is split into sample and reference beams and focused on the detector cell, as shown in Fig. 12.11. An infrared-blocking filter is used between the lamp and the cell to prevent cell heating. The sample and reference cells (e.g., 3 μL volume) are formed with a Teflon seal between a prism and a polished backing plate. Reflected light is focused on a photodiode pair for measurement. Initial setup involves adjustment of the angle of incident radiation to a subcritical angle for maximum sensitivity. A disadvantage of the Fresnel refractometer is that it requires two different prisms to cover the useful refractive-index range. This generally is not a problem, however, because one prism covers the range for most reversed-phase solvents, and the other is used for normal-phase

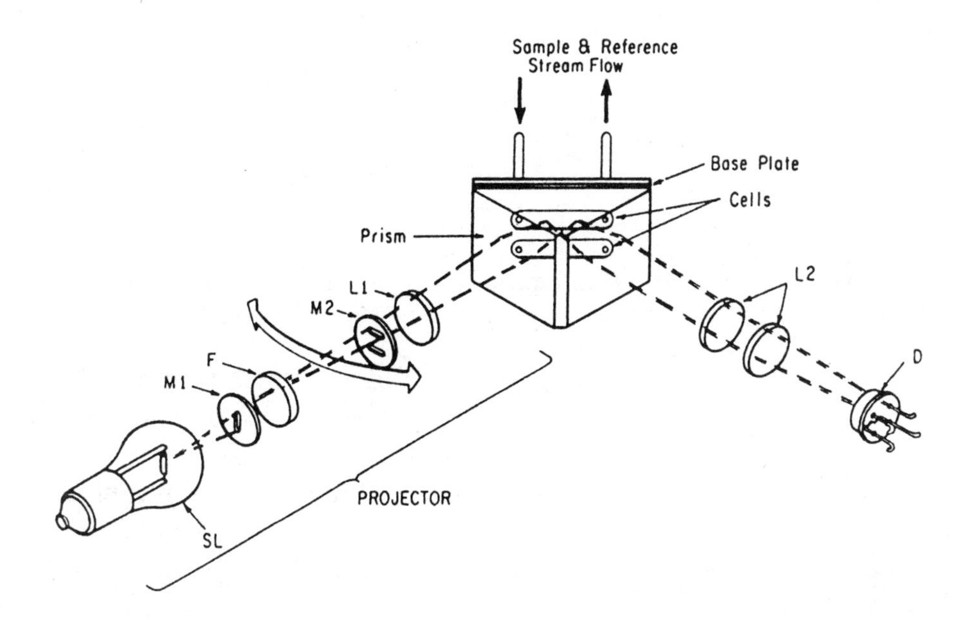


**Fig. 12.11.** Schematic diagram of a Fresnel refractometer. SL, lamp; M1, M2, masks; F, infrared blocking filter; L1, L2, lenses; D, photodetector. Courtesy of LDC/Milton Roy.

solvents. Fresnel refractometers are also more sensitive to problems resulting from a dirty flowcell than other styles.

The deflection refractometer (Fig. 12.12) uses a cell with sample and reference compartments separated by a glass divider. Light passes through the sample and reference cells, is reflected

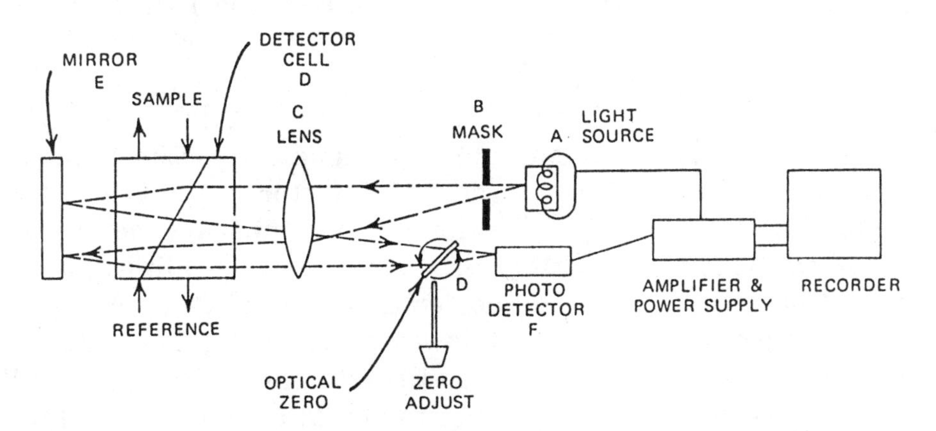


**Fig. 12.12.** Schematic diagram of a deflection refractometer. Courtesy of Waters Chromatography Division.

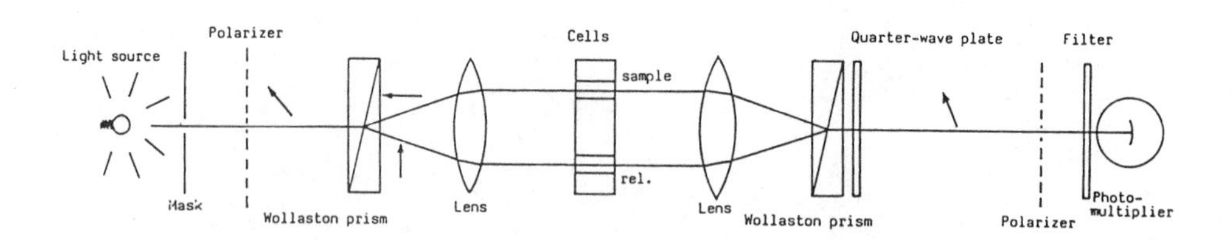


**Fig. 12.13.** Schematic diagram of an interferometric refractometer. Courtesy of Tecator Inc.

back through the cells, and is directed to a position-sensitive photodetector. Changes in the refractive index of the mobile phase (as sample flows through the detector cell) cause changes in the deflection of the light beam and thus change the output signal. The deflection refractometer has the advantage that a single cell can provide measurements over the entire refractive index range. The cell volume (e.g., 10 µL), however, is greater than the Fresnel refractometer, so more band-spreading is experienced.

The interferometric refractometer is shown in the schematic of Fig. 12.13. Light from the lamp is polarized, passes though a beam-splitter, and is directed onto the sample and reference cells. When the refractive index of the sample and reference cells is not the same, a phase shift occurs between the two beams of light. This phase shift results in a difference in the intensity of the light that reaches the photodetector, and thus changes in the detector output. The interferometric refractometer uses small-volume detector cells (e.g., 1–12 µL), so band-broadening is minimized. This RI-measurement technique can be 10–100 times more sensitive than other RI detectors.[5]

RI detectors use a liquid-filled reference cell; because of the difficulty in matching sample and reference refractive indices, gradient elution is not used with RI detectors. Whenever the mobile phase is changed, the reference cell must be thoroughly flushed with the new mobile phase. A reference-cell filling-valve (Fig. 12.14) is a labor-saving device that can be added easily to any RI detector. A three-way low-pressure valve is connected between the sample-cell outlet and the reference-cell inlet. To fill the reference cell, the valve is positioned to direct mobile phase through the sample cell and then through the reference cell. When flushing is complete, the valve is returned to the operating

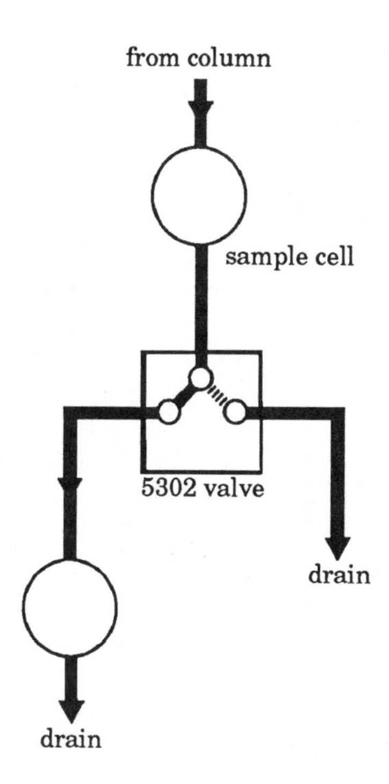


**Fig. 12.14.** Use of a three-way low-pressure valve to facilitate filling the reference cell of an RI detector with mobile phase. Courtesy of Rheodyne Inc.

position and the outlet of the sample cell is directed to waste; the reference cell inlet is blocked in this position so that liquid remains in the cell.

Refractive index detectors are very temperature-sensitive, so temperature stabilization of the mobile phase, column, and detector is necessary for maximum sensitivity. The most sensitive detectors are designed so that the injection valve, column, and detector are all mounted in a single oven with the temperature regulated to better than ±0.01°C (e.g., Tecator Model 5931).

## *Electrochemical Detectors*

Electrochemical (EC) detectors (see typical flowcells of Figs. 12.15, 12.16) measure the ability of an analyte to be oxidized or reduced in the presence of an applied potential. As the electrochemical reaction takes place at the working electrode, a potentiostat is used to maintain a constant potential across the cell.

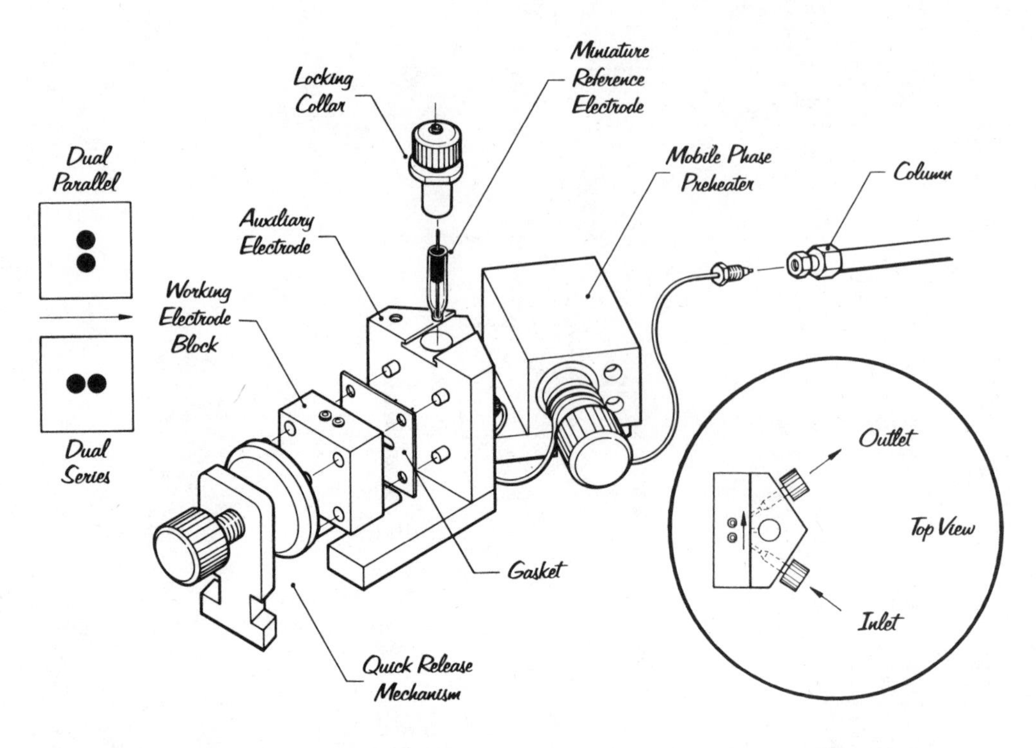


**Fig. 12.15.** Thin-film electrochemical cell. Courtesy of Bioanalytical Systems, Inc.

Because the potential across the working and auxiliary electrodes is held constant, EC detectors are more properly termed *Amperometric* detectors. The detector output tracks the adjustments made by the potentiostat. A reference electrode is used to provide a stable potential for measuring the potential at the working electrode. The two most popular configurations of the EC detector are the thin-film cell (Fig. 12.15) and the flow-through cell (Fig. 12.16), as described below.

The *Thin-Film* cell configuration is shown in Fig. 12.15. In the example shown, the working electrode is glassy carbon, the auxiliary electrode is stainless-steel, and the reference electrode is Ag-AgCl. The cell cavity is formed by a spacer-gasket mounted between the Kel-F block, which supports the working electrode, and the stainless-steel auxiliary-electrode block. The cell volume is determined by the thickness of the gasket. Working electrodes made of gold, platinum, or carbon paste also are available. The thin-film cell has a very small working surface, and as a result,

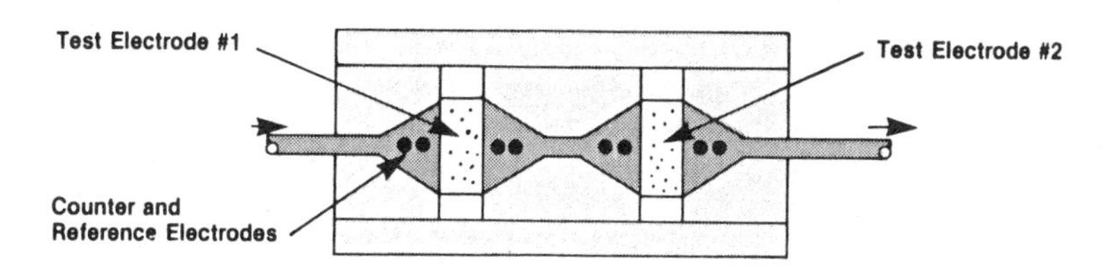


**Fig. 12.16.** Flow-through electrochemical detector cell. This cell contains two working electrodes; counter electrode is equivalent to auxiliary electrode in Fig. 12.15a. Courtesy of ESA, Inc.

contamination can be a major problem if care is not taken to keep the cell very clean. The cell can be cleaned chemically, or it may require abrasive cleaning of the working electrode. Even though the flow characteristics result in a low reaction efficiency (generally less than 5%) the EC detector is one of the most sensitive and selective detectors available.

The flow-through electrochemical detector cell is shown in Fig. 12.16. The cell consists of two fully porous graphite working electrodes placed in series. The reference and counter electrodes are made of a proprietary material and are placed in symmetrical pairs in close proximity to the working electrode. This cell design ensures that the current and potential distribution will be uniform and that the uncompensated cell resistance will be minimized. Cell volume of less than 5 μL results in minimum loss of resolution occurring within the detector.

The fully porous working electrodes allow the column eluent to flow through (rather than over) each electrode. Thus the electrode is claimed to react with 100% of an electroactive component in the chromatographic eluent as compared to 1–5% for the thin layer cell. The electrode-frit, however, is extremely susceptible to blockage, so extra care is required to assure that the column effluent is particulate-free (an in-line filter is recommended).

EC detectors are quite sensitive to bubbles, so mobile phase degassing is required. Continuous helium sparging is necessary to exclude oxygen when the reductive mode is used. Teflon tubing has sufficient permeability to oxygen to cause problems, so Teflon tubing must be replaced with stainless-steel in the reductive mode. Additionally, EC detector cells are slow to equilibrate, often requiring several hours. In certain cases, it is necessary to

Table 12.2
Common UV Detector Problems[a]

| Problem type | Problems reported, % |
|---|---|
| Noise/drift/sensitivity | 47 |
| Noise | 26 |
| Drift/stability | 12 |
| Sensitivity | 9 |
| Lamp life | 24 |
| Bubbles | 7 |
| Dirty cell | 6 |
| Other | 8 |
| No problems | 11 |

[a]Reprinted from ref. (7) with permission. Totals do not equal 100% due to rounding.

passivate all of the stainless-steel surfaces in the LC for optimum performance. In spite of these operational inconveniences, the EC detector's selectivity and extreme sensitivity make it the detector of choice for many routine LC assays.

## 12.2. Detector Problems and Solutions

The discussion in this section is limited to problems associated with UV detectors. Much of what is said can be directly applied to other detectors as well, because the problems from a troubleshooting standpoint are similar. A summary of these problems and solutions is found in Table 12.3 at the end of this chapter. Problems associated with the detector cell (bubbles, dirt, etc.) are fairly universal, so discussion of these will apply more generally than specific UV-detector problems (e.g., second-order radiation). Consult the operator's manual for the detector before attempting cleaning or adjustment, in order to determine specific recommendations for that model of detector. The operator's manual also contains exploded diagrams which will be useful for detector service procedures.

Table 12.2 lists the most common UV detector problems encountered by respondents to an *LC/GC Magazine* survey.[6,7]

Though these data are only for UV detectors, the limited number of responses for other detector types indicates that the same general trends are true for other detectors, as well. Solutions to these and other problems are covered in the following discussion. It will be seen that the same symptom can be caused by several different detector problems. Refer to the troubleshooting tree in Chapter 2 as an aid to isolating the various causes of these problems.

The survey of Table 12.2 indicated that an approximately equal number of fixed-wavelength and variable-wavelength UV detectors were in use by the respondents. However, five times as many problems were reported for variable-wavelength UV detectors. Though these data should not be interpreted quantitatively, they do suggest that fixed-wavelength detectors are more reliable, in agreement with most user's experience. Because of this, when methods are developed for the routine analytical lab, a fixed-wavelength detector is preferred from a reliablity standpoint. Maintenance is further minimized because fixed-wavelength detectors never have to be calibrated for wavelength accuracy.

### *Lamp Problems*

*Lamp failure.* With the exception of EC detectors, all the detectors discussed in this chapter use a lamp as part of the detector. Lamp failure obviously results in a total loss of detector signal. Most detectors have a lamp viewing-port or lamp-status meter which allows one to visually confirm that the lamp is operating. Note that UV radiation can cause eye damage, so do not directly view lamps which produce UV radiation (e.g., deuterium or mercury lamps). Often you can verify that the lamp is on by the purple glow reflected off other components in the detector.

*Baseline noise.* A more common problem than complete lamp failure is increased baseline noise caused by the lamp. A deuterium lamp is noisiest during the first 30 min after startup, so a minimum of 30 min should be allowed each day for warmup before the detector is used. Lamp noise also increases with lamp age. This problem most commonly is encountered with deuterium lamps, because the normal lifetime of these lamps is in the

400–1000 h range (vs 2000 h or more for mercury and tungsten lamps). Increased short-term baseline noise plus occasional noise spikes in the chromatogram are symptomatic of lamp aging. You can help to isolate other possible sources of noise by (a) rerunning a chromatogram under standard reference conditions (i.e., eliminate assay variables), (b) substituting a known good detector for the suspect one (i.e., isolate the problem to the detector), or (c) replacing the lamp. (Although lamp replacement can be expensive, it generally is easier to accomplish than isolating other problems, so many workers replace the lamp at the first indication of problems.) If noise spikes in the chromatogram are observed, isolate the source of these by turning the pump off (or set the flow rate to 0 mL/min). Noise spikes caused by bubbles will disappear when the flow is stopped, but spikes caused by the lamp will persist.

If a deuterium lamp has been in use for six months or more, it may be near the end of its useful lifetime. Therefore lamp replacement is often the simplest option. Deuterium lamps have a shelf life of 6–12 months; it is more cost-effective not to stock a spare lamp— instead order one by overnight delivery. Another option is to order a new lamp after the old one has been in use 3–4 months; in this manner a spare is available, but the shelf-life problem is reduced. If you have several detectors that use the same style of lamp, a single lamp can be stocked as a replacement part versus keeping several lamps on hand.

Keeping complete records of lamp on-time can be tedious, so replacement is best scheduled on a calendar basis (e.g., replacement every six months). Many vendors sell deuterium lamps with a built-in meter that records total on-time for the lamp. With this device, it is easy to determine when the lamp has been in use for 500–1000 h, so replacement can be scheduled on a lamp-use basis rather than a calendar basis. One vendor (ISCO) has added a proprietary electronic lamp circuit that is claimed to extend deuterium lamp life to over 5000 h.

*Lamp replacement.* Replacement of a bad lamp should eliminate the lamp problem. When replacing a detector lamp, be careful to avoid getting fingerprints on the lamp itself; when the lamp is turned on, fingerprints can result in permanent damage to the surface of the lamp. Handle the lamp only with a soft cloth or laboratory wiper (e.g., Kimwipe) and gently clean any fingerprints from the surface before turning on a new lamp. A cotton swab saturated with methanol can be used to clean the surface of the lamp. The lamp should be aligned according to the instructions in the detector operator's manual. A new lamp should be allowed to operate for at least an hour to "burn in," before using the detector for quantitative analysis.

### *Cell Problems*

*Bubbles.* The most common detector-cell problem is to have an air bubble pass through or become trapped in the cell. Transient bubbles will cause spikes in the chromatogram; trapped bubbles will cause an off-scale pen deflection. Bubbles cause similar problems in all detector types, not just UV detectors. In detectors where it is possible to view the flowcell, you can see the presence of a bubble, which appears as an annular ring rather than a clear green image at about 670 nm (be sure to wear UV-absorbing safety glasses whenever viewing such a cell with the lamp on).

Bubbles usually result from improperly degassed mobile phase, but the problem often is made worse by a dirty cell. Noise spikes in the chromatogram can be caused by lamp failure or by bubbles (see discussion above).

Bubble problems usually can be eliminated by using degassed solvents, so that the concentration of air in the final mobile phase is never greater than the saturation point. (See Chapter 6 for specific recommendations on degassing.) Sometimes it is necessary to add a restrictor to the detector waste-line in order to prevent bubble problems. Commercially available backpressure devices (e.g., SSI, Upchurch) can be attached to the detector outlet to provide a constant pressure. Alternatively, a length of small-diameter Teflon tubing (e.g., 1 m x 0.010-in. id) used as a waste line will provide sufficient backpressure to prevent bubble problems in the detector cell. Another possibility is to mount an old (short) column after the detector. In all cases, be sure not to exceed the detector manufacturer's recommended pressure limit for the cell. Finally, be aware that occasional bubbles in the detector waste-line are normal, and are not indicative of any problem.

A problem that gives the same symptoms as bubbles is the presence of an immiscible solvent in the detector cell. This can occur if a column or the LC system has been used in a reversed-phase mode and is changed improperly to the normal-phase mode (or vice versa). Once "bubbles" of immiscible solvent get into the system, they may be washed out very slowly, unless the system is flushed with a solvent miscible in both reversed- and normal-phase mobile phases. 2-Propanol is a good flushing solvent.

If a reversed-phase mobile phase is contaminated with a normal-phase solvent, flush the system with five column-volumes of unbuffered mobile phase (if buffer is used). Next, pump ten column-volumes of the strong solvent of the mobile phase (e.g., acetonitrile). Now flush the system with about 50 column-volumes of propanol to remove any remaining mobile phase plus the unwanted solvent residues. Reverse these steps to re-equilibrate the system with the reversed-phase mobile phase. If a normal-phase mobile phase is contaminated with an aqueous solvent, just flush 50 column-volumes of propanol, and then return to mobile phase. The procedure of flushing with a mutually miscible solvent (e.g., 2-propanol) can be used to change the LC system from reversed-phase to normal-phase operation (or vice versa). Be sure to pump at least ten column-volumes of the final mobile phase for equilibration, before running a sample.

*Cell blockage.* A blocked detector cell is indicated when (a) a high system-pressure is observed, and (b) the pressure drops to normal levels when the inlet fitting to the detector is loosened. The detector can become blocked in one of three places: (a) the detector inlet-line or heat exchanger, (b) the cell itself, or (c) the outlet (waste) line. If the blockage is sufficient to cause a noticeable rise in system pressure, it is unlikely that the blockage can be removed with the cell cleaning procedure descibed below, but it is worth trying. If the blockage is sufficient to prevent drawing solvent through the cell with a syringe, disassembly or replacement probably will be required. For cells with high upper pressure limits (e.g., 1000 psi), backflushing the cell using the LC pump to drive the solvent will often displace the blockage, but *Do Not* attempt to do this with normal cells (e.g., 150 psi pressure limit). It may be more cost-effective at this point to return the cell to the manufacturer for repair; consult the operator's manual for specific instructions.

If the above blockage-removal procedure is unsuccessful, the blockage must be located before it can be cleared. If the inlet and waste lines can be disconnected from the cell body, these can be tested for restrictions, independent of the cell. If both of these lines are clear, the blockage must be in the cell itself. With some cell designs, it is possible to view a blockage inside the cell before it is disassembled.

The most likely location for a cell blockage is in the inlet line, because the 0.010-in. id (or smaller) tubing will readily trap any particulate matter that enters the detector. This can be a problem when (a) home-packed columns are used and a few particles of packing are left downstream from the outlet frit, or (b) when a column is reverse-flushed for an insufficient time before reconnecting it to the detector (Section 11.5). If blockage is a recurring problem, a zero-volume in-line filter should be used between the column and the detector. Sometimes the blockage can be backflushed from the inlet line or heat exchanger if the line can be disconnected from the cell body. Just connect the cell end of the line to the pump, and pump solvent backwards through the inlet line until the blockage is cleared. For stubborn blockages, and in detectors where the line cannot be disconnected from the cell, it is often possible to cut about 1-cm of tubing from the inlet line and thus remove the blockage. Clearly, each time this procedure is performed, there is less tubing to work with, so this is not the most desirable procedure. Finally, if the line cannot be cleared, it should be replaced with the proper replacement part or returned to the factory for repair. The procedures for clearing the inlet line can also be used to clear the detector outlet (waste line).

Blockage of the cell itself usually is caused by a chemical problem (e.g., buffer precipitation); particulate matter that might block the cell generally gets trapped before the cell (in the inlet tubing). If aqueous buffers are allowed to stand in an unused system, or if an organic solvent is used to flush buffer from the system, precipitation in the detector cell can occur. For cases in which the cell is not blocked completely, it may be possible to dissolve a buffer blockage by pumping hot water through the cell. Other types of blockage sometimes can be cleared with the nitric-acid cleaning procedure discussed below. In most cases, however, the cell must be disassembled for cleaning.

*Cell contamination.* The detector cell windows can become dirty from unclean samples or from polymerization of sample components on the windows. A dirty cell gives noisier chromatograms and often increases the frequency of bubble problems.

*Cell cleaning (nitric acid).* To clean the detector cell, disconnect it from the column and waste line. Note: before starting this procedure, be sure to have proper safety protection (e.g., glasses, apron, rubber gloves, etc.) because nitric acid can cause severe burns. It is also good to know the location of an acid spill-kit in case of an accident. Place the *outlet* line from the cell in a beaker of 2-propanol. Connect a syringe (e.g., 10 mL capacity) to the cell's inlet line (a piece of small-diameter Tygon tubing is a useful coupler). Now draw about 10 mL of propanol through the cell to remove any residual mobile phase. Next, draw about 10 mL of distilled water through it and place the cell outlet line in a beaker containing ca. 6*N* nitric acid (e.g., 50 %v conc. nitric acid/water) and draw about 10 mL of this solution through the cell. Next, draw about 20 mL of distilled water through, followed by at least 100 mL of HPLC-grade water. If reversed-phase mobile phases are to be used, mobile phase may be pumped next. If normal-phase mobile phases are to be used, flush the water from the cell with about 10 mL of 2-propanol before pumping mobile phase.

Drawing the cleaning solution through the cell in the reverse direction serves three functions. First, there will be a negative pressure in the system, which greatly reduces the likelihood of spraying nitric acid on yourself or the instrument in case a connection slips. Second, by flowing in the reverse direction, any particulate matter trapped in the detector tubing is likely to be backflushed out of the system. Finally, by keeping a negative pressure on the cell, the probability of cell damage or leakage due to excessive pressure is reduced.

*Cell leakage.* The detector cell can leak (a) at the fittings, (b) at the gaskets that seal the windows on the cell, (c) between the sample and reference sides of the cell, or (d) through cracked quartz windows. Except for the case of loose or faulty fittings, cell leaks are the result of either applying too high a pressure across the cell or improper rebuilding. Excessive pressure can result from a tubing blockage, too high a flow rate, or placing too much restriction after the cell. Once the cell gaskets have leaked, it is usually necessary to replace or rebuild the detector cell to solve the problem.

*Rebuilding the flowcell.* The detector flowcell should be disassembled for cleaning only after attempts to clean the cell with solvents and/or nitric acid have failed (see cleaning procedure above). It is easier to damage the flowcell than to improve its operation when disassembly is attempted. Follow the manufacturer's directions carefully for disassembly, cleaning, and reassembly.

*Sample / reference mismatch.* UV detectors normally are operated with an air reference; i.e., the reference cell contains only air. Generally the electronics are designed with enough range to allow nulling the mobile phase background versus the air reference. In some cases (e.g., with UV-absorbing mobile phases for UV detectors, and always with refractive index detectors), the reference cell is filled with mobile phase. If the mobile phase in the reference cell is not matched with that in the sample cell, it may not be possible to zero the output signal. In this case, the reference cell should be flushed with fresh mobile phase; a valve setup such as that shown in Fig. 12.14 will facilitate the procedure. When converting from a wet-reference to an air-reference, be sure to thoroughly dry the reference cell with a stream of dry nitrogen; solvent residues can cause problems similar to bubble problems.

### *Wavelength Problems*

*Second-order radiation effects.* Some variable-wavelength UV detectors are designed to use the deuterium lamp in the visible region. An example of this is shown in Fig. 12.17a, where a deuterium lamp is being used for an absorbance measurement at 405 nm. One of the characteristics of monochromators is that they pass light at higher orders of radiation than their set values; the most important of these in LC is at half the set wavelength. In the present example, the grating would pass light at 405 (first-order) and 202.5 (second-order). The third-order radiation is well below the wavelengths of interest for LC and is absorbed by atmospheric oxygen, so it is not a problem. The second-order radiation, however, is in the normally measured UV region. This means that the detector is monitoring the absorbance at two wavelengths simultaneously (this can result in nonlinear behavior if Beer's law is violated). In the present example, the wave-

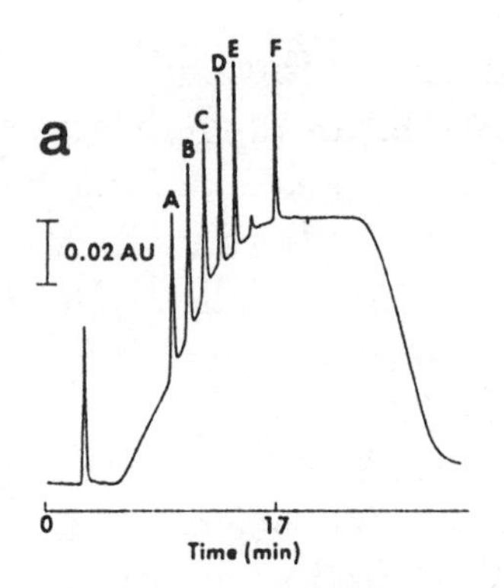


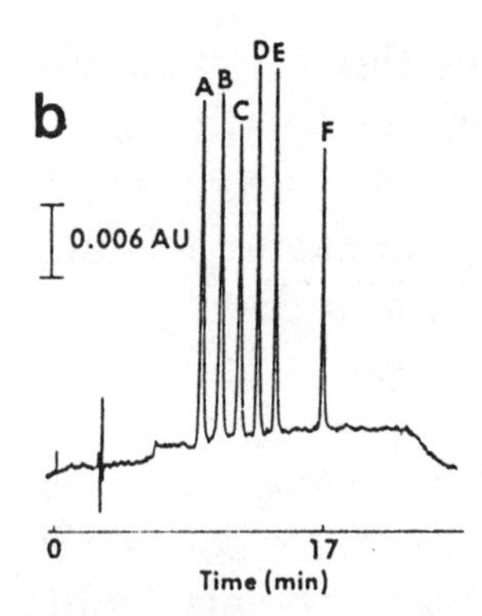


**Fig. 12.17.** Example of second-order radiation effects with UV detectors. Gradient-elution separation of urinary porphyrins with absorbance detection at 405 nm. (a) Detector using deuterium lamp; (b) Detector using tungsten lamp. Reprinted from ref. (8) with permission.

length of the second-order radiation is below the UV cutoff of the mobile phase, methanol, so when a gradient is run, the baseline shows a strong rise as the volume fraction of methanol changes. Also, the likelihood of interferences is high, because so many compounds absorb in the 200-nm region. The problem was solved in this example by using a tungsten lamp, which has no radiation in the UV region, to produce the chromatogram in Fig. 12.17b. It is wise to be aware of potential second-order radiation problems when using deuterium lamps for detection above 360 nm.

Some detectors that use a deuterium lamp for detection in the visible region (e.g., Micromeritics model 788) avoid problems with second-order radiation by using a cutoff (blocking) filter that will not transmit light below 400 nm.

The problem of second-order radiation also can cause problems with fluorescence detection, when monochromators are used to select the excitation wavelength. In this case, the second-order radiation results in the simultaneous use of two excitation wavelengths (i.e., first- and second-order), rather than a single wavelength. With filter fluorometers, diffraction gratings are not used, so the second-order problem does not exist.

*Poorly chosen detection wavelength.* The choice of the detection wavelength can affect the reproducibility of the assay results. If possible, the wavelength should be on a plateau in the absorbance spectrum for the compound of interest. A hypothetical

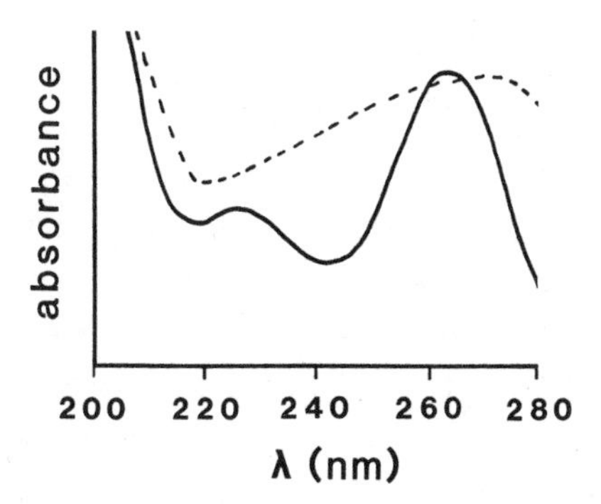


**Fig. 12.18.** Hypothetical UV absorbance spectra for two compounds, A (——) and B (- - - - -). See text for discussion. Reprinted from ref. (9) with permission.

example is seen in Fig. 12.18. If a wavelength of 260 nm is chosen, both compounds A and B of Fig. 12.18 are on a plateau. When the chosen wavelength is on a plateau, the signal is expected to be the same each time the wavelength is reset, because slight variations in the wavelength will not result in significant changes in the absorbance. This wavelength is also more likely to give the same response for more than one detector, because slight differences in wavelength calibration will give insignificant changes in sensitivity. At 254 nm, on the otherhand, the UV spectrum for A is changing rapidly. Thus, small variations in the wavelength would result in much larger changes in absorbance than at 260 nm.

When initially choosing the detection wavelength for a method, it is wise to consult the UV-absorbance spectra for the compounds of interest, so that a wavelength for maximum stability can be selected. Choosing a detection wavelength that is too close to the UV cutoff for the mobile phase can increase the baseline drift when gradient elution is used. If this is a problem, either use a higher wavelength or a different mobile phase.

*Poor wavelength-selection technique.* With many variable-wavelength UV detectors, the wavelength is selected by mechanically turning the grating until the proper wavelength is projected onto the detection photodiode. The mechanical linkage between the wavelength selection knob and the grating may consist of several gears and levers, each of which has mechanical play. With age, mechanical wear can increase the amount of play in the linkage. The net result of these variations is that the actual wavelength selected may be different than the dial setting (depending on the technique used to select the wavelength). Consider the case in which the present wavelength is 280 nm, and a new wavelength of 254 nm is required. If the wavelength selection dial is turned from 280 to 254 nm, the true wavelength will be higher than if the dial were turned to 240 and then to 254 nm. Depending on the detector design and the amount of wear, the difference between the two settings can be several nanometers.

The best way to avoid irreproducible wavelength-settings is to use the same technique each time. For example, always select a new wavelength from at least 10 nm below the desired wavelength. This would mean turning from 280 to 244 nm (or lower) to 254 nm in the above example. With all the mechanical play taken up in the same way each time, the precision of the wavelength setting should be better. Errors in the accuracy of the wavelength setting are not addressed by this technique. If the wavelength is selected electronically by entering the desired wavelength on a keypad, The same problems are potentially present unless they are compensated for instrument design.

*Wavelength calibration problems.* Miscalibration of the detection wavelength can give unexpected results for UV detection. If miscalibration is suspected, a quick check of calibration can be made using the 656 nm spectral peak of the deuterium lamp (see Fig. 12.19, note that this is a plot of spectral irradiance, not absorbance). Just turn the wavelength selection dial to about 640 nm and adjust the attenuation and detector zero so that the pen is at about 80% of full-scale on the chart recorder. Now slowly turn the wavelength dial to about 675 nm. An absorbance minimum (or maximum, depending on the how the detector is zeroed) should be seen when 656 nm is reached if the detector is calibrated properly. A similar negative peak should be seen at 486 and at 582 nm (see Fig. 12.19), but these are weaker, and often difficult to find; the observed error for all three wavelengths should be the same.

If the detector is out of calibration, it is best to consult the operator's manual for specific instructions for calibrating each model of detector. Many calibration techniques use a potassium chromate solution to calibrate both the absorbance and wavelength (275 nm). Some detectors (e.g., Hewlett-Packard diode

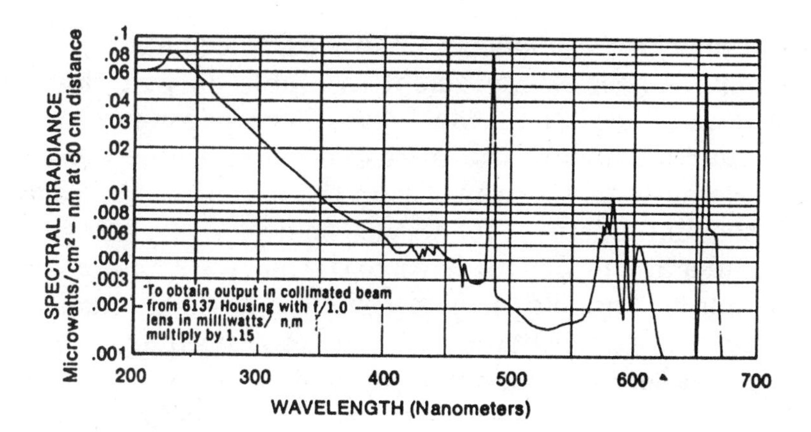


**Fig. 12.19.** Typical deuterium lamp output spectrum. Maximum irradiance (e.g., 656 nm) shows up as a minimum absorbance (negative peak) when the lamp is scanned using a UV detector. Courtesy of Oriel Corporation.

array detector) include a holmium oxide filter that is used with built-in detector software for automatic wavelength calibration.

*Low-wavelength problems.* When a low wavelength (e.g., below ca. 210 nm) is used, detector-related problems become more apparent because the detector generally is more sensitive to changes in both sample and mobile-phase composition than at higher wavelengths. The mobile-phase solvents should be chosen for low cutoff-wavelengths (e.g., acetonitrile is better than methanol); consult refs. (10) or (11) for UV-cutoff data. Detector drift is worse when a wavelength near the UV cutoff of the mobile phase is used. A buildup of ozone in the optical path of the detector can cause increased noise when detection wavelengths below 200 nm are used[12]; continuous purging of the monochromator with nitrogen or helium will solve this problem. Finally, low-wavelength detection often is chosen to enhance the detection of compounds with poor UV chromophores, but this increases the response to unwanted compounds. As a result, detection at low wavelengths can result in more complex chromatograms than for the same sample at higher wavelengths.

*Servicing the monochromator.* The monochromator typically is located in a sealed compartment, and opening the compartment will void the manufacturer's warranty. No user-servicible parts are contained in the monochromator compartment, and adjustments require special equipment available only at the factory. During years of use in a normal laboratory environment, the grating and mirrors inside the monochromator compartment may become coated with pollutants from the air. If this is suspected, consult the manufacturer for service advice.

The inlet and outlet windows to the monochromator compartment should be kept clean. These should be visibly inspected for fingerprints, dirt, or cloudiness each time the lamp is replaced. Use a methanol-saturated cotton swab to clean the windows, allowing ample time for the surfaces to dry before placing the detector back into operation. (This technique can also be used to clean the windows on the photodetector.)

### *Other Problems*

*Time constant.* An improperly selected time constant can adversely affect the chromatogram in one of two ways. First, if the time constant is too small (too fast), increased short-term baseline noise can be expected. Second, if the time constant is too large (too slow), broader, tailing, shorter bands can be expected (i.e., extra-column band broadening). Some detectors use a toggle switch on the back of the detector to select the time constant; care should be taken not to bump the switch to an unwanted postion when making cable connections at the rear of the detector.

An example of band broadening caused by an excessively large time constant is shown in Fig. 12.20. A good rule of thumb is to use a detector time-constant that is about 10% of the width of the narrowest band of interest. Thus, a setting of 0.5 or 0.1 s is suitable for most analytical columns (e.g., 15- or 25 x 0.46 cm, 5-μm particles). Shorter columns with smaller particles will produce narrower bands and thus require smaller time constant settings.

*Pump noise.* UV detectors can be affected by fluctuations in refractive index, because light scattering at the cell wall changes with refractive index. Refractive index changes with mobile-phase composition, as well as with changes in pressure and temperature. When pump pulsations occur, the pressure changes, altering the refractive index and thus slightly changing the UV transmittance of the mobile phase. Most UV detectors are rela-

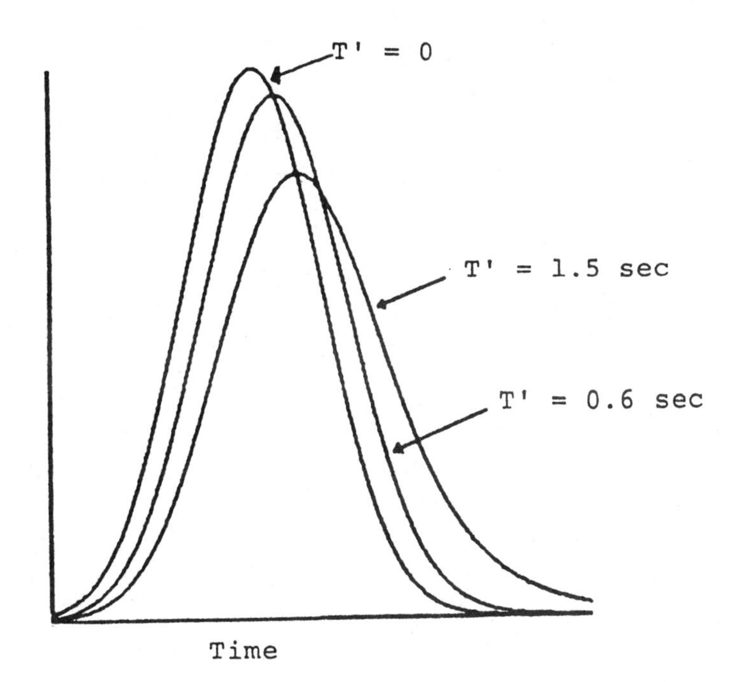


**Fig. 12.20.** Curves demonstrating peak distortion resulting from excessively large time constants $T'$. This example assumes a 25 x 0.46-cm column ($N = 12{,}000$) operated at 1mL/min. Reprinted from ref. (1) with permission.

tively insensitive to refractive-index changes in typical LC systems, because pump pulse-dampening, heat exchangers, and detector cell design (e.g., tapered cells) all work together to minimize these effects. Even so, at very sensitive settings, cyclic baseline fluctuations that correspond to the pump cycle may be seen. If this is a problem, additional pulse dampening should provide a remedy.

RI detectors are extremely sensitive to pump pulsations, so additional pulse suppression is often required if the detector is to be operated at maximum sensitivity.

*Temperature problems.* Baseline drift is often caused by environmental temperature fluctuations. A change in the temperature of the mobile phase causes a change in its refractive index, and thus a change in UV transmittance of the mobile phase (see discussion in the pump-noise section above). It is best to operate the column in a thermostated (or at least insulated) compartment for maximum baseline stability. It is important to locate the LC system out of the direct path of a draft (e.g., not in a fume hood, and away from air conditioning and heating ducts). Most detectors incorporate heat exchangers to minimize temperature problems, but because heat exchangers can add undesirable extra-column dead-volume to the system, heat exchangers may not be used in small-volume (e.g., < 8 µL) detector cells. An easy test to determine whether the heat exchanger is adequate is to run a blank baseline at maximum detector sensitivity (minimum attenuation) and then heat or cool the detector inlet-line. This is done conveniently by gripping the line between the thumb and forefingers to heat the line or by holding a piece of ice on the line to cool it. If the baseline drifts as the temperature change is made, the line should be insulated. Wrapping the transfer line with an insulating material or covering it with a piece of slit Tygon tubing is adequate for most applications.

Because refractive index detectors are extremely sensitive to temperature changes, most RI detectors include a means for thermostatting the flowcell, such as a circulating fluid heater around the detector cell. Some models (e.g., Tecator Model 5931) have the sample injector, column, and detector mounted in a single temperature-controlled cabinet for maximum temperature stability.

*Mixing problems.* Incomplete mixing of the mobile phase can cause cycling baselines, especially if a refractive index detector is used. Very slight changes in mobile-phase composition result in major changes in refractive index, which in turn can give a fluctuating baseline. This problem can be verified by changing the mobile-phase composition and seeing if the cycling baseline changes its pattern to correspond to the new mixture. It is possible that the low- or high-pressure mixing system is not working properly; consult Chapter 7 for information about troubleshooting pumps and mixing. If the mixer is working properly, additional mixing (e.g., a packed-bed mixer downstream of the standard mixer) may solve the problem. In some cases, especially where RI detection is used at the most sensitive settings, you may have to resort to using only hand-mixed mobile phases.

*Linearity problems.* A nonlinear detector response can occur because of problems with the detector or the method. Most of these can be overcome by paying careful attention to sample dilution, wavelength selection, and method design. A check of method (and detector) linearity should be made during method validation by injecting standards and samples which cover the expected range for analysis (see Section 16.1).

The most common problem is to exceed the linear range of the detector. This can happen when too large a sample mass is injected, so that the detector response is in a nonlinear region. This can also occur in an apparently "safe" region if it is possible to set the detector attenuation past the linear dynamic range. Thus, if the particular detector is linear to 1.5 AU and you set it to 2.0 AU, the sample peaks may remain on-scale, but exceed the linear dynamic range of the detector. The solution to these problems is to inject a smaller sample mass (dilute the sample and/or reduce the injection volume) and make sure that the attenuation is set within the detector's linear range.

Another source of linearity problems is observed at the low end of detector response, but this is usually caused by sample losses or interaction with the column, not a detector problem. If the method is checked for linearity during method development, this problem should seldom be encountered.

When partial-loop injection is practiced with more than about 50% of the loop volume injected, nonlinearity can occur because of the injection of different sample sizes. This is an injection-technique problem, not a detector problem (see Section 10.1 for a discussion of injection techniques).

Zeroing the detector to adjust for baseline offset (absorbance) uses up some of the linear range of the detector. With the nearly transparent solvents commonly used with LC, this loss of linearity is unnoticed. When strongly-UV-absorbing mobile phases are used, however, you can "zero away" a significant part (e.g., 1 AU) of the detector's linear range. It is best to use a non-absorbing mobile phase or operate at a wavelength at which mobile-phase absorbance is not a problem.

If too large a bandpass is used in the detector, and the detector is operated at a wavelength at which sample absorbance is changing rapidly (e.g., 275 nm in Fig. 12.18), Beer's law can be violated and nonlinearity can result. The best ways to avoid this problem are to (a) select a detection wavelength near the absorbance maximum and (b) use a small bandpass (if this is adjustable, as with a diode-array detector).

*Cabling problems.* Problems can appear at the recorder or data system, if the connecting cables to the detector are not installed properly. Most detectors provide two types of output signals: (a) recorder, and (b) computer or data system outputs.

The recorder output typically is a 0–10 mV signal intended for use with a strip-chart recorder. This signal is attenuated according to the attenuation (range) selections made on the recorder control panel. If the recorder output signal is inadvertently fed to a computer or data system, the signal generally will be too weak for adequate sample detection. Changing the cable connection to the computer output will solve this problem.

The computer or data system output generally is a 1 V/AU (volts/absorbance unit) signal, which is unaffected by the control panel settings. This allows the data system to receive the strongest possible detector signal, which then can be attenuated appropriately by the data system. If this signal is fed inadvertently to a strip-chart recorder, the recorder pen will often go off-scale in the positive direction and will not come back on scale until the cable is disconnected or the recorder range is changed. Reconnecting the cable to the recorder output on the detector will solve this problem; generally this mistake will not damage the recorder or the detector.

Finally, if the signal cable to the recorder or data system is not connected properly, undesirable baseline noise can result. A typical signal cable consists of a twisted-pair of wires to carry the signal and an outer braided shield to protect the signal from unwanted electronic noise. The shield is fastened to a ground wire at one or both ends. When properly connected, the signal wires should be connected to the + (pos.) and – (neg.) terminals on the detector and recorder (or data system). The ground wire, however, should be connected to the grounding post only at one end. If the ground wire is connected to both the detector and recorder or data system, a "ground loop" will be created, which can cause short-term noise in the chromatogram.

*Internal self-tests.* Many detectors contain internal testing routines that test the electronic and/or optical sections of the detector. These may be activated automatically upon power-up or through a special keyboard command, depending on the model. The operator's manual contains a list of error messages which may appear as a result of the self-test.

Table 12.3
Detector Problems and Solutions

| Cause of problem | Symptom | Solution |
|---|---|---|
| *Lamp problems* | | |
| Lamp failure | No signal | 1. Replace lamp |
| Insufficient warmup | Noisy baseline | 1. Allow about 30-min warmup for normal operation, 1 h for maximum sensitivity<br>2. Allow new lamps at least 1 h warmup before use |
| Lamp aging | Increased short-term noise, occasional noise spikes | 1. Replace lamp |
| *Cell problems* | | |
| Bubbles | Noise spikes in chromatogram, possible off-scale signal | 1. Degas mobile phase<br>2. Add restrictor after cell |
| Immiscible solvents | Noise spikes in chromatogram following solvent changeover | 1. Flush system with 2-propanol to remove contaminant, then with mobile phase |
| Blockage | High backpressure, leaky gaskets | 1. Clean cell (Sect. 12.2)<br>2. Replace cell if cleaning unsuccessful |
| Dirty cell | Noisy baseline, increasing bubble problems | 1. Clean with nitric acid (Section 12.2) |
| Damaged gasket | Leaks at cell body | 1. Check for blockage or excessive backpressure,<br>2. Rebuild or replace cell |
| Damaged or loose fittings | Leaks at fittings | 1. Tighten fittings<br>2. Replace fittings |
| Sample/reference mismatch | Can't zero baseline | 1. If air reference, check for fluid in reference cell, blow out with dry nitrogen |

*(continued)*

Table 12.3
Detector Problems and Solutions *(continued)*

| Cause of problem | Symptom | Solution |
|---|---|---|
| *Cell problems (continued)* | | |
| | | 2. If liquid reference, flush thoroughly with mobile phase |
| *Wavelength Problems* | | |
| Second-order wavelength problems | Very high baseline in high-wavelength region | 1. Use proper UV-blocking filter |
| Poorly chosen wavelength | Irreproducible peak heights, poor method reproducibility between detectors | 1. Choose wavelength at adsorption plateau |
| Poor wavelenth selection technique | Can't duplicate previous peak heights when wavelength is reset | 1. Dial wavelength from same direction each time (Section 12.2) |
| Calibration problems | Peak heights smaller (higher) than expected | 1. Check calibration at 486, 582, or 656 nm, recalibrate if necessary |
| Low-wavelength problems | Extra peaks in chromatogram, high baseline level | 1. Normal for low wavelength detection<br>2. Try acetonitrile instead of methanol or THF in mobile phase<br>3. Helium-sparge mobile phase to remove oxygen<br>4. Purge optical path with nitrogen to remove ozone (Section 12.2) |
| *Time constant too large* | Broadened peaks, especially for early peaks | 1. Use smaller time constant (e.g., 10% of width of first band of interest) |
| *Pump pulsations* | Cyclic baseline fluctuations, corresponding to piston cycles | 1. Check for and remove bubbles in pump<br>2. Add pulse damper to pump<br>3. Work at less-sensitive detector setting |

*(continued)*

Table 12.3 Detector Problems and Solutions *(continued)*

| Cause of problem | Symptom | Solution |
|---|---|---|
| *Temperature problems* | Excessive baseline drift, especially with RI detection | 1. Thermostat column and tubing between column and detector<br>2. Remove LC from drafts<br>3. In extreme cases, thermostat reservoir and injector in addition to step 1 |
| *Mixing problems* | Cyclic baseline fluctuations corresponding to mixer cycle | 1. Verify by changing mobile-phase composition<br>2. Add, repair, or replace mixer<br>3. Use only hand-mixed mobile phase |
| *Cable connections* | Signal too high (low) at recorder/data system, generally off in multiples of 10x expected value | 1. Use appropriate detector output for recorder/data system |
| *Ground loop* | Short-term noise in baseline | 1. Check for proper cable connections for detector output, don't ground at both ends (Section 12.2) |
| *Self-diagnostics* | Warning lamps, error messages, buzzers etc. | 1. Check detector operator's manual |

## 12.3. References


[1]Scott, R. P. W. (1986) "Liquid Chromatography Detectors," *J. Chromatogr. Library*, **33,** 2nd edn., Elsevier.

[1a]Snyder, L. R. and Dolan, J. W. (1985) *Getting Started in HPLC, User's Manual,* LC Resources Inc., Lafayette, CA.

[2]*LC User Survey IV,* Astor Publishing Corp., Springfield, OR (December 1984).

[3]Snyder, L. R. and Kirkland, J. J. (1979) *Introduction to Modern Liquid Chromatography,* Wiley-Interscience, New York, 2nd ed.

[4]Weinberger, R. and Sapp, E. (1984) *Amer. Lab.* **16(5),** 121.

[5]Berglund, R. and Thente, K. (1983) *International Lab.*, Nov/Dec.

[6]*LC User Survey VII,* Astor Publishing Corp., Springfield, OR (December 1985).

[7]Dolan, J. W. (1986) *LC, Liq. Chromatogr. HPLC Mag.* **4,** 526.

[8]Weinberger, R. and Coniglione, V. (1984) *LC, Liq. Chromatogr. HPLC Mag.* **2,** 766.

[9]Dolan, J. W. (1986) *LC, Liquid Chromatogr. HPLC Mag.* **4,** 1178.
[10]*HPLC Solvent Reference Guide* (1985) J.T. Baker Chemical Co., Phillipsburg, NJ.
[11]*Solvent Guide,* Burdick & Jackson Laboratories, Inc., Muskegan, MI, 2nd ed. (1984)
[12]van der Wal, Sj. and Snyder, L. R. (1983) *J. Chromatogr.* **255,** 463.

# Chapter 13

# RECORDERS AND DATA SYSTEMS

## Introduction

Recorders and data systems convert the electronic signal output from the LC detector into a permanent record for future use. In nearly all cases, the primary record is a chromatogram, plotted on a piece of chart paper. In addition to a chromatogram, the data may be reported as results tables and/or stored on magnetic media. A variety of recorders and data systems are available, ranging from strip-chart recorders to multisystem computer-based units that also serve as controllers for several LC systems. This chapter discusses the principles of operation, maintenance, and troubleshooting of these devices. Because of the variety and complexity of available products, you will need to refer to operation manuals for specific procedures for many systems.

## 13.1. Principles of Operation

### *System Definitions*

For the present discussion, we will divide recorders and data systems into four categories, based on their capabilities and complexity: (a) recorders, (b) integrators, (c) data systems, and (d) system controllers. One problem with this arbitrary classification of data-recording devices is that it is difficult to define a demarcation between products. Strip-chart recorders typically are analog devices, whereas integrators and data systems are digital. There is a fuzzy line between integrators and data systems as defined here. For this reason, we concentrate on data systems in this chapter; additional information on recorders and integrators is included where appropriate.

*Strip-chart recorders* historically are the primary data-gathering device for LC systems, but these are less widely used with the advent of low-cost integrators. A strip-chart recorder converts the analog detector-signal into an *X-Y* graph (chromatogram) on a continuous piece of chart paper. The signal attenuation is controlled by the detector, so the recorder acts as a transducer, converting the electronic signal into a mechanical one. Typically, input to the recorder is 0–10 mV full-scale, which is provided from the "recorder" output terminals on the LC detector. A 10–12-in.-wide chart is most popular, fed from a roll or accordian-folded pad of paper. Chart speeds of 0.5–1 cm/min give a chromatogram that has a pleasing appearance and is easy to measure. The trace is made with a ball-point or felt-tipped pen (older models may have wet-ink pens). Measurement of retention times and peak heights are made manually. If peak-area measurements are to be used, it is best to use a system with integration capabilities, because manual area-measurement from a strip chart is tedious and error-prone. Recorder sensitivity is adjusted with a potentiometer (see procedure in Section 13.2); the detector time-constant also influences the recorder sensitivity. The primary problem areas for strip-chart recorders are the ink supply and paper feed.

*Integrators* probably are the most widely used data-gathering device used in LC today. The availability of low-cost integrators, pioneered by Hewlett-Packard (e.g., model 3390), has reduced the use of strip-chart recorders to labs with low sample-loads and/or little need for quantitative analysis. Today integrators are available for about the same price that strip-chart recorders commanded ten years ago.

Integrators record a chromatogram on chart paper, just as recorders do, but they have a number of added features. Integrators use a processed signal rather than the original analog signal used by the chart recorder. The analog signal from the detector is converted to a digital signal (analog-to-digital, or A/D conversion) that is processed by the integrator. The digital signal is converted back to an analog signal (D/A conversion) for display as the chromatogram. This conversion and processing takes time, and as a result, chromatogram plotting may lag 15 s or more behind the detector output. The chart width may range from 3 to 12 in. for various models; printing is done with a dot-matrix printer using a ribbon or thermal-sensitive paper. Most units mark the time of injection and the retention time for each band next to the band on the chromatogram. This makes it easy to correlate the peaks in the chromatogram with those in the report. When the chromatogram is complete, the integrator prints a simple report including the retention time, area and/or height, and type of integration. Many integrators have the ability to do simple calibrations so that the results can be reported in calibration units (e.g., mg/mL) or as area-percent. The detector signal fed to the integrator is unattenuated, typically 1 V/AU (volt/absorbance unit) for UV detectors. The signal attenuation is selected at the control pad on the integrator. The major distinction between an integrator and a data system is that the integrator reports the results only for the current run. That is, when the next run starts, all the data for the previous run are lost. This means that summary reports, recall of data for future calculations, etc. are not possible with integrators. However, some models have the ability to store a limited amount of data for summary reports at the end of a series of runs. Integrators typically are limited to the collection of data from one or two detector channels from a single LC system. Integrators are available as stand-alone units or as add-on cards for use with personal computers (e.g., Dynamic Solutions' Baseline).

*Data systems* are the next level of sophistication for data recording in LC. Data systems have all the features of integrators plus the ability to store and process data after a run or series of runs is made. Reports may include peak names (rather than just retention times), interpretation of results (e.g., flagging results outside normal limits), statistical analysis, and customization features to make the data presentation fit the specific needs of your application. In addition to report capabilities, data systems can store raw and/or processed data on magnetic media (e.g., floppy diskettes) for archival purposes. Data systems are available that can collect and process data simultaneously from several LC systems. In the past, most data systems were designed as dedicated computers for chromatographic data processing. Today, with the wide availablility of inexpensive and powerful personal computers (PCs), PC-based data systems are offered by many vendors. These systems (e.g., PE-Nelson) have the added advantage of being useful as PCs in addition to data-system chores. PC-based data systems are the most rapidly growing area of data collection and processing in LC today.

*System controllers* are data systems that also control the operation of one or more LC systems. Thus, the flow rate, mobile-phase composition, detector parameters, and other aspects of the LC system can be dictated by the controller. This allows for unattended operation of the LC system, with method changes from one sample type to another, adjustment of system parameters based on analysis results, and automated method development. System controllers may be specially designed computers or consist of specialized software and add-on boards for a PC (e.g., Beckman's System Gold).

## *System Parameters*

There are several parameters that must be selected (either by default or by user input) for proper operation of any LC data system. These are common to different brands and models of systems, although they may differ in name. Each system will differ a little in the implementation of these measurement parameters, so you should consult your system operator's manual for specific definitions for your data system. This discussion centers on data systems, but applies equally to integrators and system controllers; exceptions will be noted. For the most part, strip-chart recorders have none of these features available.

*Data Rate.* The performance of a data system is closely linked to the rate at which data are gathered during a run. If data are gathered at too slow a rate, they do not adequately represent the peaks eluting from the column. If too high a data rate is used, an excessive amount of computer memory is used to store the data. A good rule-of-thumb is to set the data rate so that about 10 data points are collected across the narrowest band of interest. For example, a 25-cm column that generates 15,000 plates will have a peak width of about 10 s for $k' = 1$; a data rate of 1 Hz is appropriate for this setup.

*Baseline Noise.* In order for a data system to be able to detect a peak, it must be able to distinguish the peak from the background noise. This is based on measurement of the baseline noise. Generally, the system automatically measures the baseline noise for a section of baseline in which no peaks elute. This is done during setup and calibration. After the baseline has stabilized (running only mobile phase), you press the key to start collecting a background signal. After the collection period (e.g., 30 s), the system processes the data to obtain a standard baseline-noise value for use in detecting the presence of peaks.

*Peak Threshold.* The data system determines that a peak is present when certain peak-threshold criteria are met. These criteria can be user-selected or left as default values. The peak-threshold level is a signal level above which the signal is tested for being a peak. The threshold typically is expressed as a change in signal for a given period of time ($d_v/d_t$), for example microvolts per second. This can be better understood as a multiplier of the baseline noise (e.g., 3x noise). Once the signal exceeds the threshold (i.e., a certain change in signal is seen for a certain period of time), it is tested to be sure a peak (not a noise spike) is present. For example, the peak criterion might be that ten consecutive data points must exceed the threshold. When the peak criteria are met, integration begins; integration ends when the signal drops back below the threshold value.

*Attenuation.* When strip-chart recorders are used, the scale on the chromatogram is controlled by the attenuation (range) settings on the detector front-panel. Data systems, however, use the raw (unattenuated) signal from the detector, so the attenuation of the signal must take place in the data system before the chromatogram is plotted. Data systems allow you to change the attenuation on a timed basis during the plotting of a chromatogram, so that peaks don't go off scale, yet the small peaks are clearly visible. The attenuation changes affect only the plotted

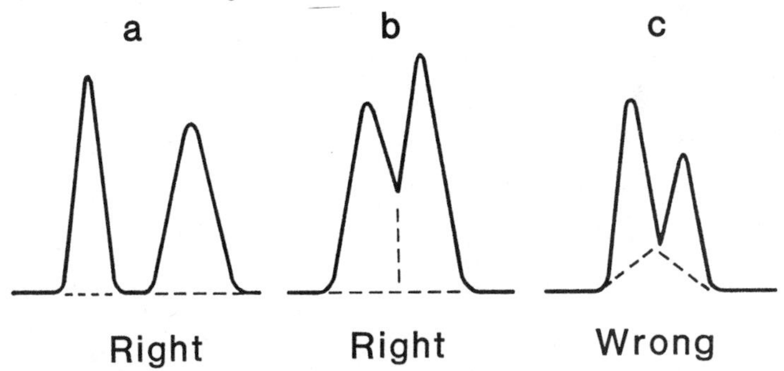


**Fig. 13.1.** Methods of drawing the baseline. (a) Baseline-baseline for well-resolved peaks; (b) Baseline-perpendicular drop, correct method for partially resolved peaks; (c) Baseline-valley, improper method in most cases. Reprinted from ref. (1) with permission.

chromatogram; that is, all the integration and processing of the data is based on the unattenuated signal.

*Peak Picking.* The retention time for a band is determined by a peak-picking algorithm. The most common method of peak picking is to take the first derivative of the signal. When the derivative of the peak signal is zero, the retention time is recorded. A signal-averaging routine may be used to make sure that a true peak maximum occurs rather than a noise spike.

*Baseline Drawing.* When a peak elutes from the LC, the baseline before and after the peak must be determined so that the data system can integrate the peak. The method that the data system used to draw the baseline often is indicated on the printout after the run. (Note: when we talk about drawing a baseline for quantitation, we are referring to the reference point(s) from which the peak height or area are measured. This is not necessarily the same as the baseline, or background, signal observed when no peaks are eluting from the column, but rather it is an average or projected value of the background signal.) When the peak is well resolved from other peaks, baseline selection is simple: just connect the baseline before and after the peak with a straight line. This type of baseline is called baseline-baseline (Fig. 13.1a).

When two peaks are partially merged (e.g., $R_s = 0.8$), the baseline can be drawn in one of two ways. The first method is to connect the baseline before and after the peaks then to make a perpendicular line from the lowest point on the valley between the peaks to the baseline, dividing the band pair roughly in half. This

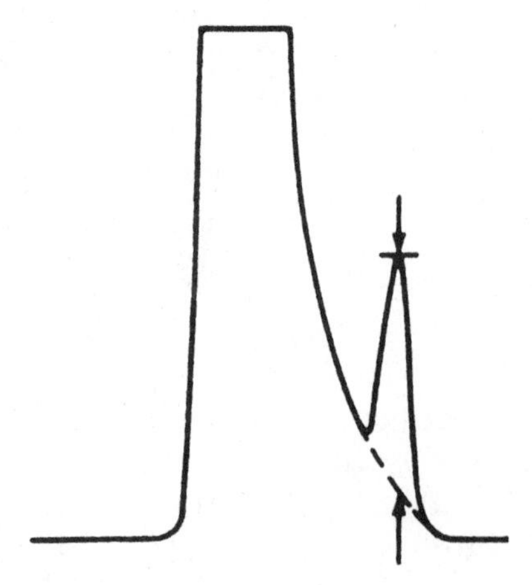

**Fig. 13.2.** Use of tangent skimming to draw the baseline for a rider peak. Arrows indicate proper points to use for measurement of peak height. Reprinted from ref. (2) with permission.

would be abbreviated as a baseline-to-perpendicular (Fig. 13.1b). Using perpendicular drop is the proper way to draw the baseline unless there is some compelling reason to draw it in another manner (e.g., a known broad peak eluting under the pair). The alternative method of drawing the baseline for the partially merged peaks is to draw the new baseline from the true baseline before the peaks to the valley between them and then to the baseline after the peaks (baseline-valley, Fig. 13.1c). The baseline-valley technique gives inaccurate results for most cases. There are a number of other baseline drawing routines available in specific data systems.

*Peak Skimming.* When a smaller peak ("rider") elutes on the tail of a larger (and broader) peak, a baseline can be obtained by drawing a tangent to the larger peak (Fig. 13.2). Depending on the system design, this baseline can be a tangent skim or an exponential skim. The particular method of peak skimming generally is fixed for a given data system.

*Peak Window.* As was discussed in Chapter 3, the retention time of a peak is characteristic of a given compound, when all other factors are held constant. Because of small changes in the mobile phase, column, flow rate, temperature, or other factors, small changes in retention time (e.g., 0.01–0.05 min) will be observed from run to run under normal conditions. Some assays will be subject to more variation than others, and some LC systems will give more consistent retention times than others. Data systems compare the retention time of an unknown peak to the retention time of a standard (held in computer memory) in order to identify the peak. The user must select a retention-time window (or use the default value) that is acceptable for peak identity. The range usually is expressed either as a percent (e.g., ±2%) or an absolute value (e.g., ±0.1 min) relative to the standard or calibrator. Depending on the data system, peaks that fall outside the peak window may be flagged or reported as unknowns. Similarly, the absence of a peak in the window may be used to alert the user that a possible malfunction has occurred.

Some data systems allow the selection of a window for the amount of the target compound in a given sample. For example, in a therapeutic drug monitoring application, the extremes for the normal blood levels of a drug might be entered for the window thresholds. When a sample is analyzed for which the drug peak either is below or exceeds the threshold, a message would signal to rerun the sample or alert the physician of abnormal results.

*Peak Group.* For some assays it may be desirable to combine the response of several peaks into one reported value. For example, because of similar biological activity, two closely eluting isomers that are present in a pharmacological preparation may be combined for reporting purposes. Most data systems have some means of allowing the user to select two or more closely eluting peaks as a peak group or peak bunch in the analysis report.

*Other Parameters.* There are many other integration and reporting parameters in use by data systems, but these are beyond the scope of this discussion. Some systems, for example, have the ability to store a blank run and then subtract it from a sample run in order to obtain a flat baseline when gradient-elution is used. Consult the operator's manual for the specific data system for directions on how to use other parameters.

## 13.2 Problem Prevention

### *Procedures*

There is little preventive maintenance required for recorders and data systems. When problems do occur, they are primarily electronic in nature, and thus beyond the capability of most users to repair. There are a few practices that fall into the preventive maintenance category. As with the previous section, the term data system will apply to integrators, data systems, and system controllers unless noted otherwise.

Make sure that all the *electrical connections* are made properly. Recorders should be connected to the "recorder" output of the detector (generally 10 mV full scale); data systems should be connected to the unattenuated "data system" or "computer" output of the detector (generally 1 V/AU for UV detectors). Use a special signal cable if this is supplied with the unit. Otherwise, use a cable with a twisted pair of signal conductors inside a grounded shield for maximum signal integrity. Connect the ground only at one end (detector or recorder/data system) in order to avoid increased noise caused by the ground loop created when both ends of the cable are grounded. If you need to use a long signal cable (e.g., <10 ft), signal losses are possible, so consult the operator's manual for recommendations for your system.

Be sure that there is an adequate supply of *chart paper* before you start each day's analyses. It is better to throw away a few feet of chart paper than to rerun samples because you ran out in the middle of a series of runs. Also, be sure to keep an adequate paper supply on hand—there are few things more frustrating than to be unable to run the LC system because you are out of chart paper.

If you are using a data system that stores data on *magnetic media* (e.g., floppy or hard disks), be sure that you have sufficient storage space for the samples that you plan to run. The operator's manual should tell you the specific memory requirements for your system so that you can calculate the total memory needed per sample. For example, if you are using a data system that requires one byte of memory per data point, and you want to run twenty 20-min assays with data collected at 1 Hz, you can calculate that 24 kbyte of storage are needed (1 byte/point x 1 point/s x 60 s/min x 20 min/assay x 20 assays = 24,000 bytes). If you are using floppy diskettes, be sure to keep an adequate supply of these on hand.

When data are stored on magnetic media, you should make *backup copies* of any data that you could not easily replace if the original records were destroyed. Backing up data files at the end of each day or week should be sufficient for most users. When lots of sample data are collected on a hard disk, you will have to make backups regularly in order to have enough room on the disk to collect more data. The backup copies should be stored where they will be protected from extreme heat, magnetic fields, and chemical or physical contamination. Backups can be made on floppy disks when a small amount of data is collected. A magnetic-tape backup can store more data much faster than can be done with floppy disks, but it is more expensive and has slower data-access time. For these reasons, tape systems are used only for backup, not for routine use.

Most strip-chart recorders use ballpoint or felt-tip *pens* to record the chromatogram on the paper. Be sure to cap felt-tip pens when the recorder is not in use, or they will dry out quickly. Keep an adequate supply of pens on hand. When wet-ink pens are used, it may be necessary to clean the pen tip occasionally with a syringe-cleaning wire to prevent blockage.

Integrators and data systems have a dot-matrix *printhead* that uses either mechanical or thermal means to record the chromatogram on the chart paper. Mechanical printheads use inked ribbons, so be sure to keep a spare ribbon or two on hand. Because the ribbons wear out gradually, there usually are several days of use left in a ribbon when you first notice it is getting faint. Reinking kits are sold, but these are messy; when the labor cost is considered, they are not very economical to use.

It is wise to check all the *system settings* before you start your first run of the day. You should make a short checklist to help ensure that all the settings are correct for your assay. Calibrate the LC system before you run your first samples; this is also a good time to check the calibration of the data system.

## *Spares*

Few spare parts are required for recorders and data systems; these are listed in Table 13.1. The primary consumable part is chart paper; an adequate supply should always be kept on hand. Users of strip-chart recorders should stock several spare pens as backups. When data are stored on magnetic media, a sufficient number of floppy diskettes or tape cartridges should be stocked for data collection as well as backup storage needs.

Table 13.1
Recorder and Data System Spare Parts

|  |
|---|
| Wet-ink recorders |
| Ink |
| Cleaning wire |
| Ball-point or felt-tip pen recorders |
| Pens in desired colors |
| Dot-matrix / ribbon data systems |
| Printer ribbon |
| All systems |
| Paper (thermal or plain) |
| Systems with magnetic data storage |
| Floppy diskettes or tape cartridges |

## 13.3. Problems and Solutions

A summary of problems and solutions can be found in Table 13.2 at the end of this chapter.

### *LC System Problems*

Many problems that show up as abnormal data-system output originate with the LC system, not the data system. Some of these problems are listed below. We have included a short discussion of the most common sources of noise originating elsewhere in the LC system, but often confused with data-system malfunction. For further information, you should refer to the chapter for the appropriate module or to Chapter 2.

If the cause of the problem is not immediately obvious, determine whether the problem is in the data system or elsewhere in the LC. First, run the self-test diagnosics for the data system (see the operator's manual for instructions). If the test fails, a fault list will be generated by the data system, and you can fix the problem as directed in the manual. If the test passes and you still suspect the data system, the most expedient way to isolate the problem is to substitute a known good data system (preferably the same model as the suspect one) for the one in question. If the problem doesn't go away, you know it is related to the LC system, not the data system; otherwise, the data system is at fault.

*Pump noise* shows up as a cyclic disturbance in the baseline, especially when the detector is run at maximum sensitivity or when refractive index (RI) detectors are used. You should be able to correlate the cycle time of the baseline noise with the piston cycle-time. When the flow rate is increased, you should see an increase in the rate of baseline cycling, proportional to the flow increase. The problem is caused by changes in pressure at the detector as a result of bubbles in the system, dirty check valves, or other pump malfunction. See Chapter 7 for pump problems.

*Mixing problems* result in a nonhomogeneous mobile phase. This in turn disturbs the detector output, and a cycling baseline is the usual result. The cycling baseline caused by mixing problems can be differentiated from pump noise because the period should correlate with the mobile-phase composition rather than the flow rate. To correct this problem, you need to fix any mixer problems, improve the mixing characteristics of the system (e.g., add a supplemental mixer), or use hand-mixed solvents. Refer to Chapter 7 for more details.

*Bubbles* passing through the detector cell will cause noise spikes in the chromatogram. These show up as abrupt changes in the chromatogram; the spikes usually go off-scale for a few seconds then return to baseline. The frequency of the spikes may or may not correlate with the pump cycle. When bubbles become lodged in the detector cell, the recorder pen will go off-scale and stay there until the bubble is dislodged. Bubble spikes are distinguished from chromatographic peaks by their abrupt onset (peaks perpendicular to the baseline rather than sloping like normal bands). Bubbles most commonly arise from insufficient solvent degassing; see Chapter 6 for more information.

*Detector lamp* problems also show up as spikes in the chromatogram, but these spikes seldom go off-scale (and stay there) as bubble spikes do. Lamp problems often are accompanied with increased baseline noise. Lamp replacement usually solves these problems. See Chapter 12 for information on troubleshooting detector problems.

### *Signal Cables*

Using the wrong signal cable or connecting it improperly between the detector and data system can cause several problems that show up in the chromatogram. When the *improper cable* is used, signal loss or increased baseline noise may result. You should use the signal cable that is supplied with the recorder or data system if the manufacturer provides one. If a cable was not supplied, use a cable consisting of a twisted pair of wires (for the signal) contained inside a woven shield that is grounded at one end. This type of cable provides maximum signal integrity with the least amount of noise. If an *excessive length* of signal cable is used, signal loss may occur. Generally, runs of 10 ft or less are no problem. If you need to run the signal a longer distance between the detector and data system, consult the operator's manual or the data system manufacturer for recommendations. When signal cables are *improperly routed,* it is possible to observe increased short-term baseline noise or noise spikes. These problems can occur when the signal cable is located too close to electronically noisy devices, such as water baths, furnaces, fluorescent lights, and air conditioners. Correct this problem by using short runs of cable, and avoid devices that switch large current-drawing motors or heaters on and off. When *poor grounding* occurs, increased baseline noise can be observed. The cable should be grounded only at one end. If the shield is grounded at both ends, a "ground loop" can be created, and a noisy baseline may be seen. If neither end of the cable is grounded, the shield is ineffective, and environmental noise may be added to the signal. *Improper connections* at the detector or data system can cause unexpected signal levels to reach the data system. Be sure that the proper "recorder" or "computer/data-system" outputs from the detector are used. Proper connections are discussed in Section 13.2. Information on identifying and eliminating common electronic noise problems can be found in ref. (3).

### *Improper Settings*

When an abnormal signal is seen at the data system, the first thing to suspect is that one of the data-system settings is incorrect. Check the settings against the list in your method sheet and make corrections as necessary. Use of the wrong *threshold value* is the most likely problem. If the threshold is too low, an excessive number of peaks will be included in the analysis report. When the threshold is too high, peaks will be left out of the report. When an automatic threshold is used (e.g., 3x noise), the threshold may be changed inadvertently when the baseline noise-level changes.

### *Paper Problems*

Problems associated with the chart-paper feed generally are obvious to correct. If the chart paper does not advance, check to be sure that there is a sufficient paper supply and that the friction or tractor-feed mechanism is engaged properly (there may be a lever or switch to disengage the paper feed). If the paper does not feed straight, it often is caused by uneven tension on the paper in-feed. Check to be sure that the feed path is clear of foreign objects (paper clips, paper scraps, etc.). When tractor feed is used, make sure that the sprockets are engaged at the same point on both edges of the paper.

### *Writing Problems*

Writing problems occur when the pen head moves across the chart paper, but no trace or printing is made. With wet-ink, ball-point, and felt-tip pens, be sure that there is a sufficient supply of ink; replace the pen or replenish the ink as needed. With thermal-pen systems, check the operator's manual for the procedure for testing and/or adjusting the pen. Often the print-head will need to be replaced when it stops writing. Be sure that you are using thermal paper, because thermal print-heads will not write on plain paper. When a dot-matrix printer is used with a ribbon cartridge, the printing gradually will become lighter as the ribbon is used up. When the print becomes too light for acceptable reading, ribbon replacement should be made. Be sure that the ribbon-feed mechanism is engaged properly, or the ribbon will not advance.

Table 13.2
Recorder & Data System Problems and Solutions

| Cause of problem | Symptom | Solution |
|---|---|---|
| *LC System (not the data system)* | | |
| Pump noise | Cyclic baseline noise correlates with flow rate | 1. Check for/remove bubbles from pump<br>2. Clean/replace check valves<br>3. See Chap. 7 for pump problems |
| Mixing problems | Cyclic baseline noise, correlates with mobile-phase composition | 1. Correct mixer problems<br>2. Add supplemental mixer<br>3. Use hand-mixed mobile phases<br>4. See Chap. 7 for mixing problems |
| Bubbles | Noise spikes, often off-scale | 1. Degas mobile phase<br>2. Add backpressure regulator<br>3. See Chap. 6 for degassing problems |
| Detector lamp | Noise spikes, increased baseline noise | 1. Replace detector lamp<br>2. See Chap. 12 for detector problems |
| *Signal cables* | | |
| Improper cable | Signal loss, baseline noise | 1. Use cable supplied with data system, or<br>2. Use shielded cable with twisted conductors |
| Excessive length | Signal loss, baseline noise | 1. Use shorter cable, or<br>2. Use signal amplifier<br>3. Consult operator's manual for recommendations |
| Improper routing | Noise spikes and/or increased short-term noise (often 60 Hz) | 1. Route cable away from electronically noisy devices (ovens, water baths, motors, air conditioners, etc.) |
| Poor grounding | Increased baseline noise | 1. Avoid ground loops by only grounding the cable at one end<br>2. Only use shielded cable |
| Improper connections | Unexpected signal levels (often off by 10- or 100-fold) | 1. Use appropriate "recorder" or "data system" output terminals at detector for data recording device |

*(continued)*

Table 13.2
Recorder & Data System Problems and Solutions *(continued)*

| Cause of problem | Symptom | Solution |
|---|---|---|
| *Signal cables (continued)* | | |
| *Improper settings* | Unexpected signal levels, extra or missing peaks in analysis report | 1. Check threshold setting<br>2. Check all other settings against method sheet |
| *Paper problems* | Paper crooked or jammed | 1. Check/replenish paper supply<br>2. Check/clear feed restrictions (paper clips, scraps, and so on)<br>3. Correct improper sprocket engagement |
| *Writing Problems* | Pen moves, but no printing occurs | 1. Make sure pen is in proper writing position<br>2. Replenish/replace ink supply, pen, or ribbon<br>3. Check/replace thermal head<br>4. Check for proper paper selection (e.g., thermal paper for thermal printers)<br>5. Check/correct proper ribbon feed |

## 13.4. References


[1]Snyder, L. R. and Dolan, J. W. (1985) *Getting Started in HPLC, User's Manual,* LC Resources Inc., Lafayette, CA.

[2]Snyder, L. R. and Kirkland, J. J. (1979) *Introduction to Modern Liquid Chromatography,* Wiley-Interscience, New York, 2nd ed.

[3]Fleming, L. H., Milsap, J. P., and Reynolds Jr., N. C. (1988) *LC/GC* **6,** 978.

# CHAPTER 14

# SEPARATION PROBLEMS

## *Band Tailing and Peak Distortion*

## Introduction

Preceding chapters examined problems that can be traced to the equipment or materials (modules, columns, solvents, etc.) used in carrying out the LC separation. In this and the next three chapters we will discuss some other kinds of system failure. Whereas our previous troubleshooting has been mainly physical or mechanical in nature, the origins of the problems we will now examine are often in the *chemistry* of the system. And because chemistry is still largely an experimental science, these kinds of problems usually are harder to solve. However, most chromatographers have a good background in chemistry or biochemistry—and all that is required in most cases is to put this training to use.

The subject of this chapter and the next—separation problems—focuses on the chromatogram. During either method devel opment or routine operation, a variety of separation artifacts can arise. These almost always degrade the quality of the final separation and its use for analysis or sample purification. For example, it may be observed that bands in the chromatogram are not symmetrical or exhibit other peculiarities in peak shape. In other cases, the bands in the chromatogram seem abnormally wide—with poor resulting separation of adjacent peaks. Sometimes extra peaks are present in the chromatogram, peaks that cannot be traced to any compound present in the sample. Occasionally negative peaks may be observed. Retention times for a given compound may change from sample to sample, without anything being wrong with the equipment or the way the procedure is being carried out. Finally, the sample may react chemically during the separation, with resulting peak distortions and low recovery of eluted material.

Everyone who has injected many samples into an LC system has experienced problems like these. Most chromatographers will recall cases in which they never really found out the cause of the problem. They tried a lot of different things, and eventually the difficulty went away. Or someone else got tired of waiting for the results, so the project was dropped. Our goal in this chapter is to get away from haphazard remedies. We will try to develop a systematic approach to these separation or chemical problems, insofar as this is possible. Fortunately a lot of work has been published in recent years that can help us in these situations.

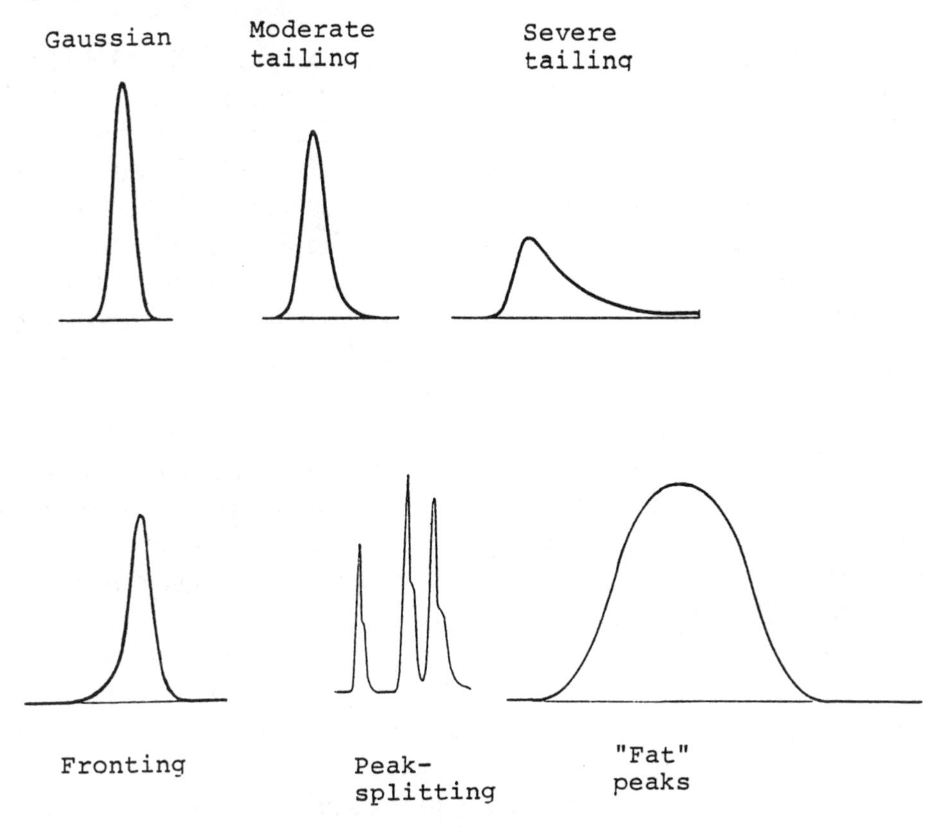


**Fig. 14.1.** Non-Gaussian band shapes.

Bands with strange shapes represent one of the most vexing problems that can arise in the LC laboratory. Figure 14.1 illustrates several examples of these non-Gaussian peaks, along with their common descriptions. As we will see, this situation can have serious chromatographic consequences. There are also about a dozen different causes of band tailing, so discovering why bands tail—and then fixing the problem—can be a difficult undertaking. Fortunately there is a systematic approach based on (a) logical analysis plus (b) practical fixes that has now been documented in numerous laboratories.

We should first consider why tailing or misshaped bands are bad. *One* reason is that such bands can be hard to quantitate. Some data systems have difficulty in measuring peak size accurately, when the band tails or its shape is otherwise peculiar. As a result, the precision and/or reliability of assay methods involving misshaped peaks is often poor when compared to good chromatography. A *second* reason for avoiding tailing bands is that resolution suffers when bands tail. As we discussed in Chapter 3, good resolution is the foundation on which we build sound LC procedures. Anything that compromises resolution will adversely affect our final result. This is especially important when a large band in the chromatogram tails, with resulting overlap of the tail onto smaller bands that elute later. A *third* negative feature of tailing bands is that columns cannot then be coupled in series to give additive plate numbers.

*Finally,* tailing bands often are a symptom of multiple contributions to the retention process—as opposed to a single retention mechanism. This in turn often means that relative retention on one column will not be the same for another identical column (see Sect. 11.4). If we can eliminate tailing bands in this situation, we will generally reduce column-to-column variability.

If tailing bands represent poor chromatographic performance, how much tailing can be tolerated? Actually, when we do everything right, LC bands can be remarkably symmetrical and free from tailing. Figure 14.2 is a good example for the separation of a dye (Red 40) by ion-pair chromatography. Here the detector attenuation has been adjusted for maximum sensitivity, which allows us to see the band edges in detail. The front and back of the band are quite similar in shape, allowing us to quantitate late-eluting, trace impurities in this sample; i.e., bands that could be lost in the tail of the major band if band-tailing were a problem.

If examined closely, most elution bands will exhibit some deviation from perfect Gaussian shapes. Here it is helpful to use the band asymmetry factor $A_s$ defined in Section 11.2 and illustrated in Fig. 14.3. Band asymmetry can be affected by many contributions: the column, the equipment, the injection procedure, the mobile-phase/sample combination, and so on.

It is useful to examine two extremes of band asymmetry. *First,* consider the case of a new column and the manufacturer's test system; for example, the elution of ethyl benzene by 50 %v methanol/water with a flow rate of 1 mL/min. This represents an ideal situation. Good columns (plus good equipment) should result in bands with $A_s$ values between 0.9 and 1.1. Some commercial columns may show $A_s$-values as large as 1.2 in this situa-

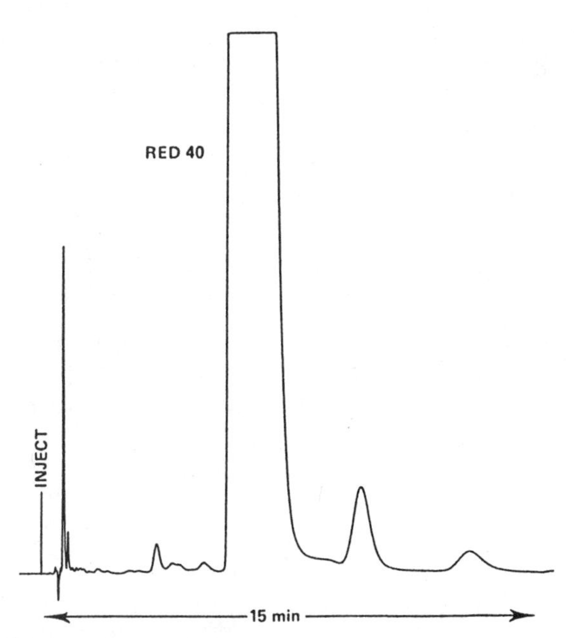


**Fig. 14.2.** Absence of band tailing for good chromatography. Separation of an impure dye compound by ion-pair chromatography. Courtesy of Waters, a division of Millipore.

tion, but the closer the $A_S$-value is to 1.00, the better. This is particularly important for avoiding problems when connecting columns in series.[1]

A *second* situation is the assay procedure for actual samples. Here it often is necessary to relax the tolerances on band symmetry. For real samples, as opposed to test compounds, we often must accept $A_s$-values of 1.3 or even greater. However, we are asking for trouble when bands show asymmetry factors greater than 1.5. A serious effort at reducing band tailing is called for in this case, before proceeding with routine analyses.

How do we solve the problem of tailing or distorted bands? The remainder of this chapter covers this topic in detail. Our approach can be summarized as follows:

1. *Possible Cause.* Consider the various general causes of band tailing and misshaped peaks as listed in Table 14.1; select a probable cause and proceed to step #2.

2. *Pattern.* Consider the nature or pattern of peak distortion; are some or all peaks tailing? does tailing change in

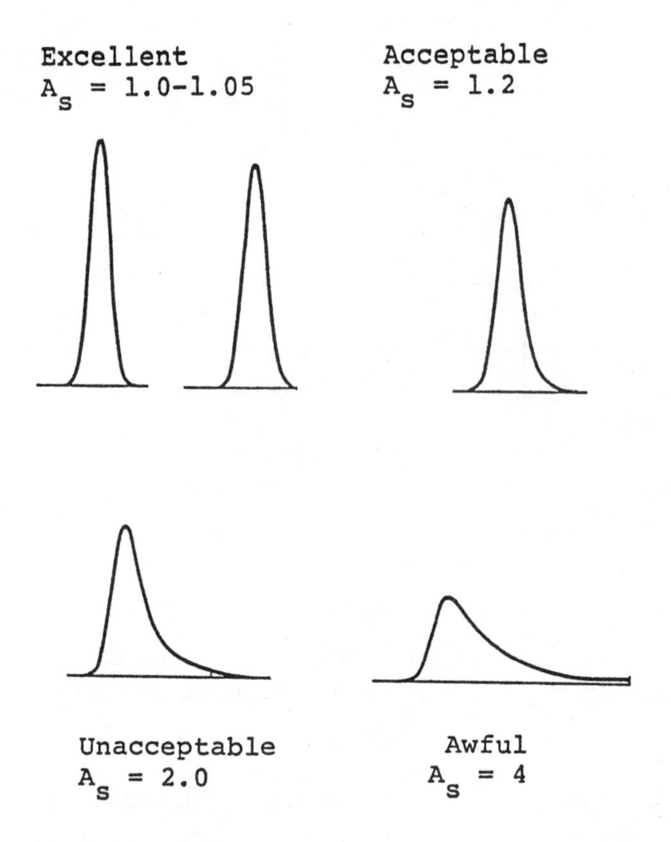


**Fig. 14.3.** Examples of band asymmetry factor $A_s$.

regular fashion from beginning to end of the chromatogram?

3. *Chemistry.* Relate the general chemistry of the separation to the specific possible cause of band-tailing under consideration (from step #1); what LC method are we using? what is the structure of the sample? what is the pH of the mobile phase? is the sample compound(s) ionized under the conditions of separation?
4. *Try Fixes.* Try various specific fixes as described below for each kind of band tailing.
5. *Is It Logical?* When you have arrived at a solution to the problem, ask yourself if everything about the problem and your fix make sense; only then proceed with the separation.

Table 14.1
Different Contributions to Band Tailing

| | |
|---|---|
| 1. | Bad column (blocked frit or void) |
| 2. | Sample overload |
| 3. | Wrong solvent for sample |
| 4. | Extra-column effects |
| 5. | Band fronting |
| 6. | Strong retention sites (normal-phase or ion-exchange) |
| 7. | Secondary-retention effects |
| 8. | Inadequate buffering |
| 9. | Miscellaneous other effects |
| 10. | Pseudo-tailing |

Our starting point in efforts aimed at reducing band tailing is to select a probable cause from Table 14.1. The possibilities in Table 14.1 are listed in order of decreasing promise. That is, we generally should begin with cause #1 (bad column) and proceed downward. By decreasing promise, we don't just mean decreasing probability. Rather we mean a combination of (a) ease in confirming or eliminating a possible cause of band tailing, and (b) the frequency of a particular cause. Often it is possible to quickly rule out a potential cause of band tailing and go on to the next possibility. Following sections (14.1–14.10) cover the various band-tailing problems of Table 14.1.

## 14.1. Bad Column (Blocked Frit or Void)

By a "bad column" we mean a column with some defect in its flow characteristics. This can have many causes. The packing material may have settled, causing a void at the column inlet. Or the inlet frit may be partially blocked by particulates. There are other things that can be wrong with a column, as discussed in Chapter 11: (a) active sites that strongly retain acids, bases or ionized molecules (see Section 14.7 below), (b) contamination of the stationary phase by strongly-held compounds from previous samples, (c) mismatch with the LC equipment (small-volume or microbore columns, see Section 14.4), (d) a low plate-number (Section 15.1), (e) change in retention characteristics, etc. All of these latter effects can be associated with tailing bands. Blocked frits and voids at the column inlet, however, are the most common

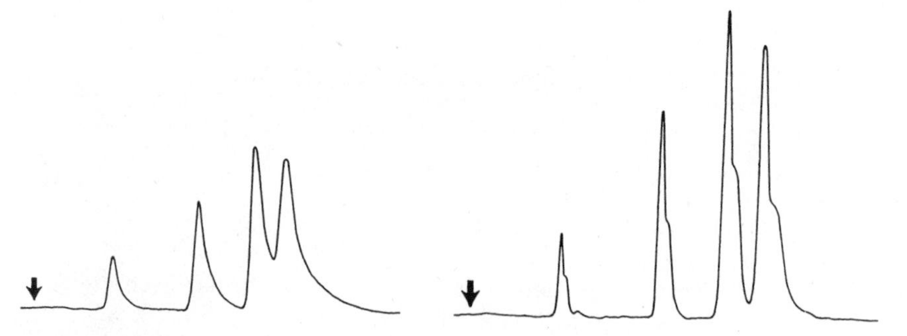

**Fig. 14.4.** Tailing and peak-splitting (double-peaking) caused by a bad column. Courtesy of Spectra Physics.

problems. Fortunately, they exhibit clear-cut symptoms for easy recognition, and they are the easiest problems to resolve.

Typically a column becomes bad with use; new columns have been tested by the manufacturer, and it is rare that such columns will have voids or blocked frits. The first sign of a bad column is band tailing or distortion that is similar for every peak in the chromatogram, as illustrated by the two examples of Fig. 14.4. This condition can arise suddenly, or develop gradually over time, but it tends to become worse with further injections of sample. A bad column can be confirmed simply by running the manufacturer's test system (e.g., ethyl benzene). If the same (distorted) band shape is seen in this case, then a blocked frit or void is likely. Our general approach is then as follows:

1. *Reverse and Flush.* Remove the column from the system and reconnect it with the ends reversed; i.e., the original column outlet is now the inlet, and the new inlet goes directly to waste (not to the detector—the flowcell could be fouled by dirt from the column). Flush the column with mobile phase for 30 min (1–2 mL/min) and rerun the sample that gave tailing peaks (leave the column reversed). Some columns may not be stable to reversal; if there is a question about this, contact the column manufacturer for more information.

   About one time out of three, the tailing will disappear, and often the pressure will drop as well. In this case, the problem is solved; continue operation in the reverse direction. Note that some users prefer step #2 (replace frit) before reversing the column; since either is acceptable, this is a matter of personal choice.

2. *Replace Frit.* If column reversal plus backflushing does not solve the problem, remove the column from the system again. Carefully open the original inlet-end of the column and remove the frit (Section 11.5). Examine the column packing for any settling (1 mm or more) or voids (holes in the column surface). If the column packing is smooth and level with the top of the column, insert a new frit and reconnect the column endfitting (discard the old frit, do not clean and reuse). Reinstall the column in the system (normal direction), flush with mobile phase for 30 min (1–2 mL/min), and reinject the sample. About one time out of three, this will solve the problem, and again the column pressure may drop.
3. *Fill Void and Reverse.* If there is any evidence of settling of the column packing or void formation, it may be necessary to try to extend the column life rather than discard the column. If this choice is determined to be cost-effective, see the discussion of Section 11.5 and ref. (1a) for instructions on filling the void. Reference (1a) indicates that when the void is filled, the column bed will be more stable if the column is operated with the direction of flow *reversed* from its original direction.
4. *Discard.* If (a) replacing the frit (step 2) did not reduce the pressure to acceptable levels, and (b) filling the void (step 3) did not restore acceptable column performance, then further repair of the column cannot be justified; discard the column and replace it with a new one.

## 14.2. Sample Overload

When one or more bands within the chromatogram tail, and these bands are larger than normal, the column may be overloaded. That is, if the total mass of solute is large enough, the linear capacity of the column is exceeded. The result is a decrease in retention time and change in band shape. This situation is illustrated by the hypothetical example of Fig. 14.5b, compared

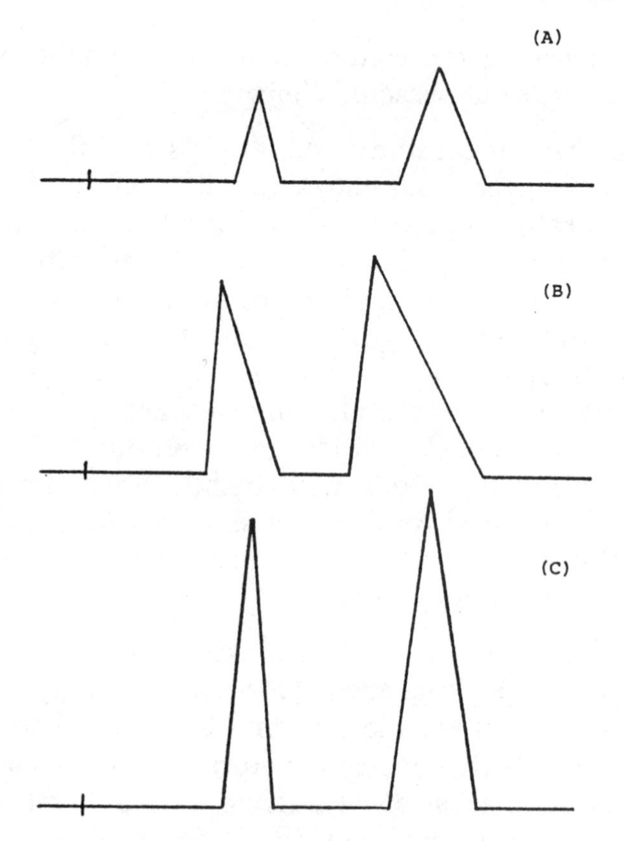


**Fig. 14.5.** Sample overload and band tailing for hypothetical sample. (a) Normal sample; (b) tailing sample; (c) sample of (b) diluted 1/4 and rerun with detector set 4x more sensitive.

with a normal chromatogram in Fig. 14.5a. In this case, the simplest check for sample overload is to dilute the sample four-fold and reinject it—after decreasing the detector attenuation four-fold. The result is illustrated in Fig. 14.5c. Now the originally tailing peaks of run (b) are more symmetrical, and their retention times have increased to the values for the normal sample of Fig. 14.5a. We have confirmed that Fig. 14.5b involves sample overload.

Sample overload normally should not be a problem in a routine LC method, because an important part of method development is to test for the effects of increasing sample size or concentration. When sample overload is seen (broader and/or tailing bands), maximum limits can be set on reported concentration values. For example, it might be found for a 50 μL sample that concentrations of compound X greater than 1 mg/mL cause the band for X to broaden, tail, and/or change its retention time significantly. The method therefore should specify that any samples having apparent concentrations of X greater than 0.5 mg/mL must be diluted and rerun. Just as the column can be overloaded, it is possible to overload the detector. UV detectors should be linear to about 1 absorbance unit, but this is not always the case—particularly when measuring on the side of an absorption band, and at lower wavelengths. A nonlinear detector response can lead to band distortion (usually a fat peak) and to other problems. When detector overloading is suspected, a calibration plot (Chapter 16) should be determined for a range of sample concentrations that overlap that of the problem sample. Detector overloading will be evident as nonlinearity of the peak-height/concentration plot, in the concentration region of the problem sample.

## 14.3. Wrong Solvent for Sample

Good chromatographic results require that the volume and kind of solvent used to dissolve the sample (before injection) fall within certain limits. Ideally a small volume of sample (dissolved in the mobile phase) will be injected. In many cases, a larger volume of sample dissolved in a weaker solvent can be used (see discussion of Section 3.4). Thus 100–500 μL (or more) of sample dissolved in water often is injected for reversed-phase separations. When a relatively large volume of a solvent stronger than the mobile phase is injected, however, there usually is a severe degradation in the quality of the chromatogram. This is illustrated in the chromatogram of Fig. 14.6, for the separation of a two-component sample. In this normal-phase separation on silica, the mobile phase was 0.5% dioxane/isooctane. The sample was injected in a 100-μL volume of pure dioxane (a much stronger solvent than the mobile phase). The result is the obviously odd chromatogram of Fig. 14.6, where three (not two) bands are observed, and the widths of the bands change in abnormal fashion (second band widest, last band narrowest). When the sample was reinjected in a 100-μL volume of mobile phase, two normal-looking bands were observed.

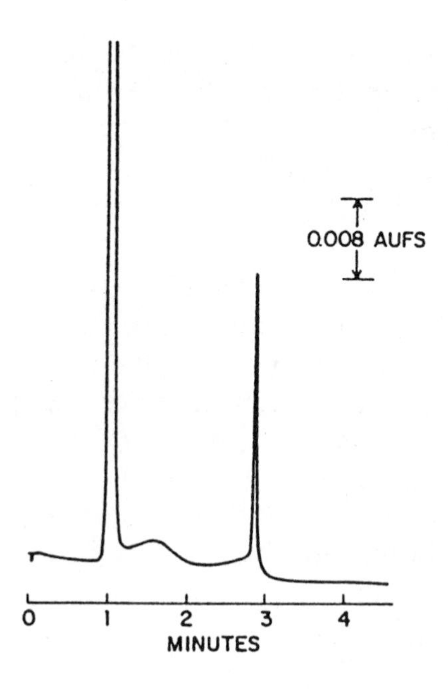


**Fig. 14.6.** Injection of sample in too large a volume of too strong a solvent. Silica column, 0.5 % v dioxane/heptane mobile phase. A 100-µL sample is injected as solution in pure dioxane. Courtesy of Du Pont.

Additional examples of sample-solvent peak distortion are presented in Fig. 14.7. In Figs. 14.7a,b, 30 µL of the same sample was injected onto a reversed-phase column. In Fig. 14.7a, the sample was dissolved in a strong solvent: pure acetonitrile. In Fig. 14.7b, the sample was injected as a solution in the mobile phase. Peak distortion and increased band-broadening are evident in the chromatogram of Fig. 14.7a, whereas the separation of Fig. 14.7b is quite normal. Figures 14.7c,d show a normal-phase separation of a single compound (decamethrin), with 30 µL of sample injected in Fig. 14.7c in a solution of strong solvent (methylene chloride), and in Fig. 14.7d dissolved in the mobile phase. Here peak-splitting is seen in Fig. 14.7c for the injection of the strong-solvent solution.

Sample-solvent problems should be suspected whenever a large volume of a solvent other than the mobile phase is injected, particularly when the sample-solvent is obviously stronger than the mobile phase. For example, you should question the injection of 100 µL of sample dissolved in methanol for a reversed-phase

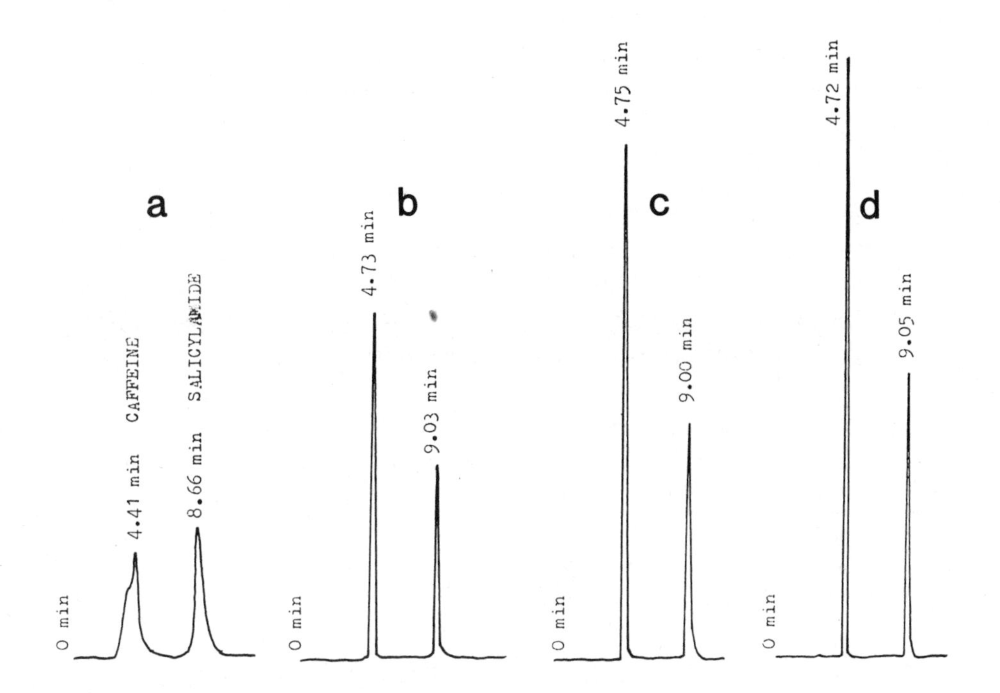


**Fig. 14.7.** Other examples of a sample-solvent that is too strong: (a) 30 μL injection of sample in acetonitrile; reversed-phase separation with 18% acetonitrile/water as mobile phase; (b) same as (a), sample dissolved in mobile phase; (c) 30 μL injection of sample in methylene chloride; silica column, 8% methylene chloride/hexane mobile phase; (d) same as (c), sample dissolved in mobile phase. Reproduced from ref. (3) with permission.

separation on a C8 column with 30 %v methanol/water as mobile phase. The general rule is that up to 25 μL of strong solvent can often be injected without problems, for columns of the usual size (15–25 x 0.46 cm); smaller-volume columns require proportionately smaller sample volumes. However, whenever a stronger solvent is used to dissolve the sample, the effect of sample volume on the chromatogram should be tested. If it is suspected that a 25 μL volume of methanol as solvent is causing problems, try one of two alternatives:

1. *Reinject a Smaller Volume.* Reinject 5–10 μL of sample and look for changes in the chromatogram (change detector attenuation to obtain similar peak heights); if the two

Table 14.2
Choosing the Solvent for Dissolving the Sample

1. *Ideal.* The best option is to dissolve the sample in the mobile phase and inject 10–50 μL.
2. *Practical.* Alternatively, a larger volume of weaker solvent can be injected; e.g., 100–500 μL (or more) of sample dissolved in water for the case of reversed-phase LC. The main disadvantage is a larger baseline upset at the beginning of the chromatogram in some cases.

   Larger volumes (100–500 μL) of sample dissolved in the mobile phase are also OK, except that resolution may suffer for early-eluting bands and/or smaller-volume columns.
3. *When Necessary.* When convenient or required, 10–25 μL of a stronger solvent can be injected; e.g., 25 μL of methanol solution for reversed-phase LC.

chromatograms (e.g., 5 and 25 μL injections) are similar, then sample-solvent effects are absent

2. *Dilute the Sample with a Weaker Solvent.* Dilute the sample (dissolved in strong solvent) with weaker solvent in 4:1 ratio, then inject a fivefold greater sample volume; again examine the chromatogram for any differences caused by sample-solvent effects. For the preceding example of 30%v methanol/water as mobile phase and a 25 μL methanol solution of sample, this would mean injecting 125 μL of a sample dissolved in 20%v methanol/water (instead of 25 μL in pure methanol)

Depending on the outcome of the above experiments, an alternative means of injecting the sample usually can be found that avoids sample-solvent problems. One other sample-solvent problem is the use of too large a volume of sample. The effect of large sample volumes on band width is discussed in Section 15.2 (Anomalous Band Broadening). Sample volumes that are too large can also result in band distortion, with the appearance of fat or flat-topped peaks as in Fig. 14.1.

Table 14.2 summarizes the preferred approach to selecting the kind and volume of solvent used for injected samples.

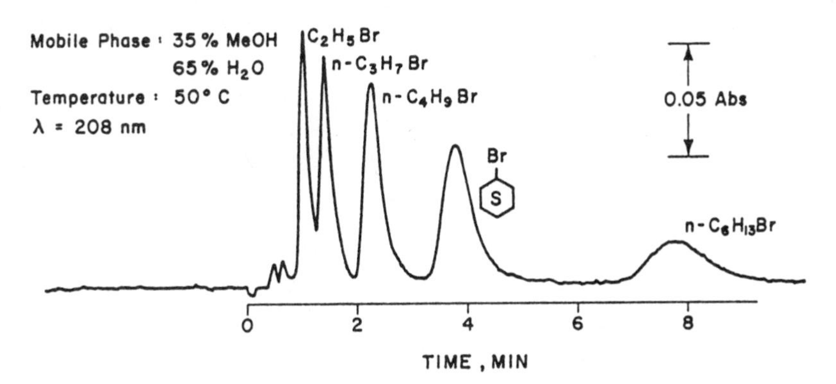


**Fig. 14.8.** Band-tailing caused by extra-column band broadening. Mixture of alkyl bromides separated by reversed-phase with 35 %v methanol/water as the mobile phase. Reprinted from ref. (5a) with permission.

## 14.4. Extra-Column Effects

Some of the LC equipment now in use was designed to handle columns that were first introduced a decade ago: 30 x 0.46-cm columns with 10-µm particles. Since that time, columns with smaller dimensions and smaller particles have become more popular. This means that extra-column band-broadening (see Sections 3.4 and 15.2) often contributes adversely to the plate number and shape of early bands in the chromatogram. Tailing is then seen for bands with small $k'$-values, with asymmetry factors decreasing as $k'$ increases. An example is given in Fig. 14.8. This generally is not a serious problem, because we attempt to adjust $k'$-values for bands of interest into the range $1 < k' < 10$, and band asymmetry often is not a problem for $k' > 1$. That is, the (common) tailing of bands near $k' = 0$ can be ignored.

For the case where extra-column band-broadening significantly affects the shape of peaks of interest, either another LC system (with less extra-column volume) must be used, or the first system must be replumbed to reduce its extra-column volume. For well-designed LC equipment, the major contribution to extra-column volume is usually the detector flowcell. Some detectors allow the substitution of smaller-volume flowcells (2 µL or smaller), which reduces extra-column effects, but usually also reduces detector sensitivity. In some cases, the connecting tubing may be

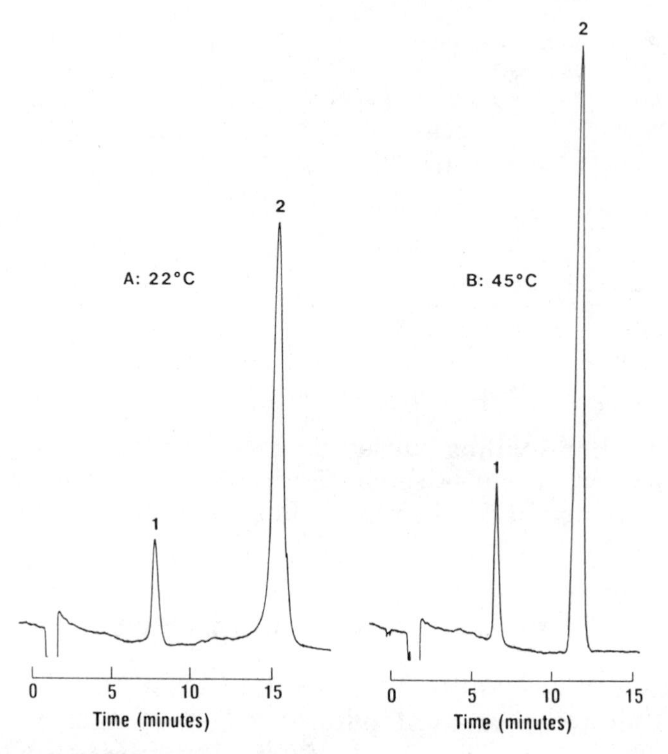


**Fig. 14.9.** Peak fronting in ion-pair chromatography as a function of separation temperature. Column, Zorbax C8; mobile phase, 33 %v acetonitrile/water plus 10-mM sodium dodecyl sulfate, pH 6; 50-μL sample. (a) Ambient temperature; (b) 45°C. Reprinted from ref. (4) with permission.

too long and/or have an internal diameter that is too large. Minimum lengths of 0.007- or 0.010-in. id tubing can be substituted in these cases. Finally, a major contribution to extra-column volume often is found in the autoinjector (see Chapter 10).

## 14.5. Band Fronting

Fronting bands (Fig. 14.1) are less commonly observed in LC, but they are readily distinguished from other band-shape problems. Column temperature problems can cause fronting bands in ion-pair chromatography (Fig. 14.9a shows) the separation of an antibiotic amine at ambient temperature. In this case repeating the separation at 45°C (Fig. 14.9b) eliminated the fronting problem. It generally is good practice to run ion-pair separations under thermostatted conditions, because relative retention tends to vary with temperature in ion-pair chromatography. Temperatures of 40–50°C are also generally favored for ion-pair chromatography, because narrower bands and better separation result.

Another source of fronting bands in ion-pair chromatography is the use of a sample-solvent other than the mobile phase. For a variety of reasons (Section 15.3), in ion-pair chromatography the sample should only be injected as a solution in the mobile phase; no more than 25–50 µL of sample should be injected, if possible. Failure to follow these recommendations can lead to fronting bands and other problems.

Still another cause of band fronting[5] is for the case of anionic sample molecules separated with higher-pH mobile phases. For silica-based packings, the packing has an increasingly negative charge as pH increases, and this results in repulsion of anionic sample molecules from the pores of the packing. However, with larger sample sizes, this effect is overcome by the corresponding increase in ionic strength caused by the sample. For this case, a more logical remedy for the problem is to increase the ionic strength of the mobile phase, by increasing mobile-phase buffer concentration to the range of 25–100 mM. Note that ionic or ionizable samples should never be separated with unbuffered mobile phases (see Section 14.8)

Column voids and blocked frits can also cause band fronting. See Section 15.1 for information on these problems.

Finally, fronting bands are the opposite of tailing bands (which are far more common). Whereas tailing bands suggest that sample retention decreases with increasing sample size or concentration, fronting bands suggest the opposite: increasing retention for larger samples. In both cases, a decrease in sample size may eliminate peak distortion—but this is often not practical, because some minimum sample size is required for good detectability. In the case of band tailing, it is believed that peak distortion often arises because too large a sample uses up some part of the stationary phase. For fronting bands, however, the cause is seldom well understood.

## 14.6. Strong-Retention Sites

Separations by normal-phase (adsorption) or ion-exchange chromatography involve the binding of sample molecules to spe-

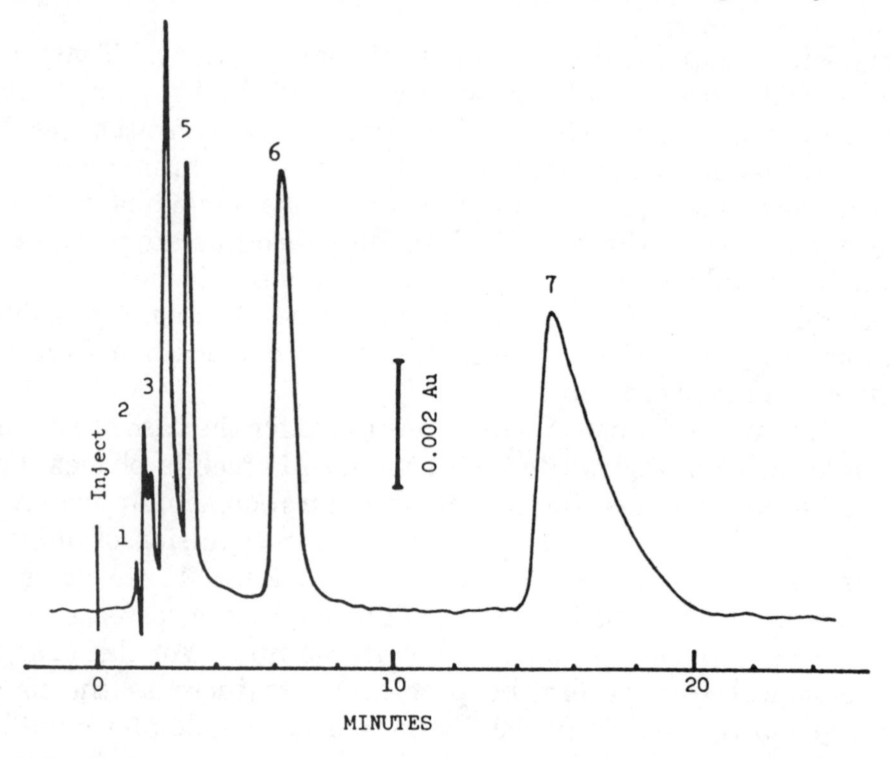


**Fig. 14.10.** Tailing of late-eluting bands in the cation-exchange separation of some aniline derivatives. Mobile phase, 0.2 mM phosphate, pH 2.9, 25°C; 10-µL sample. Reprinted from ref. (6) with permission.

cific sites on the surface of the column packing. For example, the silanol (≡Si—OH) groups on silica, or the sulfonate ($—SO_3^-$) groups on a cation exchanger. Often these sites are not all equivalent; some sites are favorably situated for particularly strong interaction with sample molecules. The strongest retention sites will be preferred by sample molecules, and these sites will be used up first (by retained sample molecules). Because these strong sites often are present in low concentration (constitute only a small fraction of all retention sites on the packing), they are quickly depleted by the sample, so that only weaker sites are then available for further retention of sample molecules. This means that such packings (for normal-phase or ion-exchange LC) may overload more quickly than is the case for other LC methods (SEC, ion-pair, or reversed-phase LC).

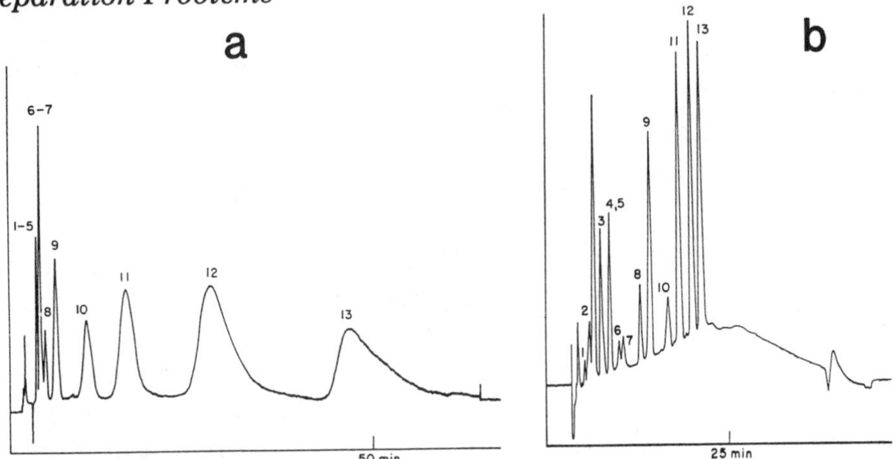


**Fig. 14.11.** Elimination of tailing in late-eluting bands from isocratic anion-exchange separation of carboxylic acids, by using gradient elution: (a) 55 m*M* sodium nitrate; (b) gradient from 10 to 100 m*M* nitrate in 20 min. Reprinted from ref. (7) with permisssion.

Another feature of strong retention sites is that they generally attract sample molecules with larger $k'$-values; i.e., strong sites interact particularly strongly with strongly retained molecules. This means that premature column overloading occurs mainly for later-eluting bands in the chromatogram, usually those with $k' > 10$. Consequently it is the last bands in the chromatogram that overload first, and exhibit band-tailing as a result. This is illustrated in Fig. 14.10 for the cation-exchange separation of some aniline derivatives. Here only the last band ($k' = 10$) tails, with earlier bands having acceptable peak shapes. Band-tailing in ion-exchange or normal-phase LC as a result of strong retention sites can sometimes be reduced by using less sample, but this compromises detection sensitivity. A more effective approach is to increase the strength of the mobile phase; i.e., reduce the $k'$ value of the last (tailing) band in the chromatogram to a value well below 10. If lowering $k'$ results in unacceptable loss in resolution at the beginning of the chromatogram, then the only alternative is the use of gradient elution—this is shown by the ion-exchange separations of Fig. 14.11. Here bands #12,13 tail in the isocratic separation of Fig. 14.11a, whereas tailing is eliminated in the gradient separation of Fig. 14.11b.

For normal-phase separation with silica columns, strong-retention sites can be suppressed by adding water to the mobile phase. Water-deactivation of the column in this way can in turn improve the symmetry of late-eluting bands. This is further discussed in Section 15.4.

## 14.7. Secondary-Retention Effects

In a well-designed LC separation, sample molecules will be retained by a single retention process. For example, in reversed-phase LC, the solute will interact hydrophobically with the nonpolar alkyl chains of the column packing. However, for silica-based packings, interaction of some sample compounds with the silanol groups is also possible. This secondary-retention process (silanol interaction) can lead to band-tailing. Two reasons for this kind of band-tailing exist:

1. Secondary retention often involves a limited number of retention sites, and these sites are used up quickly; i.e., the column quickly overloads for those compounds that interact strongly with these secondary sites
2. In some cases the interaction of sample molecules and secondary sites is quite strong, and the sorption-desorption kinetics are slow; this can result in severely tailing bands.

Band-tailing caused by secondary retention probably is the most common and serious example of misshaped peaks. The presence of this kind of band-tailing is also an indication that sample retention (and separation) is likely to vary from column to column (column irreproducibility, see Sections 11.4 and 15.4). For these reasons we will look closely at this problem, and consider how it affects the different LC methods.

### *Reversed-Phase LC*

Secondary retention in reversed-phase separations has been variously attributed to silanol groups of different kinds and to the presence of trace metal impurities in the column packing. We will consider silanol groups first. A good example of these silanol effects is given by the examples of Fig. 14.12. In each case, a standard mixture of compounds that are acidic (VMA, HVA, SAL), basic (PA, NAPA), or neutral (CAF) was injected. Two different reversed-phase columns were used: (a) Supelcosil LC-18 and (b)

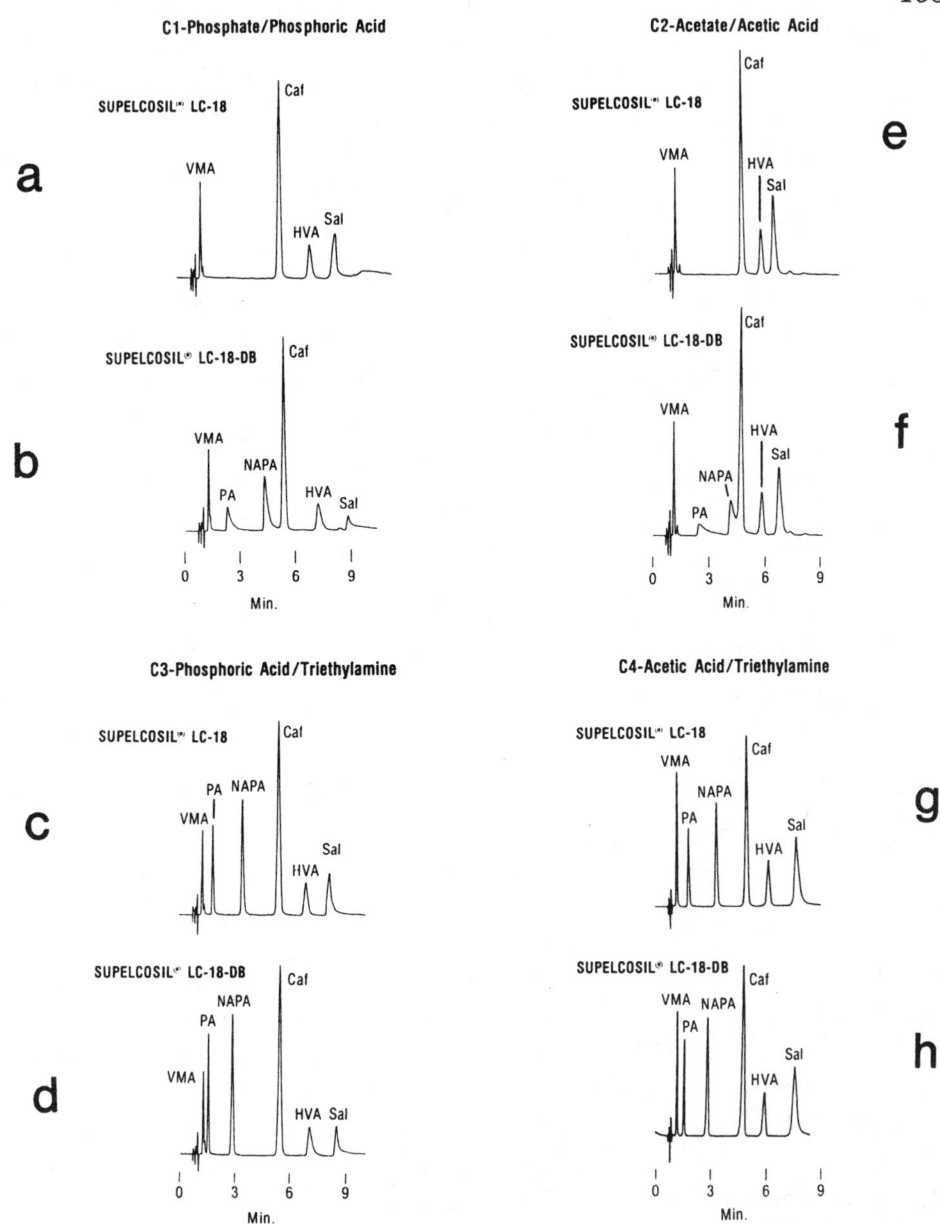


**Fig. 14.12.** Secondary-retention effects in the reversed-phase separation of a mixture of neutral, acidic and basic compounds. Solutes are caffeine: (CAF) (neutral); homovanillylmandelic acid (HVA), vanillylmandelic acid (VMA), and salicylic acid (SAL) (acids); procainamide (PA) and *N*-acetyl procainamide (NAPA) (bases); Columns are Supelcosil LC-18 (Figs. a,c,e,g) and Supelcosil LC-18 DB (Figs. b,d,f,h). Mobile phase is 7 %v acetonitrile/water, pH 3.5; 10 m*M* triethylamine added in (c,d,g, h); 1% acetate added to (e–h). Courtesy of Supelco, Inc.

Supelcosil LC-18DB. The DB column has been specially processed to minimize the retention (and tailing) of basic compounds.

Figures 14.12a, b show the separation on each column using a mobile phase that is 7%v acetonitrile/water (pH 3.5, phosphate buffer). The basic compounds PA and NAPA are retained strongly and tail badly on the LC-18 column (bump after SAL in Fig. 14.12a), which is typical of most reversed-phase LC packings. On the LC-18 DB column (Fig. 14.12b) these two basic compounds now elute much earlier, with improved (but inadequate) peak shapes. It is believed that this strong retention and tailing of amines such as PA and NAPA in most reversed-phase systems is due to two kinds of silanol interactions:

Hydrogen bonding,

$$R_3N: + HO\text{-}Si\text{-} \Leftrightarrow R_3N: \cdot\cdot H \cdot\cdot\text{-}Si\text{-} \qquad (14.1)$$

Ion-exchange

$$R_3NH^+ + Na^+ {}^-O\text{-}Si\text{-} \Leftrightarrow R_3NH^+ {}^-O\text{-}Si\text{-} + Na^+ \qquad (14.2)$$

The greater retention and tailing of the basic compounds PA and NAPA in Fig. 14.12a suggests that the silanols of the LC-18 column are more acidic, vs the case of the LC-18 DB column (Fig. 14.12b).

When secondary-retention effects such as those of Fig. 14.12 are present, the most effective solution generally is the addition of some mobile phase modifier that will preferentially interact with (and block) these secondary-retention sites (silanol groups in the present example). Amine-additives such as triethylamine (TEA) are commonly used for this purpose, with small concentrations of the amine (1–20 m*M*) usually being effective. Figures 14.12c,d show the effect of adding 10 mM TEA to the mobile phase for the LC-18 and LC-18 DB columns, respectively. The result for each column is a pronounced sharpening of the bands for PA and NAPA, with the elimination of any band-tailing. This is typical of many such reversed-phase separations of basic compounds. Presumably the amine-modifier (TEA) replaces the solute $R_3N^+$ in Reactions 14.1 and 14.2, thereby effectively eliminating the silanol groups as sorption sites for sample molecules. Typically 30–50 m*M* of TEA is adequate for this purpose, with little increased effect for higher concentrations. The bands for acidic compounds HVA and SAL in Figs. 14.12a–d show considerable variation in bandwidth and peak asymmetry, and tailing of SAL is particularly pronounced on the LC-18 DB column. This tailing of SAL is not improved by adding TEA (Fig. 14.12d vs 14.12b), which presumably neutralizes the effect of acidic silanol groups. This suggests that different kinds of silanol groups are present on the silica surface, some of which preferentially bind bases—while others bind acids. One possibility is:

$$\text{R-COO}^- + \text{HO–Si*-} \Leftrightarrow \text{R-COO}^- \cdots \text{H} \cdot \text{O-Si*–} \quad (14.3)$$

where ionized acids $R\text{–}COO^-$ hydrogen-bond with this special kind of silanol group (-Si*-OH), or alternatively,

$$\text{R-COOH} + {}^-\text{O-Si*-} \Leftrightarrow \text{R-COO}^- \cdots \text{H} \cdot \text{O-Si*-} \quad (14.4)$$

There is substantial evidence for the existence of different kinds of silanol groups on the silica surface.[8a] If the previous hypothesis is correct, then the addition to the mobile phase of a carboxylic-acid modifier should reduce the secondary-retention of acidic samples as in Eqs. (14.3, 4). This possibility is tested in the examples of Fig. 14.12 by adding 1% acetic acid to the mobile phase, while holding pH constant: Fig. 14.12e (LC-18) and Fig. 14.12f (LC-18 DB). The acidic compounds HVA and SAL now are seen to give quite sharp bands with no discernible tailing. So the addition of a carboxylic acid (e.g., acetic acid) to the mobile phase appears to improve band shape for carboxylic-acid samples, whenever the latter compounds show peak tailing.

The examples of Figs. 14.12c-f can be generalized as follows:

1. Whenever band-tailing caused by secondary retention is observed, the addition of modifiers of related structure to the mobile phase usually will result in a reduction of band tailing. As the structure of the added modifier more closely approximates that of the sample compound, the effectiveness of the modifier in reducing tailing should increase correspondingly. This suggests that very severe cases of band tailing not resolved by the usual modifiers may require special modifiers, selected for their structural similarity to the sample compounds of interest.
2. Samples that contain both acidic and basic compounds (as in Fig. 14.12) present a special problem. This is seen in

Figs. 14.12e,f, where addition of acetate to the mobile phase improves the band shape of the acidic components of the sample, but worsens the shape of the basic compounds (and increases their retention). That is, acidic modifiers can intensify the secondary retention of basic sample compounds. The solution is to add both an acidic and a basic modifier to the mobile phase, as illustrated in Figs. 14.12g,h (1% acetate, 10 mM TEA). Now all the bands in chromatograms for either column (LC-18, Fig. 14.12g; LC-18 DB, Fig. 14.12h) are quite symmetrical. It should also be noted that retention for each compound in the sample now is almost identical for the two columns. That is, the suppression of secondary retention effects in this case has also eliminated retention variability (Figs. 14.12a vs 14.12b). We will comment on this in Section 15.4.

Because basic compounds more often present band-tailing problems, special interest exists in the selection of the best amine modifier to minimize secondary retention. Recent systematic studies[9] have shown that tertiary-amines with short alkyl groups (methyl or ethyl) are most effective in this regard. The modifier triethylamine (TEA) has proven broadly useful as an antitailing modifier for basic sample compounds, confirming the general applicability of this conclusion. However, even TEA sometimes fails to yield symmetrical bands, especially for some columns and for tertiary-amine samples. In this case, modifiers of the general structure $R(CH_3)_2N$ have proven more effective, where R is either hexyl or octyl.[10,11] This is illustrated by the data of Table 14.3.

Though dimethylalkylamine modifiers such as DMOA and DMHA (Table 14.3) often are more effective in suppressing the secondary retention of amine samples, TEA is a better first choice in this regard for several reasons. *First,* DMOA and DMHA are retained more strongly and may prove difficult to remove from the column when changing mobile phases (for another assay procedure). This can result in retention variability for repeated injections of the same sample (Section 15.4). *Second,* equilibrating the column with DMOA or DMHA can require a longer time than in the case of more weakly held amine modifiers. *Third,* some workers have observed that the use of DMOA or DMHA improved the

Table 14.3.
Optimum Amine Modifier for Difficult Samples
($R[CH_3]_2N$ Structures)

| Amine modifier | Compound | Column | Asymmetry factor, $A_s$ |
|---|---|---|---|
| None | Amphetamine[a] | ODS-Hypersil | 5.6–6.2 |
| 50 mM TEA | | | 4.0 |
| 50 mM DMHA[c] | | | 1.5 |
| None | Imipramine[a] | Lichrosorb RP-18 | 9 |
| 10 mM octyl amine | | | 3 |
| 10 mM DMOA[c] | | | 2.5 |
| None | Imipramine[b] | Polygosil | 2.0 |
| 10 mM octyl amine | | | 1.3 |
| 10 mM DMOA[c] | | | 1.0 |

[a] (10); [b](11); [c]DMHA, dimethyl hexyl amine; DMOA, dimethyl octyl amine.

shape of tailing bands in a particular chromatogram, but at the same time worsened the shapes of other bands. Although this behavior may not be typical, it suggests that these modifiers resemble powerful drugs with serious side effects. Like such drugs, DMOA and DMHA should not be used indiscriminately, but only when the safer modifier TEA proves ineffective. The data of Table 14.3 also suggest that changing to another (less acidic) column may prove to be a more effective approach in some cases; for example, Polygosil column (less acidic) vs Lichrosorb column (more acidic).

*Ion-Exchange Effects.* Detailed studies by Vigh et al.[12,13] have shown conclusively that secondary retention in reversed-phase LC can arise from ion-exchange of protonated sample bases with ionized silanols [Eq. (14.2)]. Ion-exchange effects have been observed even for low pH mobile phases (e.g., pH = 2.0), suggesting that some silanol groups are quite acidic. When secondary retention involving ion-exchange is suspected, it can often be controlled by increasing the buffer concentration so as to increase the ionic strength of the buffer and drive Eq. (14.2) to the left. An example is shown in Fig. 14.13, for the reversed-phase separation of several anilines. In Fig. 14.13a, a buffer concentration of 0.4 m*M*

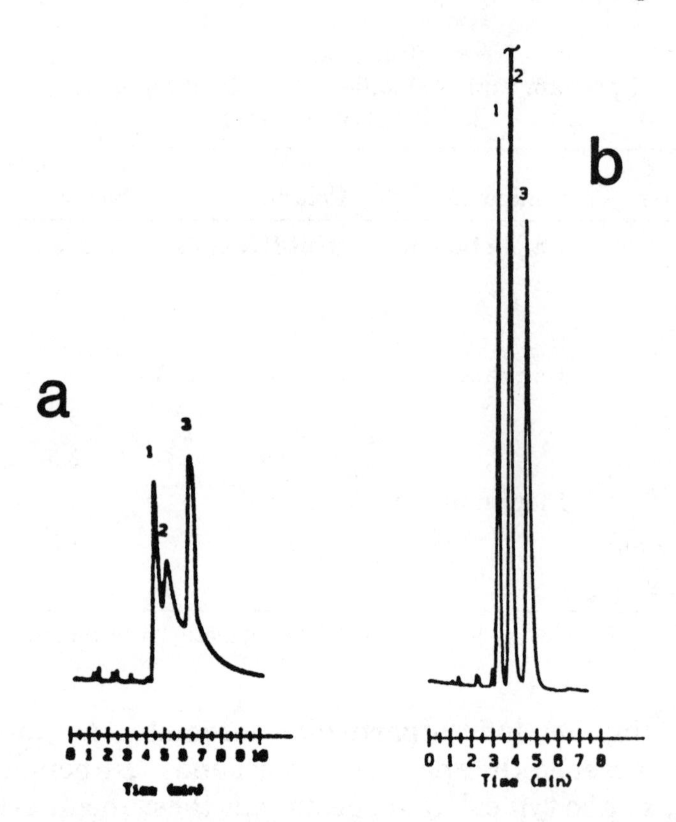


**Fig. 14.13.** Secondary-retention and tailing of anilines as a result of ion-exchange in a reversed-phase system; decrease in tailing with increase in mobile-phase buffer concentration. Zorbax C8 column; mobile phase 45% methanol/water, pH 3.6; 0.4 m*M* acetate in (a), 2.7 m*M* in (b). Courtesy of Du Pont.

phosphate results in severe tailing for all three aniline bands. Increasing the buffer concentration to 2.7 m*M* (Fig. 14.13b, no change in pH) results in a dramatic improvement in peak shape.

*Trace Metals.* A few studies suggest that some reversed-phase packings are contaminated with heavy metals (iron, copper, etc.), and that these metals give rise to secondary-retention effects that simulate those attributed above to silanols.[14,15] An example is given in Fig. 14.14a, for the elution of two dihydroxynaphthalene (DHN) derivatives from a particular C18 packing. The 2,3-DHN isomer is capable of chelation with heavy metals, and is seen to give a broader, more tailing band than the 1,7-DHN isomer, which is incapable of chelation. In Fig. 14.14b the separa-

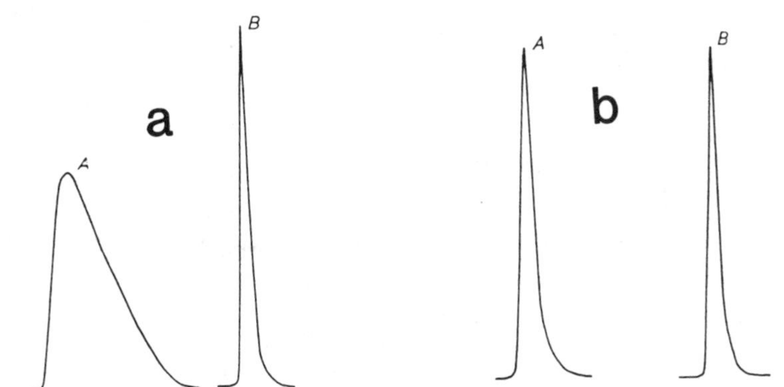


**Fig. 14.14.** Heavy-metal contamination of the column packing and resulting tailing of chelating sample compounds. C18 column. Mobile phase is 30 %v acetonitrile/water at pH 6.7. Bands are 2,3-dihydroxynaphthalene (A) (a metal chelator) and 1,7-dihydroxynaphthalene (B) (nonchelator); (a) original column; (b) EDTA-washed column. Reprinted from ref. (15) with permission.

tion was repeated, after washing the column with the metal-chelator EDTA. A marked change in band shape is noted for the 2,3-DHN isomer, suggesting that the EDTA wash has removed or masked metal contaminants that were responsible for this problem. Whenever it is known that a particular sample compound complexes with heavy metals, and where the compound shows pronounced tailing that is not affected by TEA as modifier, EDTA can be used to wash the column (or the EDTA can be added to the mobile phase). Although other studies[16,17] suggest that metal contamination of the column packing is not a general problem, the chelation and retention of sample compounds by the stainless steel frit of the column can be significant in some cases. The use of stainless-steel screens (in place of frits) has been shown to be effective in reducing this metal-related problem.[17]

### *Normal-Phase LC*

Secondary retention in normal-phase chromatography can arise from:

1. Ion-exchange of sample molecules with ionized silanols on silica.
2. Interaction of sample molecules with silanols, in the case of polar-bonded-phase packings.

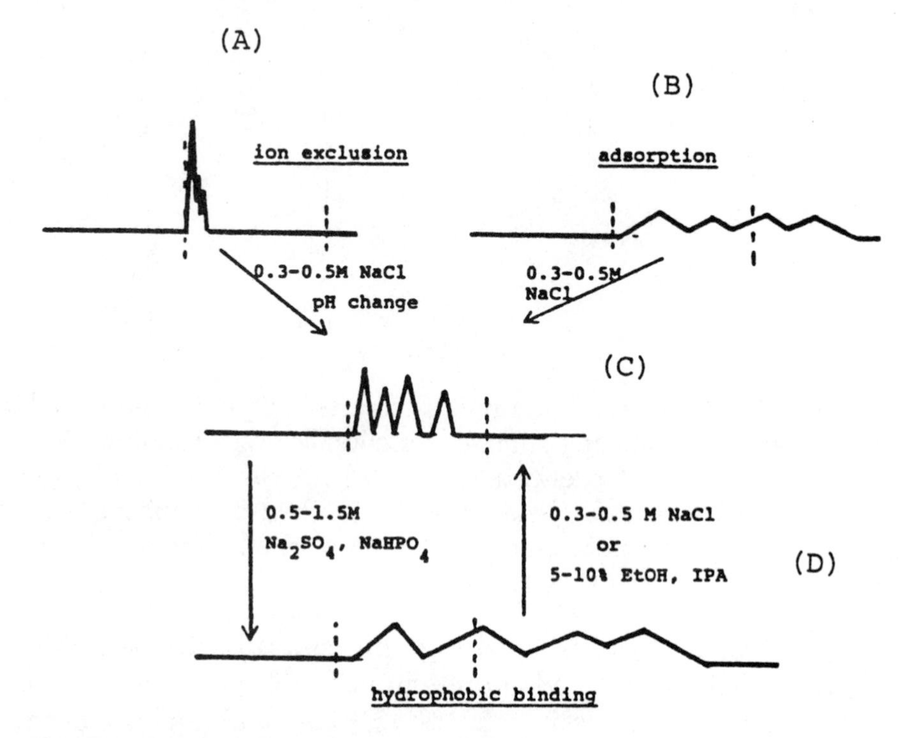


**Fig. 14.15.** Secondary-retention in the gel filtration (SEC) separation of proteins and other polyelectrolytes: (a) ion-exclusion; (b) ion-exchange retention; (c) normal retention; (d) hydrophobic retention.

In practice, these effects have been minimized simply by adding about 0.1% of an amine modifier (TEA) for the separation of basic compounds, and 0.1% acetic acid for the separation of acidic compounds. Neutral compounds such as steroids and prostaglandins tend to tail on cyano-silica columns, and addition of either TEA or 0.2% water is recommended for such cases.[19,20]

## *Size-Exclusion Chromatography*

Secondary-retention effects are noted most often on silica-based packings used with aqueous mobile phases (gel filtration). These can arise from a variety of secondary-retention effects summarized in Fig. 14.15:

1. Ion exclusion
2. Ion exchange
3. Hydrophobic (reversed-phase) binding

These secondary-retention effects may or may not be accompanied by peak tailing (see discussion of ref. 20).

*Ion Exclusion.* At higher pH values (>4) silica-based SEC packings carry a pronounced negative charge, caused by the ionization of silanol groups. If the sample molecule is also negatively charged, it will be repelled from the particle pores, and sample retention will be decreased as in Fig. 14.15a (vs normal retention in Fig. 14.15c). There are two solutions to this problem:

1. Increase the buffer concentration or add salt, to increase total salt concentration to 0.3–0.5 *M*. This swamps the charge on the silica surface, and eliminates ion exclusion.
2. Reduce the mobile phase pH so as to reduce the negative charge on the sample molecule, thereby reducing the ionic interaction with the silica surface.

*Ion Exchange.* If the mobile-phase pH is above 4, and the silica-based SEC particles carry a negative charge as above, positively charged sample molecules can be attracted to the particle surface by ion exchange [Eq. (14.2)]. This then results in sample bands eluting *after* the total-permeation limit of the column (Fig. 14.15b). The solution to this problem is similar to that for ion exclusion:

1. Increase the salt or buffer concentration so as to drive the reaction of Eq. (14.2) to the left, thereby decreasing the ion-exchange retention of the sample.
2. Increase the mobile phase pH so as to reduce the positive charge on sample molecules.

*Hydrophobic Binding.* Silica-based SEC packings for use with aqueous mobile phases (gel filtration) generally have some hydrophobic character ($-CH_2-$ groups in the bonded phase). This weakly hydrophobic surface can bind sample molecules that are especially hydrophobic. Hydrophobic binding leads to excessive retention of sample bands, as illustrated in Figure 14.15d. Two solutions to this problem exist:

1. Add 5–10 %v of ethanol or isopropanol to the mobile phase, thus increasing the strength of the mobile phase with respect to hydrophobic (reversed-phase) binding.

2. Replace polyvalent salts or buffers in the mobile phase with monovalent salts or buffers, choose salts that have a lesser salting-out effect, and/or reduce the total salt concentration; this reduces salting out of the sample.

### *Ion-Pair Chromatography*

This LC method commonly involves two major types of retention: ion-exchange and reversed-phase LC.[21] By changing the concentration of the ion-pair reagent, more or less of the reversed-phase packing is covered with the ion-pair reagent and converted into a dynamic ion-exchanger. However, this does not result in tailing bands (caused by secondary retention) in most cases, rather the opposite. The reason is that the stationary phase for both retention processes is present in significant amounts, so that there is no overloading of one retention process by the sample. Therefore band-tailing (for the same sample) is less common in ion-pair LC than for other LC methods. Likewise, it is observed that there is less column-to-column variability (Section 11.4) for ion-pair chromatography compared to reversed-phase LC.

Although band tailing generally is reduced in ion-pair chromatography, it is nevertheless desirable to add TEA as an anti-tailing modifier when separating basic sample compounds. Usually alkyl sulfonates are used as ion-pairing reagents in this case, and TEA should be added at a concentration that is either (a) 10–20% of the concentration of ion-pairing reagent, or (b) 20–30 m*M*.[22]

### *Diagnosing Secondary-Retention Problems*

In most cases, secondary-retention problems can be anticipated from the chemistry of the LC system, as outlined in the above discussion. We also can obtain additional confirmation of secondary retention by inspecting the chromatogram. For compounds of similar structure, secondary-retention effects generally increase with increasing $k'$ of the sample compound. Secondary-retention then appears as bands that tail increasingly (larger asymmetry factors) from beginning to end of the chromatogram. This is illustrated by the normal-phase separation of Fig. 14.16 for a bonded-ether-phase column. For samples containing compounds of quite different functionality, a regular increase of band tailing with $k'$ may not be observed. This is illustrated by the

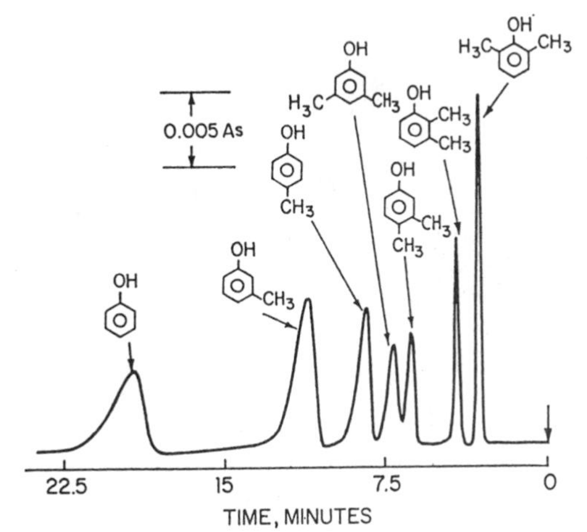


**Fig. 14.16.** Secondary-retention and tailing of aromatic alcohols in normal-phase separation on an ether-phase column. Reprinted from ref. (23) with permission.

tailing bands of Fig. 14.12f. In this case, only the basic compounds are subject to secondary-retention effects with resulting severe tailing, while more strongly retained acids are free from such effects.

## 14.8. Inadequate Buffering

Buffered mobile phases are a "must" whenever acidic or basic samples are being separated. In the absence of any buffering of the mobile phase, the sample compounds themselves will cause the mobile phase pH to vary in that part of the column where a sample band is found. For example, if carboxylic acids are separated in an unbuffered reversed-phase system, ionization of these compounds will lower the pH at that point within the column where the bands are located. This change in pH will vary with sample concentration, resulting in varying ionization of the sample compound; the result is a tailing band.

Buffer concentrations generally should exceed 10 m*M*, with values of 50–100 m*M* being advisable. Figure 14.17 shows the effect of increasing buffer concentration from 10 to 100 m*M* in a typical ion-pair LC separation. The severely tailing bands in Fig. 14.17a are much improved in Fig. 14.17b.

Many workers prefer lower buffer concentrations (e.g., 5–25 m*M*), especially in the case of phosphate or other inorganics. The

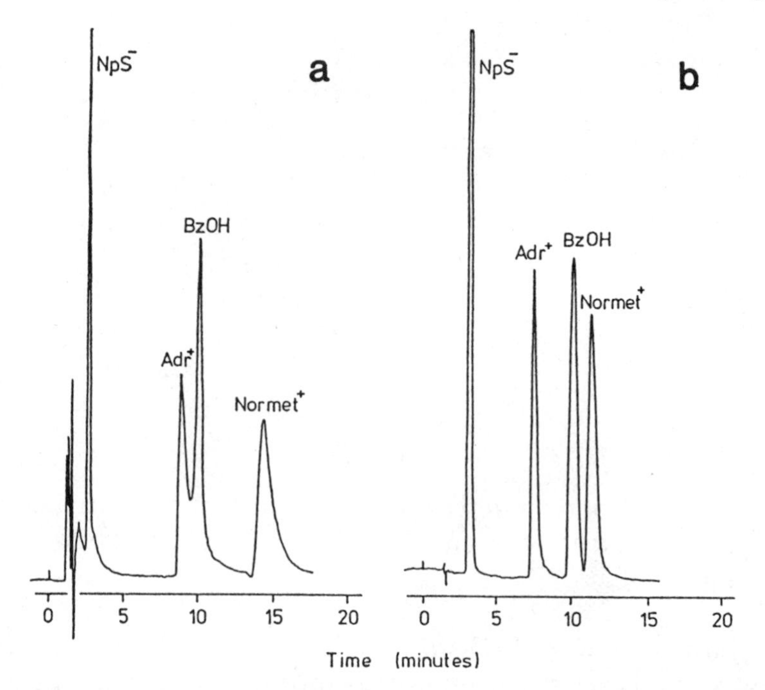


**Fig. 14.17.** Tailing of bands in ion-pair chromatography caused by too low a buffer concentration. Solutes are naphthalene sulfonate (NpS), adrenaline (Adr), benzyl alcohol (BzOH), and normetanephrine (Normet). Mobile phase is 15 %v methanol/water, 10 m*M* octyl sulfate; (a) 5 m*M* buffer concentration; (b) 100 m*M* buffer. Reprinted from ref. (24) with permission.

reason is that higher concentrations of buffers which form hard crystals are abrasive to pump seals (Chapter 7). Higher buffer concentrations for reversed-phase gradient elution also should be avoided, because of the possibility that the buffer might precipitate out during the gradient. The choice of buffer concentration in these cases therefore represents a compromise, but the buffer concentration should always be high enough to avoid tailing bands as in Fig. 14.17.

## 14.9. Miscellaneous Other Effects

Artifactual peaks (ghost peaks, negative peaks) may not correspond to any compound present in the sample, yet if they overlap a real peak, the result is a distorted sample band. An example is given in Fig. 14.18a for nitrobenzene as sample. Further discussion of this problem is given in Section 15.2.

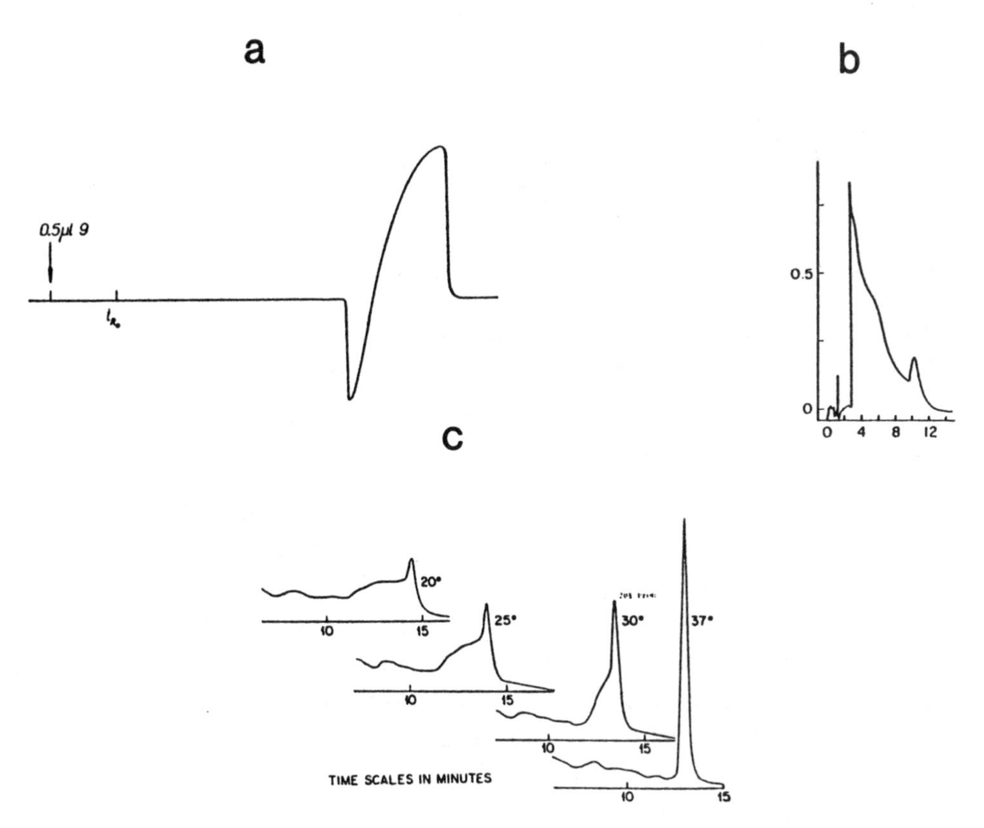


**Fig. 14.18.** Miscellaneous other examples of band tailing: (a) ghost peak overlapping band of nitrobenzene[25]; (b) reaction of desferoxamine peak with iron in LC system during separation[26]; (c) denaturation of ribonuclease during reversed-phase separation.[27] Reprinted with permission.

Reaction of a sample band during its passage through the column can yield a badly distorted peak, as in the example of Fig. 14.18b. Here a sample of desferoxamine has undergone partial complexation with iron (picked up from the LC system). Some compounds can undergo a slow interconversion between two forms of the same compound, with resulting band tailing. Fig. 14.18c shows an example of this, for the reversed-phase separation of the protein ribonuclease. In Fig. 14.18c, the protein exists in both native and denatured forms during separation at 25°C, with resulting severe distortion of the elution band. The same separation at a higher temperature (37°C) involves only the dena-

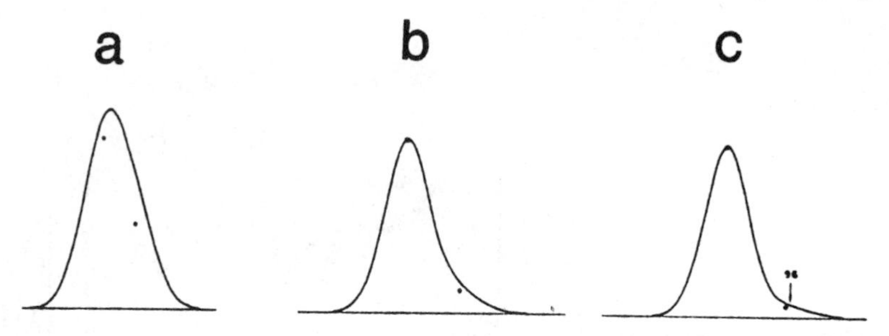


**Fig. 14.19.** Partially resolved bands that appear to tail: (a) $R_s = 0.4$, 2/1 ratio of band heights (dots show band centers); (b) $R_s = 0.6$, 8/1 ratio of band heights; (c) $R_s = 0.7$, 16/1 ratio of band heights. Reprinted from ref. (2) with permission.

tured compound, with a much improved band shape. Section 15.3 provides a more detailed discussion of problems due to sample reaction or interconversion.

The separation of samples that are barely soluble in the mobile phase, or of compounds that tend to form aggregates in solution, can lead to band tailing. For example, the separation of ionic surfactants in size-exclusion chromatography with organic mobile phases can lead to micelle-formation for larger sample sizes, with resulting decrease in retention times caused by an increase in molecular weight of the resulting sample aggregates. Here a knowledge of the properties of the sample is the key to figuring out what is going on—and then solving the problem.

## 14.10. Pseudo-Tailing

In some cases, tailing bands may not actually represent a problem—or the problem may not fit into the above band-tailing categories. One example is an apparently tailing band that is actually two partially resolved sample bands. For example, an unknown interference might partially overlap a compound known to be present in the sample. This is illustrated in the examples of Fig. 14.19, showing overlapping bands that masquerade as tailing bands. Partly resolved bands should be suspected whenever a single tailing band is seen in the chromatogram, and when other bands of similar molecular structure and concentration do not tail; e.g., the "anti" band of Fig. 14.20.

Another kind of pseudo-tailing is often observed for polymer bands that elute near the exclusion limit in size-exclusion chro-

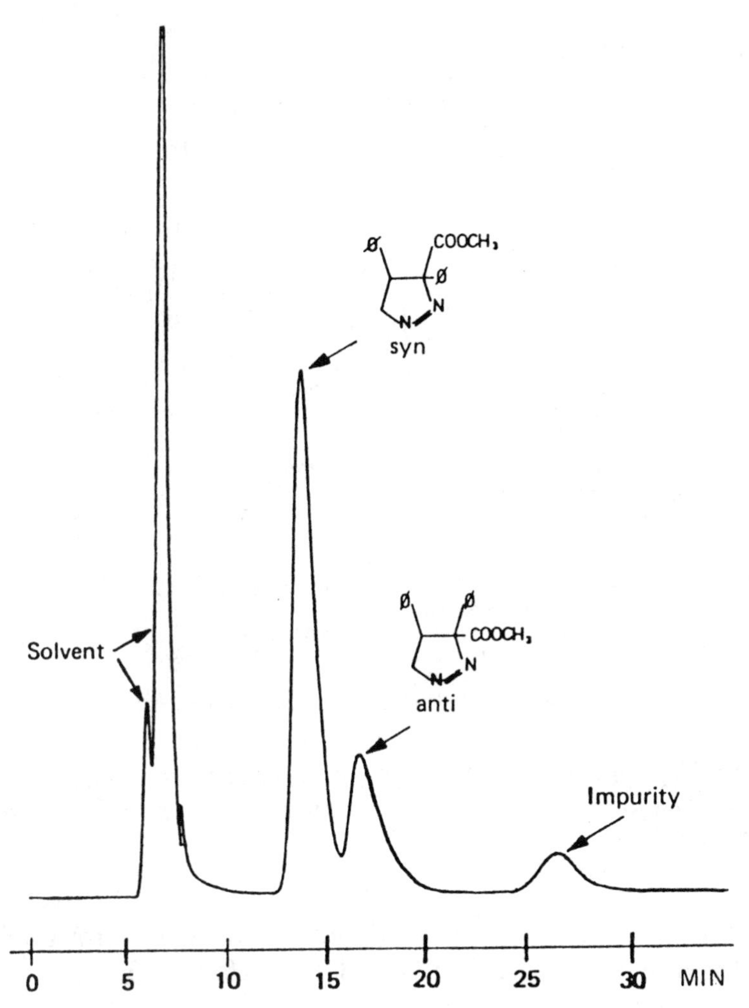


**Fig. 14.20.** Presumed tailing of anti-band caused by band overlap. Courtesy of Hewlett Packard.

matography. An example is shown in Fig. 14.21 for the separation of a mixture of polyvinylpyrrolidone (PVP) polymer and ethanol. The PVP peak tails, though the later ethanol peak does not. In this and similar cases, the molecular-weight distribution of the polymer determines band shape. None of the previously discussed cures for band tailing would change the shape of the PVP band in this separation.

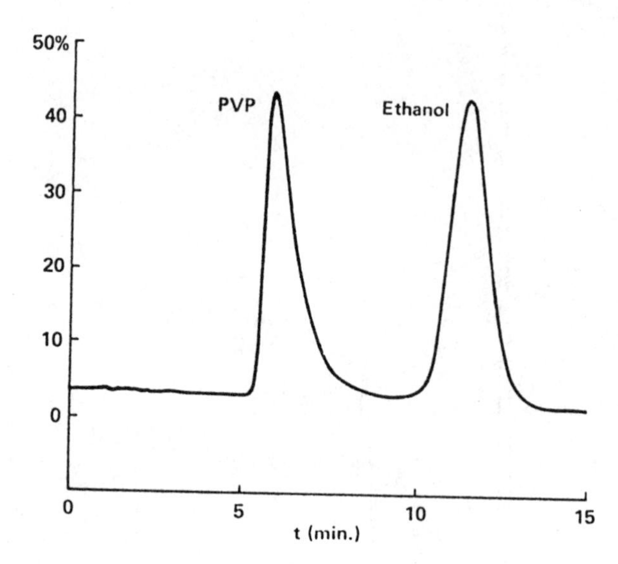


**Fig. 14.21.** Apparent tailing of polymer band eluted at exclusion limit in size-exclusion chromatography. Reprinted from ref. (2) with permission.

Our preceding discussion of band-tailing is summarized in Table 14.4, where symptoms and fixes for each type of band tailing are given.

Table 14.4
Band-Tailing Problems and Solutions

| Cause of problem | Symptom | Solution |
|---|---|---|
| *Bad column* | | |
| Blocked frit | High pressure; split or distorted peaks | 1. Reverse and flush<br>2. Replace frit<br>3. Replace column |
| Void | Peak splitting or distortion; depression at top of column | 1. Fill void and reverse column<br>2. Replace column |
| *Sample overload* | Band tailing that is eliminated by dilution and reinjection of sample | 1. Use smaller sample size (mass)<br>2. Use higher-capacity column (larger i.d.) |
| *Wrong sample solvent* | Peak-shape distortion, eliminated by smaller injection volume or weaker injection solvent | 1. Use smaller sample volume<br>2. Use weaker injection injection solvent |
| *Extra-column effects* | Band broadening; increased $A_s$-values, especially for early bands | 1. Replumb or change LC<br>2. Also see Section 3.4 |
| *Band fronting* | | |
| Low temperature | Especially in ion-pair | 1. Increase column temperature |
| Wrong sample solvent | Fronting bands | 1. Use only mobile phase for injection solvent<br>2. Keep injection volume no more than 25–50 μL |
| Sample overload | Fronting bands | 1. Decrease sample size (weight) if possible<br>2. Try another LC mode |
| Bad column | See above | See above |
| *Strong-retention sites* | Particularly with normal-phase or ion-exchange; change in retention with sample size, especially later bands | 1. Reduce $k'$ of last tailing band to $k' < 10$; use gradient elution<br>2. Water-deactivation of normal-phase columns |

*(continued)*

Table 14.4
Band-Tailing Problems and Solutions *(continued)*

| Cause of problem | Symptom | Solution |
|---|---|---|
| *Scondary-retention effects* | | |
| Reversed-phase | Strong tailing of polar compounds on reversed-phase columns | 1. Add TEA for basic compounds and/or<br>2. Add acetate for acidic compounds<br>3. Adjust pH of mobile phase; use pH ≤ 3.5<br>4. Add EDTA if chelators present in sample<br>5. Use deactivated bonded phase (Chapter 11) |
| Normal phase | Tailing bands on normal-phase columns | 1. Add TEA for basic compounds and/or<br>2. Add acetate for acidic compounds<br>3. Add TEA or water with cyano columns |
| Size exclusion | Tailing bands on size-exclusion columns | 1. Increase buffer concentration or add salt<br>2. Adjust mobile-phase pH<br>3. Add ethanol or isopropanol to mobile phase |
| Ion pair | Tailing bands in ion-pair chromatography | 1. Add TEA to mobile phase |
| *Inadequate buffering* | Band tailing that varies with sample concentration | 1. Use buffered mobile phases<br>2. Increase buffer concentration (to 50–100 m*M*) |
| *Artifactual peaks* | Ghost peaks, negative peaks | 1. See Section 15.2 |
| *Inadequate resolution (hidden peaks)* | Tailing or double peaks | 1. Change selectivity of mobile phase or column<br>2. Increase *N* |

## 14.11. References

[1]Snyder, L. R., Dolan, J. W., and van der Wal, Sj. (1981) *J. Chromatogr.* **203,** 3.
[1a]Vendrell, J. and Aviles, F. X. (1986) *J. Chromatogr.* **356,** 420.
[2]Snyder, L. R. and Kirkland, J. J. (1979) *Introduction to Modern Liquid Chromatography,* 2nd ed., Wiley-Interscience, New York.
[3]Ng, T.-L. and Ng, S. (1985) *J. Chromatogr.* **329,** 13.
[4]Asmus, P. A., Landis, J. B., and Vila, C. L. (1983) *J. Chromatogr.* **264,** 241.
[5]Sadek, P. C., Carr, P. W., and Bowers, L. D. (1985) *LC, Liq. Chromatogr. HPLC Mag.* **3,** 590.
[5a]Majors, R. E. (1974) *Bonded Stationary Phases in Chromatography* (E. Grushka, ed.) Ann Arbor Science Publishers, Ann Arbor, MI.
[6]Sakurai, H. and Ogawa, S. (1976) *J. Chromatogr. Sci.* **14,** 499.
[7]Aurenge, J. (1973) *J. Chromatogr.* **84,** 285.
[8]Eksteen, R., Supelco Inc.
[8a]Iler, R. K. (1979) *The Chemistry of Silica,* Wiley-Interscience, New York.
[9]Kiel, J. S., Morgan, S. L., and Abramson, R. K. (1985) *J. Chromatogr.* **320,** 313.
[10]Gill, R., Alexander, S. P., and Moffat, A. C. (1982) *J. Chromatogr.* **247,** 39.
[11]Lagerstrom, P. -O., Marble, I., and Persson, B.-A. (1983) *J. Chromatogr.* **273,** 151.
[12]Papp, E. and Vigh, Gy. (1983) *J. Chromatogr.* **259,** 49.
[13]Papp, E. and Vigh, Gy. (1983) *J. Chromatogr.* **282,** 59.
[14]Verzele, M., *LC, Liq.Chromatogr. HPLC Mag.* **1,**217.
[15]Vespalec, R. and Neca, J. (1983) *J. Chromatogr.* **281,**35.
[16]Kohler, J., personal communication.
[17]Carr, P. W., personal communication.
[18]De Smet, M., Hoogewijs, G., Puttemans, M., and Massart, D. L. (1984) *Anal. Chem.* **56,** 2662.
[19]Plaisted, S. M., Zwier, T. A., and Snider, B. G. (1983) *J. Chromatogr.* **281,**151.
[20]Kato, Y. (1983) *LC, Liq. Chromatogr. HPLC Mag.* **1,**540.
[21]Hearn, M. T. W. ed., *Ion-Pair Chromatography: Theory and Biological and Pharmaceutical Applications,* Dekker, New York, 1985.
[22]Goldberg, A. P., Nowakowska, E., Antle, P. E. and Snyder, L. R. (1984) *J. Chromatogr.* **316,**241.
[23]Kirkland, J. J., unreported data.
[24]Knox, J. H. and Hartwick, R. A. (1981) *J. Chromatogr.* 204, 3.
[25]Slais, K. and Krecji, M. (1974) *J. Chromatogr.* **91,**161.
[26]Cramer, S. M., Nathanael, B., and Horvath, Cs. (1984) *J. Chromatogr.* **295,** 405.
[27]Cohen, S. A., Benedek, K., Tapui, Y., Ford, J. C., and Karger, B. L. (1985) *Anal. Biochem.* **144,**275.

## Chapter 15

# SEPARATION PROBLEMS

## *Other Changes in the Appearance of the Chromatogram*

## Introduction

In Chapter 14 we discussed the varying causes of misshaped bands: tailing bands, fronting bands, and other kinds of non-Gaussian bands. Odd-shaped peaks are easy to spot by simply glancing at the chromatogram. Closer study of the chromatogram may reveal other kinds of anomalies. For example, we normally expect bandwidth to increase from the beginning to the end of the chromatogram. In other words, the plate numbers of different bands in the chromatogram should be roughly constant. Not infrequently, however, one or more bands within the chromatogram will be obviously wider than bands on either side. Sometimes such a band also exhibits tailing, but not always. At other times we may see unexpected peaks in the chromatogram—peaks that do not correspond to any compound known to be both present in the sample and responsive to the detector. These extra peaks often deflect downward from the baseline (negative peaks), rather than upward as in the normal case.

Occasionally the sample will react during its passage through the column, leading to conversion of one or more components into other compounds. This can lead to peak distortion, as in Fig. 14.18b. In other cases, however, a single well-shaped peak may be observed—corresponding to the reaction product of the original sample compound. A final separation anomaly is a change in retention time for a given compound from run to run—without any intentional change in separation conditions.

## 15.1. Wide Bands (Small *N*-Values)

The theory of band width and compound *N*-values has been reviewed in Section 3.3. Average *N*-values calculated from a particular chromatogram will vary, depending on the column and separation conditions used. Further variation in plate number (and resulting bandwidth) is possible for one or more compounds in a sample, as a result of differences in molecular weight or retention *(k')*—or because of certain problems with the separation. In this section we will discuss separation problems that result in wide bands (and poor resolution).

Many chromatographers can readily recognize a band that is unusually wide by comparing it with adjacent bands. In general, the width of various bands in the chromatogram will increase with increasing *k'*-value of the band. So if a band is wider than the one following it, we would conclude that the earlier band is too wide (its *N*-value is too small). Wide bands often tail noticeably. In these cases, Chapter 14 should be consulted for possible remedies. Not infrequently, however, a band will be wider than expected without much tailing. An example of this is seen in Fig. 15.1, for the separation of several basic drugs by ion-pair chromatography. These four bands elute with *k'*-values that are fairly close to each other. However, their band widths vary by almost sixfold. In this case, it is obvious that band A and especially band D are too wide.

Excess band broadening can easily go unnoticed for various reasons, but it can have a serious detrimental effect on sample resolution. Whenever a band appears too wide, its plate number should be measured and checked against expected values.[1,2] A plate number that is low by 50% or more is an indication of something wrong with the separation. Commercial software (DryLab™, LC Resources, Lafayette, CA) is available to make such comparisons more conveniently than manual methods.

Consider next some of the possible causes of anomalous band broadening:

1. *Column deterioration with use.* Every column shows a more or less gradual decrease in efficiency (*N*-values) with continued use.
2. *Extra-column band broadening.* Use of a particular column with a different LC system can lead to a lowering of the plate number, because of the greater extra-column band broadening with the new system.
3. *Chemistry effects.* The column plate-number is determined by both chromatographic and chemical contributions. The chro-

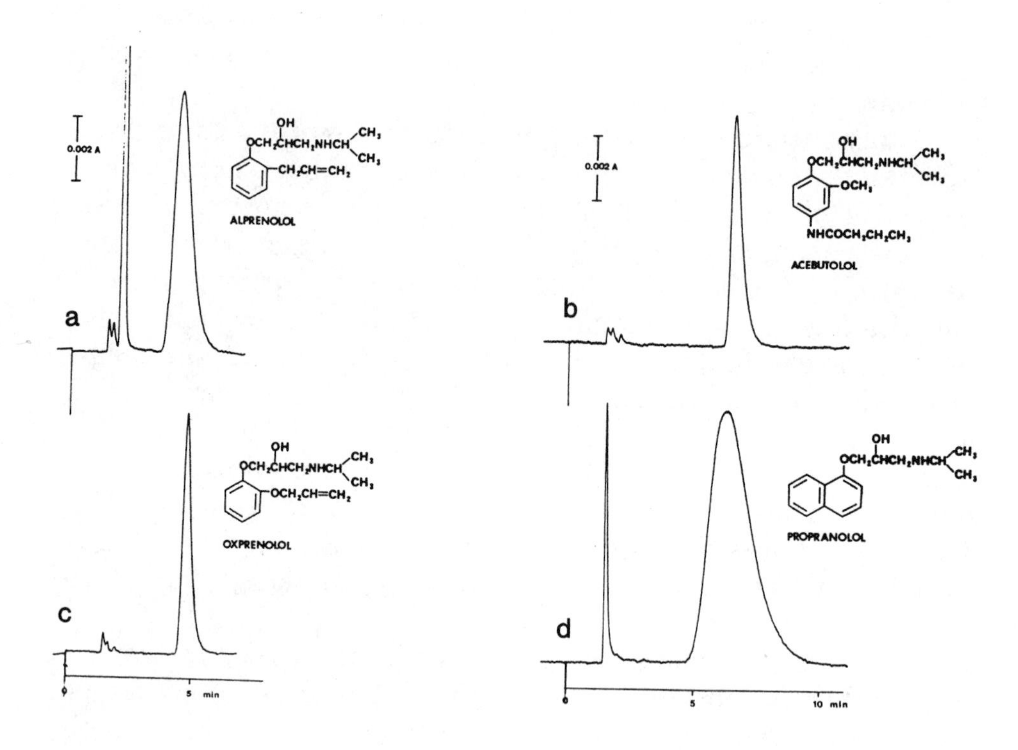


**Fig. 15.1.** Excess band broadening without severe tailing. Separation of basic drugs by ion-pair chromatography; same conditions for each compound. See text. Reprinted from ref. (4) with permission.

matographic effects are well understood, and these are predictable (DryLab software). Chemical effects are not as well defined, but can be reduced (resulting in an increase in $N$) by various changes in the mobile phase.

Loss of column efficiency with continued use has been discussed previously (Chapter 11). This problem can be confirmed by re-running a standard chromatogram and comparing the current value of $N$ with the value for the column when it was new. Alternatively, the DryLab software allows this comparison to be made for any chromatogram, eliminating the need for a second confirmatory run. Whenever a column has shown a 50% or greater loss in $N$ since its initial use, in most cases it is time to discard the column.

We already have considered extra-column band broadening (Section 3.4), which can markedly lower the $N$-values of early-eluting bands. A change in column plate-number caused by this effect can result from using another LC system for the same column. The DryLab software allows a determination of the extra-column band broadening of any LC system, by carrying out systematic measurements on the system.

What if anomalously wide bands (a) are observed in the chromatogram, (b) are confirmed by DryLab, and (c) are found not to arise from column age or system extra-column effects? In this case we may be dealing with "chemical" effects—usually associated with secondary-retention phenomena (Section 14.7). Though such secondary-retention effects often result in band tailing as well as band-broadening, this is not always the case.[3,4] Chemical effects usually will be encountered at the time of method development, most commonly for acidic, basic, or ionized sample compounds. The same procedures used for minimizing secondary retention (Section 14.7) can be employed to reduce band broadening arising from chemical effects. Changes in mobile-phase pH may also be helpful.

## 15.2. Artifactual Peaks

Figure 15.2a shows the chromatogram that results from injecting a mixture of four alkyl sulfonates (ion-pair chromatography). Detection is at 254 nm, where the sample compounds do not absorb. Therefore a flat baseline (no sample peaks) would be expected. Instead we see one negative peak (at 6 min), and four normal bands at 7–27 min. Because no peaks should have appeared in the chromatogram, we refer to these anomalous bands as artifactual peaks. Negative peaks of this type are referred to as *Vacancy peaks,* and positive peaks are called *Ghost peaks.*

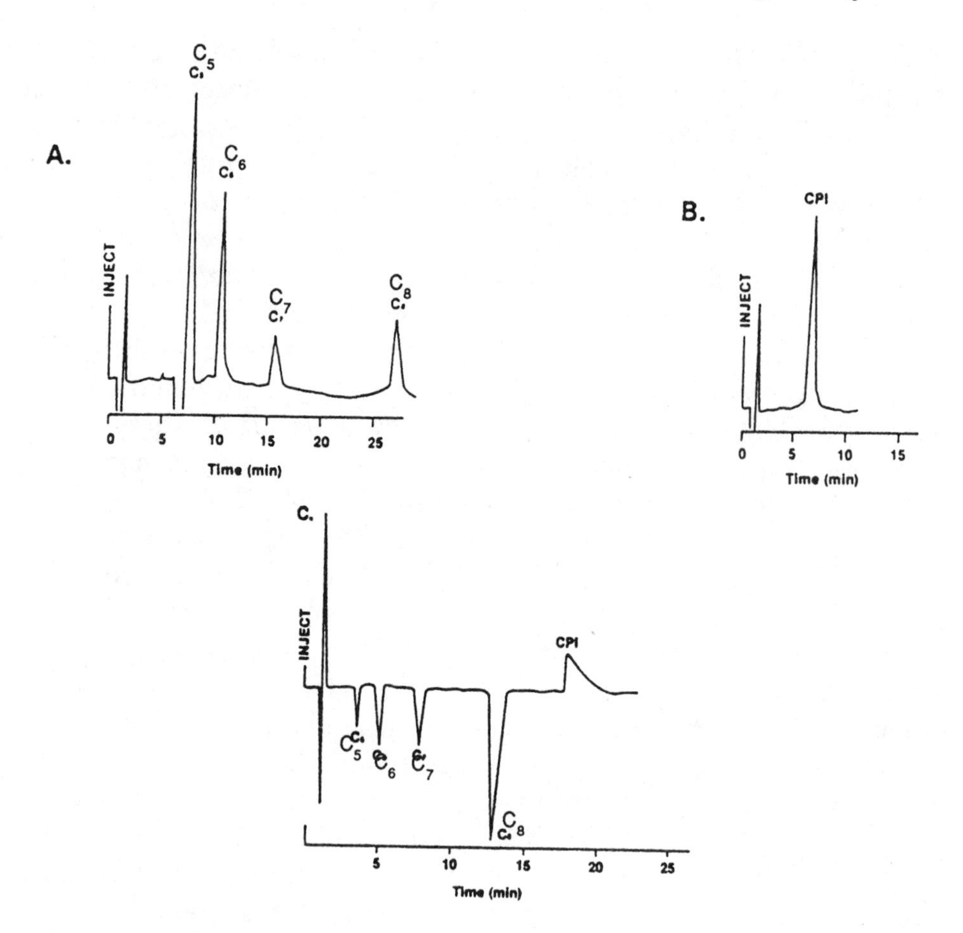


**Fig. 15.2.** Artifactual peaks in ion-pair chromatography. Mobile phase contains a UV-absorbing ion-pair reagent (cetyl pyridinium chloride, CPI); sample is a mixture of C5-C8 alkyl sulfonates (non-UV-absorbing). Buffer concentration 0.0 m*M* in (a,b), 10 m*M* in (c). (a), injection of alkyl sulfonates; (b), injection of CPI; (c), injection of alkyl sulfonates. Reprinted from ref. (4a) with permission.

## *A Simple Picture of Ghost and Vacancy Peaks*

In some cases, artifactual peaks are deliberately produced by selecting favorable chromatographic conditions. This can allow the visualization of bands that would otherwise not be detected (as in Fig. 15.2a). However, in other cases artifactual peaks can cause problems in interpretation and calculations based on the chromatogram. Therefore we need to understand how these peaks arise and how they can be avoided.

In the simplest case, we might have a mobile phase that absorbs light at the wavelength used for detection. Injecting a solvent that does not absorb at this wavelength then creates a "hole" (i.e., a mobile-phase segment or volume with lower UV absorbance) in the mobile phase, that moves through the column and gives a negative peak at $t_0$. In a slightly more complicated case, the mobile phase might have a UV-absorbing component (or impurity) M that has a retention time of $t_M$; injecting a solution of M onto the column then gives a positive peak at $t_M$ minutes. Now if pure solvent is injected instead, again a "hole" will be created in the mobile phase, but now the "hole" moves through the column at the same speed as the compound M. So a negative peak appears at $t_M$ minutes.

A still more complicated picture is created when the mobile-phase component M can interact with some non-UV-absorbing compound S in the sample. Suppose first that S and M interact to form a complex S-M. Now if S is injected onto the column, the complex S-M is formed, depleting the mobile phase of M. The resulting "hole" now moves through the column, appearing as a negative band at $t_M$ minutes. Likewise the complex S-M will have a retention time $t_S$, and the complex will elute at $t_S$ minutes as a positive band—because an excess of (UV-absorbing) M is associated with molecules of S as they leave the column (even though S by itself does not absorb light). In this way we see that both (a) ghost peaks (for S-M) and (b) vacancy peaks (for M) are possible.

### *Some Examples of Ghost and Vacancy Peaks*

Returning to Fig. 15.2, the ion-pairing reagent in this case is the UV-absorbing quaternary ammonium compound, cetyl pyridinium chloride (CPI). When an alkyl sulfonate $RSO_3^-$ is injected onto the column, the following equilibrium can be hypothesized:

$$RSO_3^- + CPI^+ \Leftrightarrow (RSO_3^- CPI^+) \tag{15.1}$$

where molecules of $RSO_3^-$ and $CPI^+$ in the mobile phase combine to form an ion-pair ($RSO_3^- CPI^+$) that is retained by the column packing (stationary phase). So the initial injection of alkyl sulfonate sample results in the removal of some CPI from the mobile phase. The presence of CPI (UV-absorbing) in the mobile phase leads to a higher baseline, and when CPI is removed, the absorbance of the mobile phase (and the baseline) will decrease. This decrease in concentration of CPI upon injection of a CPI-free sample behaves in the same way as injecting CPI into the column; that is, the negative concentration change moves through the column as a negative peak (7 min in Fig. 15.2a), appearing in the chromatogram at the same retention time as for injection of CPI (Fig. 15.2b).

The CPI that was removed from the mobile phase by reaction (15.1) then moves through the column in association with each alkyl sulfonate injected. When these non-UV-absorbing compounds leave the column, they carry with them the associated CPI molecules, which yield UV absorbance and a positive peak for each alkyl sulfonate (the four peaks in Fig. 15.2a). Similar (unintended) effects in a routine LC procedure can result in the formation of artifactual peaks that can confuse and complicate the chromatogram. While we normally avoid adding UV-absorbing components to the mobile phase, these can be present as impurities in the various mobile phase components: solvents, buffers, ion-pairing reagents, modifiers to reduce tailing, etc. Therefore artifactual peaks often are caused by impure reagents used to formulate the mobile phase. Artifactual peaks also require components in the sample that either promote or hinder the retention of the UV-absorbing mobile-phase impurities (cooperative or competitive sorption). These latter sample components often are impurities or components of the solvent used to dissolve the sample. This means that large volumes of injected sample generally lead to more pronounced artifactual peaks. It is therefore good practice to inject a smaller volume of a more concentrated sample solution—whenever artifactual peaks are observed or expected (as in ion-pair chromatography).

### *The Model of Stranahan and Deming*

An excellent general discussion of these artifactual peaks has been given by Stranahan and Deming.[5] We will review this treatment briefly so as to illustrate the principles of artifactual

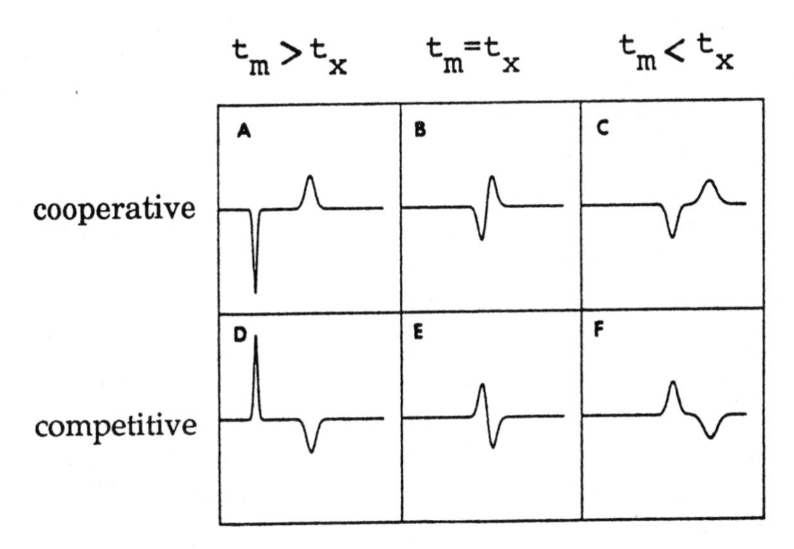


**Fig. 15.3.** Computer simulations of artifactual peaks. See text for details. Reprinted from ref. (5) with permission.

peaks. Two types of sample (X)-mobile phase (M) interaction are possible:

*Cooperative sorption* [as in Eq. 15.1]

$$X_m + M_m \Leftrightarrow XM_s \tag{15.2}$$

*Competitive sorption*

$$X_m + M_s \Leftrightarrow X_s + M_m \tag{15.3}$$

Here subscripts *m* and *s* refer to molecules in the mobile phase or stationary phase, respectively. We will assume as in Fig. 15.2 that the sample molecule X is not detectable (e.g., non-UV-absorbing), and that the mobile-phase component M is detectable. The mobile-phase component M often will be some impurity in the mobile phase, while the sample component X can be either (a) a component or impurity in the solvent used to dissolve the sample, or (b) an actual component of the sample. Stranahan and Deming[5] have modeled the cases of Eqs. (15.2) and (15.3) by computer simulation, and their results are summarized in Fig. 15.3. Examples 15.3A–C are for the case of cooperative sorption [Eq. 15.2], with three different possibilities for the relative retention times of X ($t_x$) and M ($t_m$)—indicated at the top of Fig. 15.3.

Consider first the case where the mobile-phase component leaves the column first ($t_m < t_x$): Fig. 15.3C. This is the situation for Fig. 15.2a that we just discussed: cooperative sorption of M and X by ion-pairing, with earlier elution of M vs different sample compounds X (sulfonates in Fig. 15.2). In Fig. 15.3C (or 15.2a) we see that M leaves the column as an initial negative band, followed by a positive band for X. If, on the other hand, the retention times for M and X are reversed ($t_m > t_x$ in Fig. 15.3A), X now leaves the column first as a negative band, followed later by elution of a positive band for M. This is seen in Fig. 15.2c, for the ion-pair separation of these same sulfonates under different conditions, such that CPI leaves the column after the sample ions X ($t_m > t_x$). Now the sample bands appear as initial negative peaks, followed by a positive band for CPI.

The pattern of Figs. 15.2 and 15.3A–C normally will be observed for ion-pair chromatography, which can be regarded as a cooperative sorption process. In the case of reversed-phase or normal-phase separations (e.g., 6,7), retention is competetive rather than cooperative [Eq. 15.3]. This results in a different pattern for artifactual peaks, as summarized in Fig. 15.3D–F. Now the first artifactual peak is positive, and following peaks are negative.

### *Remedies for Artifactual Peaks*

Once the presence of artifactual peaks is suspected, what can we do to minimize their interference with other (real) bands in the chromatogram? The following recommendations can be made:

1. *Avoid impure mobile-phase reagents.* In order to avoid artifactual peaks in ion-pairing, high-quality buffers and ion-pair reagents must be used to formulate the mobile phase; alternatively, try different lots of these reagents to determine which combination of reagents (in the mobile phase) minimizes artifactual peaks when injections of the sample-solvent are made (no sample dissolved in the solvent).

2. *Use the mobile phase as a sample solvent.* Impurities in the sample solvent (which also can generate artifactual peaks) can be avoided by using the mobile phase to dissolve the sample. The use of a sample-solvent other than the mobile phase (particularly in ion-pair LC) also can lead to artifactual peaks.

3. *Inject a minimum volume of the sample solution.* Because artifactual peaks often are proportional to the volume of solvent injected, minimizing this volume is recommended. For the case of ion-pair chromatography, keep the volume of injected sample solution to less than 50 µL.
4. *Pretreat the sample.* When impurities or interferences in the sample matrix contribute to artifactual peaks, removing these interferences can reduce artifactual peaks.

When artifactual peaks are present and cannot be removed using the above procedures, this problem can be treated in the same manner as for any other interfering peaks. Remember that artifactual peaks correspond (indirectly) to specific compounds and behave just like other bands in the chromatogram. Therefore chromatographic conditions often can be varied so as to move these interfering artifactual peaks away from peaks of interest.

## 15.3. Reaction of the Sample

Though it is relatively uncommon, it is possible for sample reaction to occur—either before or during LC separation. At a minimum, reaction of a sample component results in loss of that compound, with a reduction in peak size and inaccurate measurement of the component. In some cases this also results in extra bands (reaction products). In other cases, a distorted sample-band is observed (Section 14.9). Three separate cases of sample reaction can be defined, each with its own peculiar consequences and remedies:

1. *Reaction prior to injection.* This can occur during pretreatment of a sample, or during storage of sample solutions after they are made up (for example, during extended standing on an autosampler tray). The result is a loss in the concentration of reacted sample, and low results in the LC assay. Reaction products may also appear in the chromatogram, confusing its interpretation and interfering with peaks of interest.
2. *Reaction during separation.* In this case the reacted sample component is continuously converted to product as it moves through the column. The result usually is a badly distorted band, as in the example of Fig. 15.4a.
3. *Reaction in gradient elution.* This case is intermediate between the previous two, because in gradient elution the sample

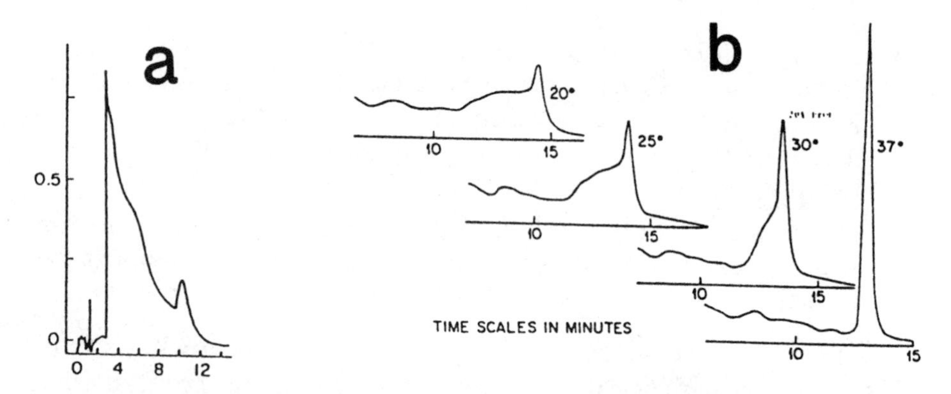


**Fig. 15.4.** Some examples of sample reaction during separation: (a) reaction of desferoxamine peak with iron in LC system during separation; (b) denaturation of ribonuclease during reversed-phase separation. Repeated from Figs. 14.18b,c.

initially sits at the column inlet (and reacts), then at a later time (when the right point in the gradient is reached) moves through the column. In some cases, two or more separate peaks are observed; in other cases, only a single distorted peak is seen (Fig. 15.4b).

### *Reaction Prior to Injection*

The main observation that suggests sample reaction prior to injection, is incomplete recovery of the assayed compound (low LC results). Whenever it is known that the compound in question is unstable, sample reaction should be suspected. Reaction may occur during sample workup, where the sample may be exposed to ambient (or higher) temperature and unfavorable conditions of pH or contact with the air (oxidation). If calibrators or standards are not treated in the same way as samples, and especially if extra peaks show up in samples, consider speeding up the pretreatment procedure (and using lower temperatures, if possible) to see whether this reduces the problem. Alternatively, in order to confirm sample reaction, allow more time (and more extreme conditions of temperature, pH, exposure to oxygen, etc.) to see whether sample loss increases, and extra peaks increase in size. The final remedy involves adjusting the chemistry of the sample pretreatment procedure to minimize sample reaction.

In other cases, reaction of the sample may occur as a result of dissolving the sample in the sample solvent and storing the sample solution prior to injection. This is sometimes a problem in the case of batching samples for injection by an autosampler. Two approaches exist for minimizing this problem:

1. Use a low-temperature autosampler (e.g., Perkin-Elmer ISS-100)
2. Run standards more frequently, and correct for sample loss by plotting the calibration factor vs position on the autosampler sample tray. In this way, a separate calibration factor is used for each sample.

### *Reaction During Separation*

In the general case, some sample component A reacts during separation to form some product B. If the reaction is fast, and the LC separation is slow, complete conversion of A to B may occur before significant sample migration occurs. In this case, a normal-appearing band corresponding to B will result. At the other extreme, the reaction is slow and the separation is fast (i.e., the usual case). In this situation, a typical band for compound A results. In between these two extremes, the ratio ($r$) of retention time $t_R$ to reaction half-life $t_{1/2}$, is neither very large or very small:

$$r = t_R/t_{1/2} \tag{15.4}$$

Then both A and B will be present in the final chromatogram. In this case, a severely distorted band (as in Figs. 15.4a,b) results. Some additional examples of peak distortion as a result of sample reaction during the LC separation are shown in Fig. 15.5a, for 2,6-dimethylhydroquinone as sample (oxidation of this compound yields the corresponding quinone).

When distorted peaks are suspected to arise from sample reaction during separation, the cause of the problem can be confirmed by increasing or decreasing flow rate by e.g., fourfold. This will in turn change the value of $r$ (Eq. 15.4), and cause a significant change in band shape. An example of this is shown in Fig. 15.6, again for the elution of 2,6-dimethylhydroquinone under reversed-phase conditions. The solution to the problem is to identify the nature of the reaction and select separation condi-

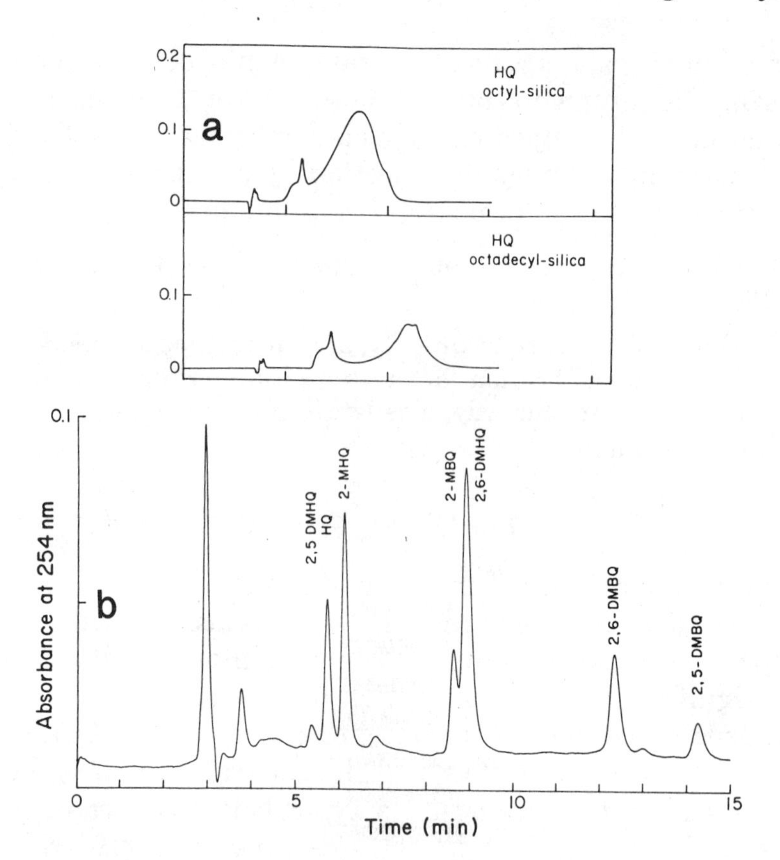


**Fig. 15.5.** Peaks for 2,6-dimethylhydroquinone that have undergone oxidation during reversed-phase separation: (a) peaks from different columns; (b) separation of mixture of hydroquinones (HQ) and benzoquinones (BQ) with reducing mobile phase (hydroxylamine + EDTA added). Reprinted from ref. (11a, 11b) with permission.

tions that minimize sample reaction. This is illustrated in the separation of Fig. 15.5b, where a mixture of hydroquinones and quinones are separated under similar conditions as in Fig. 15.5a for 2,6-dimethylhydroquinone. Now normal chromatography is observed, because of the addition to the mobile phase of a reducing mixture (hydroxylamine + EDTA) that suppresses sample oxidation. Specifically, the 2,6-dimethylhydroquinone peak at 11 min is perfectly normal. Alternatively, as in Fig. 15.4b, it may be desirable to alter separation conditions to provide for complete conversion of starting compound A to reaction product B. In Fig.

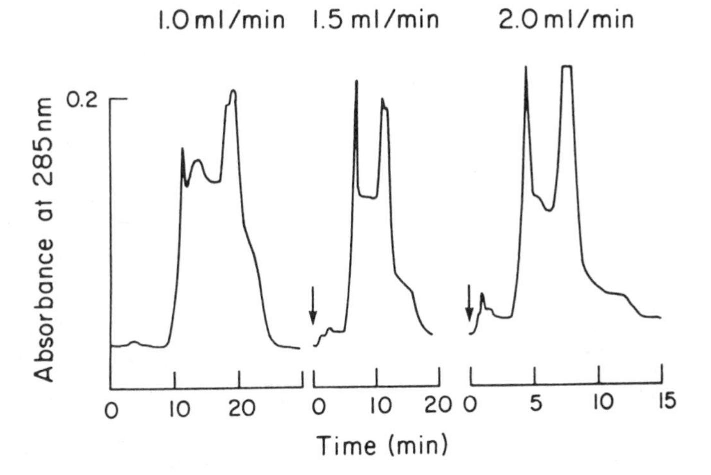


**Fig. 15.6.** Change in peak shape with flow rate for 2,6-dimethyl-hydroqinone as a result of oxidation of sample during separation. Reprinted from ref. (11a) with permission.

15.4b, A is native ribonuclease, and B is denatured ribonuclease. By selecting conditions (35°C in the latter example) that favor complete conversion of A to B, a well-shaped peak for B results; this allows easy measurement of B, from which the starting concentration of A can be inferred.

## *Reaction in Gradient Elution*

Reaction of the sample may occur after sample injection, giving two well-shaped peaks for the case of gradient elution. The reason is that during the initial development of the gradient, the compound (A) may simply remain on the column at the inlet end. At some later point the compound will elute rapidly from the column. Under these conditions it is possible to have partial reaction of A to B, followed by separate elution of A and B from the column. This is illustrated in Fig. 15.7 for the case of the protein papain as sample. Here column temperature and mobile pH are varied so as to change the reaction rate of papain (conversion of native to denatured protein). At low temperatures and higher pH values, the native protein is favored (slower reaction rate), and a single peak for the native protein is seen in Fig. 15.7a (21°C, pH = 4.8). As the mobile phase pH is lowered, a second peak for denatured papain is seen at 23 min (Fig. 15.7b, 21°C, pH = 3.0). Finally, at a still lower pH and higher temperature, only a single peak for the

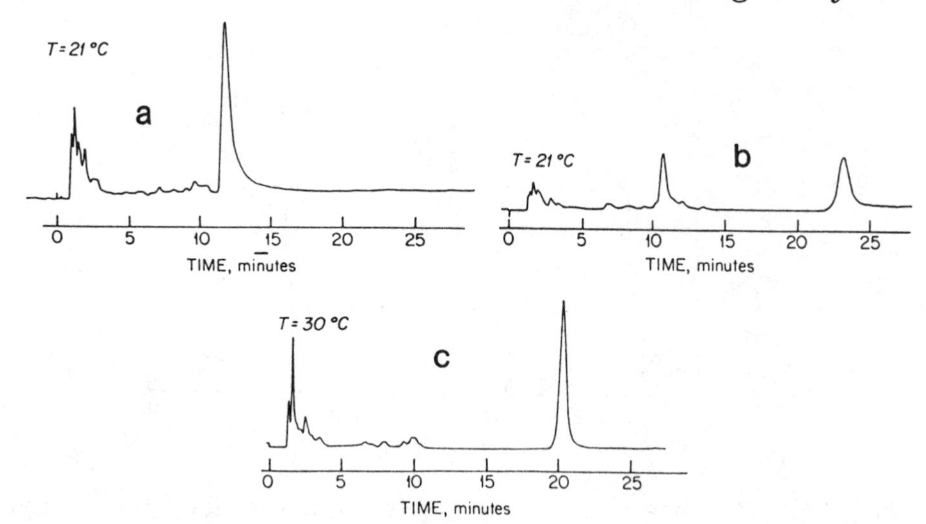


**Fig. 15.7.** Reaction of protein (papain) during gradient elution. Reversed-phase column; mobile phase is water/propanol: (a) 21°C, pH = 4.8 (native protein); (b) 21°C, pH = 3.0; (c) 30°C, pH = 2.2 (denatured protein). See text for details; reprinted from ref. (11c) with permission.

denatured papain is observed (Fig. 15.7c, 30°C, pH = 2.2). Problems with reaction during gradient elution are treated in the same way as other cases of reaction during separation (see above).

## 15.4. Varying Retention Time

Much of chromatographic analysis is based on the assumption that the retention for a given compound remains constant when separation conditions are unchanged. For a standard LC procedure, this allows us to identify what compound corresponds to each peak, by matching retention times of standards and samples for analysis. Constant retention times also are important in scheduling samples for analysis, particularly when using autosamplers. Thus, if retention times increase, one chromatogram may overlap another—with resulting problems in both separation and in data handling. Finally, if retention changes for any reason, this can adversely affect resolution. In this section we will examine the various causes of change in sample retention. Changes in retention can be classified as follows:

1. *Column-to-column.* It is often found that a new column of nominally the same type does not give the same retention times for

a given LC procedure (same conditions and sample). This column-variability problem was discussed in Section 11.4.

2. *Day-to-day or within-day.* Sometimes retention times change with time. Generally there is an increase or decrease in retention for all bands.
3. *Sample-to-sample.* Occasionally it is found that retention times for samples and calibrators (standards) are not the same; i.e., the sample matrix causes a change in retention.

We will be concerned mainly with day-to-day and within-day changes in retention (#2).

The causes of retention change include the following:

1. *Poor control of separation conditions.* Retention is a function of temperature, mobile-phase composition, and flow rate. If any of these conditions vary, retention also will vary.
2. *Slow equilibration of the column.* When the mobile-phase composition or temperature are changed, a certain time is required for the column to become re-equilibrated for the new conditions. Samples should not be injected for analysis until the column is equilibrated fully.
3. *Deactivation of the column.* Over a long period of running samples on a given column, the column can be degraded in various ways. Strongly retained components (or contaminants in the samples or mobile phase) can be retained irreversibly by the column, leading to generally decreased retention for all compounds. Bonded-phase packings can be stripped from the packing particles, leading to increased retention for some compounds, and decreased retention for others.
4. *Column-overload and sample-matrix-interaction effects.* Exceeding the linear capacity of the column (too large a sample mass) generally leads to a decrease in retention. This can result from overload by the compound itself, or by other components of the sample (matrix effects). The sample matrix can also affect retention in other ways.

These various causes of change in sample retention are discussed below.

### *Variation in Separation Conditions*

Modern LC equipment has been designed for the precise control of most separation conditions. Thus flow rates usually are

Table 15.1
Effect of Change in Separation Conditions on Sample Retention

| Variable | Method | Change in variable | Average change in $t_R$ |
|---|---|---|---|
| Flow rate | All methods | +1% | -1% |
| Temperature | All methods except SEC | +1°C | -(1–2%) |
| *Mobile phase composition:* | | | |
| Organic solvent | Reversed phase[a] | +1 %v | -(5–10%) |
| pH | Reversed phase | +0.01 unit | ± (0–1%) |
| Strong solvent | Normal-phase | +1% | - (1–2%) |
| Buffer, organic solvent | Size exclusion | +1% | 0% |

[a]Including ion-pair LC.

within better than ±0.5°C, although often the column is not thermostatted. Pressure is not directly controlled, but pressure variations usually have a negligible effect on sample retention (e.g., 9,10). On the other hand, equipment malfunction is possible and should not be ruled out as a source of retention variation.

The approximate dependence of retention on separation conditions is summarized in Table 15.1; this will aid the operator in assessing the likelihood that a particular variable is causing varying retention.

*Varying Flow Rate.* Two kinds of flow-rate variability can be encountered. *First,* a sudden change in flow rate, leading to a discontinuous change in retention times for all bands, may simply be caused by operator error. The wrong flow rate could have been entered into the pump setting or system controller. *Second,* random fluctuations in retention and flow rate from run to run may arise from malfunction of the pump; e.g., a bubble problem (Section 7.6). In the second case, retention-time ratios, as in Table 15.2, may not be constant from run to run. Pressure fluctuations often are noted in conjunction with pump malfunctions.

Variation in mobile-phase flow rate usually is recognizable from corresponding variations in $t_0$—if a $t_0$-peak is visible within the chromatogram. Alternatively, check the ratio of retention times for each compound in the two runs under comparison.

Table 15.2.
Hypothetical Change in Retention Time as a Result of Change in Flow Rate

| | Retention times | | |
|---|---|---|---|
| Compound | Run A | Run B | Retention time ratio, A/B |
| 1 ($t_0$) | 2.50 | 3.00 | 1.20 |
| 2 | 5.00 | 6.00 | 1.20 |
| 3 | 6.00 | 7.20 | 1.20 |
| 4 | 8.00 | 9.60 | 1.20 |

the chromatogram. Alternatively, check the ratio of retention times for each compound in the two runs under comparison. When the retention-time ratios are identical for every compound, the most probable cause of retention variability is a change in flow rate. This is illustrated in the hypothetical data of Table 15.2.

*Varying Mobile-Phase Composition.* Changes in composition of the mobile phase can arise from various causes: (a) error in manually preparing the mobile phase; (b) proportioning errors from pump malfunction in on-line mixing; (c) changes in composition on standing (e.g., selective evaporation of a solvent, uptake of carbon dioxide from the air with change in pH). A change in mobile-phase composition often will cause a larger percentage-change in $t_R$-values for later bands (larger $k'$). Table 15.1 indicates that small changes in the concentration of a reversed-phase solvent can lead to rather large changes in retention times.

Often a small change in the %v of some reversed-phase solvent has little effect on band spacing. For this reason, the concentration of organic solvent sometimes is varied to adjust separation time, when changes in retention (without band-spacing changes) have occurred for any reason.

When error in the make-up of the mobile phase is suspected, reformulate a new batch of mobile phase and repeat the run. If on-line mixing is being used to blend the mobile phase, replace the solvents in the pump reservoirs with manually blended mobile phase; a properly working pumping-system should give the same retention times as those of the preceding run with on-line mixing.

*Varying Temperature.* When the column is not thermostatted, a common cause of varying retention times is change in column temperature. When a continuous change in retention times for all bands is observed, and a column thermostat is not being used, consider first whether changes in laboratory temperature correlate with the observed change in retention time. As the laboratory cooled off during the overnight shift, did retention times for all samples increase? The magnitude of these retention-time changes also should be considered. Do they correlate with the usual 1–2% change in retention per 1°C change in temperature? If temperature variation is confirmed as the cause of changes in retention, then three possible remedies exist:

1. Thermostat the LC column
2. Minimize temperature variation within the laboratory
3. Insulate the column so as to minimize the effect of changing laboratory temperature on the temperature of the column.

Thermostatting the column is the most straightforward fix, but sometimes is not possible with the equipment at hand. Covering the column with insulating material, and keeping the column out of any drafts will minimize short-term temperature-related retention shifts. (A piece of 0.5-in. id Tygon tubing split lengthwise will slip conveniently over a column and provide protection from drafts.) If you insulate or thermostat the column, however, be sure also to insulate the connecting line between the column and detector, in order to minimize baseline drift.

### *Slow Equilibration of the Column*

When separation conditions are changed for a new LC procedure, the column usually equilibrates to the new conditions fairly quickly. Flushing the column with 10–20 column volumes of the new mobile phase and/or running for 30–60 min usually is sufficient. Table 15.3 summarizes the void-volumes of various columns. Alternatively, you can estimate the void volume of 0.46-cm id columns as 0.1 mL/cm (e.g., 2.5 mL for a 25-cm long column).

Column equilibration can be checked by simply re-injecting the same sample every 10–20 min. When no change in retention time occurs between successive runs, the column is equilibrated. Note, however, that a column may be equilibrated for one sample-compound and not for another. Therefore, it is important to establish that retention is constant for *all* sample components of interest, before injecting samples for analysis.

Table 15.3.
Approximate Void-Volumes of Various Columns and
Usual Volumes of Mobile Phase Required for Equilibration

| Column dimensions, cm | Dead volume, mL[a] | Volume of new mobile phase to equilibrate column, mL |
|---|---|---|
| 25 x 0.46 | 2.5 | 25–50 |
| 15 x 0.46 | 1.5 | 15–30 |
| 5 x 0.46 | 0.5 | 5–10 |

[a]Equal to 0.5 $L\,dc^2$, where $L$ is the column length (cm) and dc is the column diameter (cm).

There are some important exceptions to the usual rapid equilibration of the column. In these cases, column equilibration may require several hours or longer. These cases are summarized below:

1. Mobile phase containing an amine modifier
2. Mobile phase containing ion-pairing reagent
3. Silica columns
4. Mobile phase containing tetrahydrofuran

A common reason for slow column equilibration is that some component of either the old or new mobile phase is strongly sorbed to the column, and is present in the new mobile phase in small or zero concentrations. If a particular LC procedure requires extensive column equilibration, or if changing from one LC procedure to another is excessively time-consuming, consider dedicating a particular column to the slow-equilibration procedure (i.e., falling into one of the above four categories). This also requires that the column be stored filled with mobile phase when not in use.

*Amine Modifiers.* Amine modifiers, such as triethylamine (TEA), commonly are added to reversed-phase and ion-pairing mobile phases, to minimize the tailing of basic compounds in the sample (Section 14.7). Often no more than 1–10 m*M* concentrations are used, and this may result in longer equilibration times. When this is the case, initial injections of higher concentrations of the amine modifier (e.g., 100 µL of 0.5–1*M* amine modifier in the mobile phase) can speed up column equilibration.

A more difficult problem is changing from an amine-containing mobile phase to one that does not contain the amine additive. Here it is often observed that certain sample-compounds (generally bases) show shifting retention times for a long time after beginning the new mobile phase. This is particularly true for very strongly retained amine modifiers, such as dimethylhexyl and dimethyloctylamine. Washing the amine modifier from the column can be expedited by using 20–50 column-volumes of a strong solvent (e.g., methanol, acetonitrile) at elevated temperatures (50–60°C) before switching over to the new mobile phase.

*Ion-Pairing Reagents.* Quaternary-ammonium ion-pairing reagents, such as tetrabutyl ammonium ion (TBA), usually are washed from reversed-phase columns very slowly. As a result, the retention of acidic compounds continues to change during passage of 50–100 column-volumes of mobile phase through the column.[12] The best remedy is to use an intermediate wash solution: 50 %v methanol–water containing 100–200 m*M* of buffer or salt. Apparently, removing the last traces of TBA requires a strong mobile phase (high percent organic), *plus* buffer to compete with the ion-exchange retention of TBA onto acidic silanol sites.

Long-chain anionic ion-pair reagents such as C10–C12 sulfonate or sulfate are washed from reversed-phase columns quite slowly,[12a] and in some cases complete removal of the ion-pairing reagent is not possible—even with 100% methanol as wash-solvent. This in turn will result in continuing shifts with time in the retention of cationic solutes (e.g., protonated amines). When long-chain ion-pair reagents are specified for a given assay, it is best to devote a particular column to that separation only.

*Silica Columns.* LC procedures using silica columns (normal-phase LC) generally exhibit random variations in retention—sometimes increasing and sometimes decreasing. The usual reason is varying mobile-phase water content, with resulting variable uptake of water by the column. Silica columns are deactivated by water, with a large resulting decrease in retention times. If water-free mobile phases are used, it is almost impossible to avoid uptake of water from the air by the mobile phase. The best solution to this problem is to deliberately add a certain quantity of water to the mobile phase, such that the uptake of water by the column remains constant for different mobile phases (so-called isohydric mobile phases). Procedures for preparing isohydric mobile phases have been described.[13] However, these are somewhat tedious, and many workers prefer to use polar-bonded-phase columns in the normal-phase mode as alternatives to silica columns.

*Tetrahydrofuran as Solvent.* Reversed-phase separation with tetrahydrofuran (THF) in the mobile phase normally is not a problem as far as column equilibration is concerned. That is, THF behaves similarly to methanol or acetonitrile, and 10–20 column-volumes of mobile phase are sufficient to equilibrate the column. Occasional exceptions have been noted, however, for certain kinds of samples. In one case[14] for the separation of water-soluble vitamins, it was noted that changing to a THF-free mobile phase (from a THF-containing one) required about half a day to equilibrate the column (i.e., before the retention times of the vitamins became constant). Some workers have confirmed this effect for other samples. In such cases, there is nothing that can be done to accelerate column equilibration, and the best solution is to dedicate a particular column to assays that are sensitive to residual quantities of THF.

### *Deactivation of the Column*

With continued use, the column packing generally changes in its composition. Strongly held sample constituents may be irreversibly bound to the packing material, or the surface of the packing may be chemically attacked; e.g., the bonded-phase may be partially removed. As a result, changes in retention result.

*Sorption of Chemical "Garbage."* When the surface of the packing becomes blocked by the sorption of strongly-bound compounds from previous samples, the usual result is a decrease in retention for all compounds, with a concommitant drop in plate number. The process can be visualized as a simple loss (covering up) of stationary phase of the front end of the column. There are two remedies for this problem. First, the problem can be prevented by the use of a guard column (Section 11.3). Second, in some cases the column contaminants can be removed by washing the column with a strong mobile phase. Table 15.4 summarizes two wash protocols, one for reversed-phase columns and one for normal-phase columns. Cleaning the column in this way is more effective when a reverse-flow procedure is used, in which case it is

Table 15.4
Cleaning Columns That Have Been Contaminated with Strongly Bound Substances[a]

*Reversed-phase columns*

1. 10 column-volumes of methanol (be sure mobile-phase is miscible with methanol; e.g., no salt precipitation)
2. 10 column-volumes of methylene chloride
3. 10 column-volumes of hexane
4. 10 column-volumes of methylene chloride
5. 10 column-volumes of methanol
6. 20 column-volumes of mobile phase (*see* miscibility note on #1)

*Normal-phase columns*

1. 20 column-volumes of 50 %v methanol/chloroform
2. 20 column-volumes ethyl acetate
3. 20 column-volumes of mobile phase

[a]Flush the indicated volumes of solvent through the column in sequence.

also a good idea to vent the column effluent directly to waste, rather than through the detector (to keep the detector clean).

Keep in mind that compounds bound onto the packing surface may eventually react to form polymeric products that cannot be washed from the column. This suggests that the column should be cleaned frequently, when it is exposed to dirty samples (e.g., biological extracts, environmental samples, etc.). It is recommended that reversed-phase columns be flushed daily with methanol or acetonitrile. Whenever sorption of unwanted material is suspected, a good column should be used (Sect. 11.3)

*Loss of Bonded Phase.* Bonded-phase columns slowly lose their organic layer during use of the column. This process is favored by pH extremes (pH < 2.5 or pH > 7.0), and such conditions should be avoided if possible. Water and methanol seem to accelerate this process, and reversed-phase columns are therefore particularly subject to loss of bonded phase. The symptoms of loss of bonded phase in reversed-phase LC are:

1. Nonbasic compounds show decreased retention
2. Basic compounds show either decreased, increased, or similar retention.

Loss of bonded phase can be slowed down by using a silica precolumn (Section 11.3), although this can be inconvenient (slow equilibration follwing change of mobile phase). Procedures for restoring the bonded phase to the column have been described,[15] but are not recommended in most cases. Too much time is required to carry out the column re-bonding procedure, equilibrate the column, check its performance, and so on.

### *Column Overload and Matrix Effects*

Column overload has been discussed in Section 14.2. The usual symptoms are decreased retention times for samples that are more concentrated (give larger bands). Reducing the amount of sample injected is the usual solution.

The column also can become overloaded by sample constituents that are (a) not of interest, (b) come out near $t_0$, or (c) are not detected by the detector. In this case it is observed that retention times are different for calibrators (that do not contain these sample ingredients) and actual samples. The problem can be confirmed by the method of standard additions, where the sample is supplemented with an added amount of compound exhibiting retention variability. If the retention time remains the same for the band believed to be the compound of interest, it indicates that sample matrix effects are present. The best solution to the problem is to pretreat the sample to remove these interfering species. Alternatively, the method of standard additions can be used to calibrate the method for matrix effects (Section 16.3).

## 15.5. Varying Band Spacing

Variations in band spacing are a special case of retention variation. Therefore the general discussion of Section 15.4 applies to this problem as well. However, changes in band spacing can be much more serious than a simple increase or decrease in the retention times of every band in the chromatogram. Usually the latter problem can be resolved by changing the solvent strength of the mobile phase (Section 3.2), as by changing mobile-phase water content in reversed-phase separations. In other cases, changes in flow rate can be equally effective in restoring the original retention-time values. However, neither of these expedients is likely to work when peaks have changed relative positions within the chromatogram, or when two previously resolved bands become totally overlapped (changes in separation factor A).

Poor resolution as a result of changes in band spacing usually is associated with one of two possibilities:

1. *Column degradation (same column).* With continued use, the column either picks up chemical garbage or loses bonded phase; the resulting change in the composition of the stationary phase often leads to changes in band spacing.
2. *Column-to-column variation.* We have discussed this problem in Section 11.4.

In either case, what can be done when the column no longer provides adequate spacing and resolution of adjacent bands? First, check for any mistake in separation conditions (wrong mobile phase, wrong temperature, wrong column). Once the correct separation conditions are confirmed, however, we are faced with a more difficult situation—one that sometimes requires complete redevelopment of the LC procedure.

The best solution to band-spacing problems is to anticipate them before the LC procedure is first developed. Method development in LC usually is done by systematically varying separation conditions,[16] so as to optimize the spacing of all sample bands within the chromatogram. For example, in the reversed-phase separation of ionizable compounds, the solvent might be changed (methanol, acetonitrile, tetrahydrofuran, or mixtures thereof), or pH or buffer concentration might be varied. Changes in band spacing can be noted during this process, as a function of separation conditions. Then at some later time, these same changes in band spacing with conditions can be deliberately invoked to restore acceptable band spacing—when necessary.

This approach is illustrated for the reversed-phase separation of the 20 phenylthiohydantoin (PTH) amino acids in Fig. 15.8a. Adequate resolution of all 20 compounds is seen as a result of the excellent spacing of bands within this chromatogram. Figs. 15.8b–d show examples of this same separation in which some part of the chromatogram is no longer well resolved—as a result of changes in band spacing. However, in each case it is possible to restore the original (good) band spacing by some change in sepa-

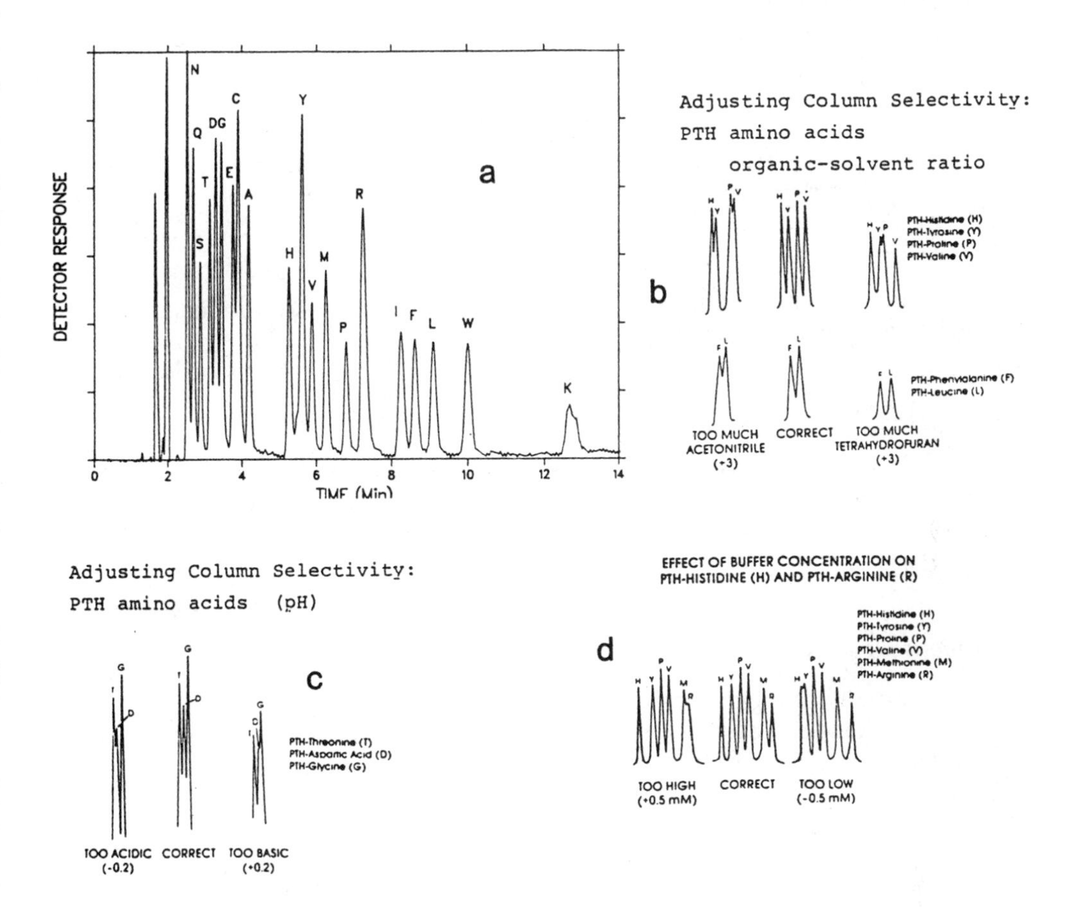


**Fig. 15.8.** Systematic correction of poor band spacing (change in retention) for separation of PTH amino acids by reversed-phase LC: (a) Separation of 20 PTH amino acids ("good" chromatogram); (b) poor resolution of the histidine, tyrosine, proline, and valine peaks—correction by change in %v of acetonitrile; (c) poor separation of the threonine, aspartate, and glycine peaks—correction by change in pH; (d) poor separation of the methionine, and arginine peaks—correction by change in buffer concentration. Reprinted from ref. (18) with permission.

ration conditions: changing the %v acetonitrile in the mobile phase (Fig. 15.8b), changing pH (Fig. 15.8c), or changing buffer concentration (Fig. 15.8d). When method development is done with the aid of DryLab computer programs,[19] it is easy to simulate the effect of minor changes in mobile-phase conditions.

Table 15.5 summarizes the separation problems that we have discussed in this chapter, their symptoms, and possible remedies for each problem.

Table 15.5
Problems and Solutions for Changes in Appearance of the Chromatogram

| Cause of problem | Symptom | Solution |
|---|---|---|
| *Wide bands* | | |
| Column deterioration with use | Gradual decrease in $N$ with use | 1. Flush with strong solvent<br>2. Replace column<br>3. Use guard column/or better sample prep to extend column life |
| Extra-column band broadening | Lower-than expected $N$-values, especially for early bands | 1. Replumb system<br>2. Adjust mobile phase for larger $k'$-values |
| Chemical effects | Wide bands not due to above | 1. See Section 14.7<br>2. Change mobile-phase pH |
| *Artifactual peaks* | | |
| Contaminants in mobile phase or sample | Ghost or negative peaks, especially with ion-pair | 1. Use pure reagents<br>2. Dissolve sample in mobile phase for injection<br>3. Inject minimum sample volume<br>4. Improve sample pretreatment |
| *Sample reaction* | | |
| Reaction prior to injection | Loss of sample component(s); extra bands | 1. Improve pretreatment<br>2. Protect samples from degradation (e.g., refrigerated sample tray)<br>3. Standardize more frequently |
| Reaction during separation | Often badly distorted | 1. Use mobile-phase additives to reduce reactivity<br>2. Use lower temperature<br>3. Change to more stable column and/or mobile phase<br>4. Derivatize sample during pretreatment |

*(continued)*

Table 15.5 *(continued)*
Problems and Solutions for Changes in Appearance of the Chromatogram

| Cause of problem | Symptom | Solution |
|---|---|---|
| *Sample reaction (continued)* | | |
| Reaction in gradient elution | Multiple peak or distorted peaks for single compound | 1. Same as for reaction during separation, above |
| *Varying retention time* | | |
| Column-to-column variations | Retention different between columns | 1. Normal to a certain extent<br>2. See Section 11.4 |
| Poor control of separation conditions | Retention time changes for all bands, drifting or cyclic | 1. Better control of temperature, mobile-phase composition, and/or flow rate |
| Slow column equilibration | Retention time changes slowly in one direction, then steady | 1. Allow longer equililibration (more flushing volume)<br>2. Inject "slug" of minor mobile-phase components (e.g., TEA) to speed equilibration |
| Column deactivation | Slow change in retention, often a decrease for all compounds | 1. Flush with strong solvent<br>2. Protect column with guard and/or precolumn<br>3. Use better sample preparation<br>4. Use more rugged mobile phase (e.g., water deactivation in normal phase)<br>5. Replace column |
| Overload or sample-matrix interactions | Distorted peaks | 1. Improve sample cleanup<br>2. Inject smaller sample volume<br>3. Use method of standard additions |
| *Band-spacing variation* | | |
| Column degradation | Band spacing changes with use of column | 1. More frequent column washing<br>2. Use precolumn and/or guard column to extend column life<br>3. Improve sample cleanup<br>4. Rework method for more rugged conditions |
| Column-to-column variations | Band spacing changes when new column is installed | 1. Normal to certain extent; see Section 11.4 |

## 15.6. References

[1]Snyder, L. R. and Antle, P. E. (1985) *LC, Liq. Chromatogr. HPLC Mag.* 3, 98.
[2]Stout, R. W., DeStefano, J. J., and Snyder, L. R. (1983) *J.Chromator.* **282,** 263.
[3]Bowers, L. D. and Mathews, S. E. (1985) *J. Chromatogr.* **333,** 231.
[4]Persson, B.-A., Jansson, S.-O., Johansson, M.-L., and Lagerstrom, P.-O. (1984) *J. Chromatogr.* **316,** 291.
[4a]Bidlingmeyer, B. A. and Warren, Jr., F. V. (1982) *Anal. Chem.* **54,** 2351.
[5]Stranahan, J. J. and Deming, S. N. (1982) *Anal. Chem.* **54,** 1540.
[6]Vigh, G. and Leitold, A. (1984) *J. Chromatogr.* **312,** 345.
[7]Snyder, L. R. (1983) *High-Performance Liquid Chromatography. Advances and Perspectives,* Vol. 3, (Cs. Horvath, ed.) Academic, New York, p. 157.
[8]Simpson, R. C. and Brown, P. R. (1985) *LC, Liquid Chromatogr, HPLC Mag.* **3,** 537.
[9]Prukop, G. and Rogers, L. B. (1978) *Sep. Sci.* **13,** 59.
[10]Quarry, M. A., Grob, R. L., and Snyder, L. R. (1984) *J. Chromatogr.* **285,** 19.
[11]Schoenmakers, P. J., Billiet, H.A.H., and de Galan, L. (1979) *J. Chromatogr.* **185,** 179.
[11a]Huang, J.-X., Stuart, J. D., Melander, W. R., and Horvath, Cs. (1985) *J. Chromatogr.* **316,** 151.
[11b]Huang, J.-X., Bouvier, E. S., Stuart, J.D., Melander, W.R., and Horvath, Cs. (1984) *J. Chromatogr.* **330,** 181.
[11c]Cohen, S.A., Benedek, K.P., Dong, S., Tapui, Y., and Karger, B. L. (1984) *Anal. Chem.* **56,** 217.
[11d]Dolan, J. W. (1988) *LC/GC* **6,** 20.
[12]Eble, J.E. (1986) Ph.D. Thesis, Villanova Univ.
[12a]Knox, J. H. and Hartwick, R. A. (1981) *J. Chromatogr.* **204,** 3.
[13]Snyder, L. R. and Kirkland, J. J. (1979) *Introduction to Modern Liquid Chromatography,* 2nd edn., Wiley-Interscience, New York, Chapter 9.
[14]Antle, P. E., Du Pont, personal communication.
[15]Verzele, M. (1985) *Chromatographia,* **19,** 443.
[16]Snyder, L. R., Glajch, J. L. and Kirkland, J. J. (1988) *Practical HPLC Method Development,* Wiley, New York.
[17]Glajch, J. L., Gluckman, J. C., Charikofsky, J. G., Minor, J. M., and Kirkland, J. J. (1985) *J. Chromatogr.* **318,** 23.
[18]Du Pont application sheet.
[19]Snyder, L. R. and Dolan, J. W. (1987), *LC, Liq. Chromatogr. HPLC Mag.* **5,** 970.

## Chapter 16

# PROBLEMS IN QUANTITATION

## Introduction

Previous chapters have dealt with individual parts of the LC system and with various symptoms that require troubleshooting: leaks, abnormal pressure, changes in the chromatogram, and so on. The various underlying problems suggested by these symptoms can, in most cases, affect quantitative analysis. That is, calculated concentrations for various sample components can be in error as a result of system malfunction of various kinds. In this chapter we will examine the relationship between various LC-system problems and quantitative analysis. We will also look at errors in reported results as another symptom of some underlying problem. This will in turn allow us to use the assay results from a particular LC procedure as a further diagnostic aid in troubleshooting. Before continuing in this chapter, review the basics of LC quantitation in reference (1).

## 16.1. General Considerations

Quantitative analysis in liquid chromatography is comprised of the following steps:

1. Run a calibration sample that contains known concentrations of the compounds to be assayed (C1, C2, etc.); the calibrator may also contain one or more internal standards (whose concentrations are Cs1, Cs2, etc.)
2. Measure the size (height or area) of the bands for calibrator compounds (A1, A2, etc.) and internal standards (As1, As2, etc.) from the run of step #1
3. Run a sample to be assayed

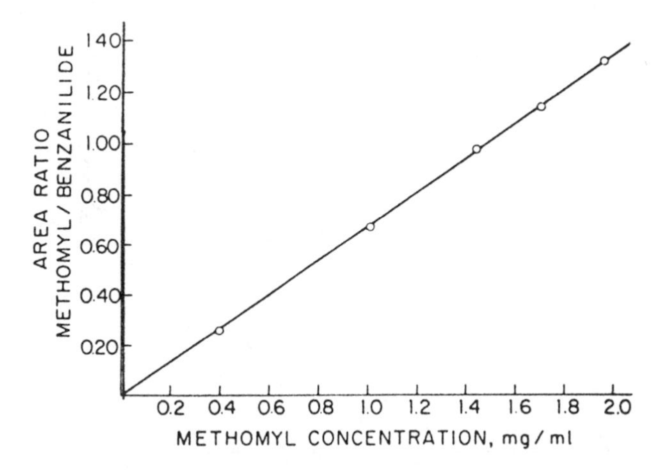


**Fig. 16.1.** Calibration plot for LC assay of methomyl in crude samples. Internal-standard method; peak area used for quantitation. Reprinted from ref. (1) with permission.

4. Measure the size (height or area) of the bands for compounds to be assayed (B1, B2, etc.) and corresponding internal standards (Bs1, Bs2, etc.)

In most LC procedures it will be found that peak size and concentration are exactly proportional; i.e., a calibration plot or linearity check is linear and passes through the origin ($x = 0$, $y = 0$). This is illustrated in Fig. 16.1 for the assay of the compound methomyl by an internal standard procedure. Methods that do not give plots of this type (linear and through the origin) generally will show reduced precision, and should be avoided if possible. Usually this means that an LC assay procedure will be refined until plots such as those of Fig. 16.1 are obtained. In this chapter we will assume that the calibration plot is linear and passes through the origin (but see Sect. 16.3).

The calculation of assay results will be illustrated for a sample-compound "X" (Fig. 16.2) in a chromatogram, using either external-standard or internal-standard procedures. For the case of *External standardization,* we have (see Fig. 16.2a):

$$\text{Concentration compound X} = (\text{Cs/As})\ \text{B1}$$
$$= (\text{CF})\ \text{B1} \qquad (16.1)$$

Here the factor (Cs/As) = (calibrator concentration)/(calibra-

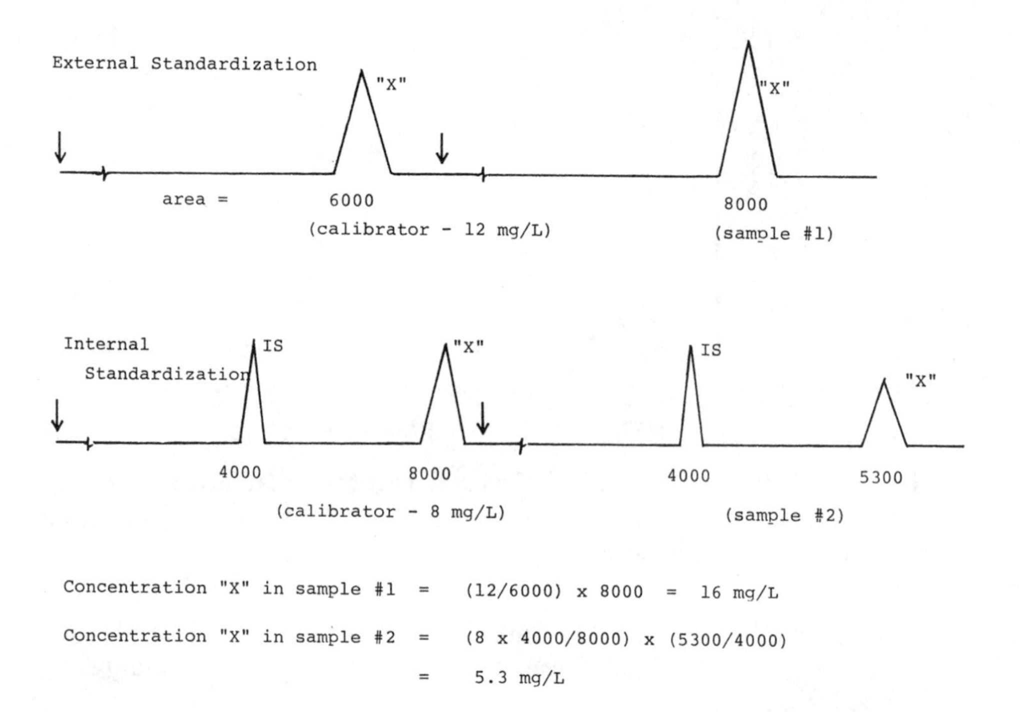


**Fig. 16.2.** Hypothetical chromatograms illustrating LC assay calculations: (a) External-standard method; (b) internal-standard method.

tor band size) = CF is the calibration factor for the method. This usually will be determined one or more times during the assay of a set of samples for compound X.

In the case of *Internal standardization,* a constant volume of the same internal-standard solution will be added to all samples and calibrators. The calculation of the concentration of compound #1 is then carried out (see Fig. 16.2b) using

$$\text{Concentration compound X} = (\text{Cs As1/A1})\,(\text{B1/Bs1})$$

$$= (\text{CF})\,(\text{B1/Bs1}) \qquad (16.2)$$

Note that the peak-size ratio (A1/As1, or B1/Bs1) replaces peak size (A1 or B1) of Eq. 16.1 for the internal-standard method.

The assay results generally will deviate from the correct sample concentrations by some amount. These deviations can be classified according to precision and accuracy. *Precision* refers to the variability of results when a single sample is assayed several times; it measures the variation of individual concentrations from the mean or average value for the set of replicates.* *Accuracy* refers to the similar comparison of experimental values with the "true" concentration of the sample—assuming this can be known. If a control sample containing a known (e.g., weighed-in or independently measured) concentration of compound X is analyzed several times by an LC procedure, and the different assay values are averaged (to give 16.45 g/L of X), the result might be:

$$\text{Concentration X} = 16.45 \pm 0.13 \ \ (1\ \text{SD})$$

Let us assume that the concentration of compound X is known to be 16.9 g/L (vs the measured value of 16.45 g/L). The precision of the method is ±0.13 g/L (1 standard deviation), as determined from the usual formula for standard deviation. The *Bias* of the method is (16.45 – 16.9) = -0.45 mg/L, which is equivalent to its lack of accuracy.

We should note that precision can be defined for different situations:

1. Within-day precision, for a single laboratory
2. Day-to-day precision, for a single laboratory
3. Interlaboratory precision (day-to-day)

Generally it is found that imprecision increases in the sequence $1 < 2 < 3$. It is desirable to define assay precision in terms of these different situations, because this helps to pinpoint precision problems and can help to solve them.

## *System Suitability*

The precision and accuracy of an LC method should be determined at the time of method development (during method validation). Simulated samples (in the same sample matrix as the samples for assay) must be formulated, containing known concentrations of the analyte over the range to be reported by the method. For example, if the method is to be applied to samples containing 1–100 mg/L of compound X, then validation samples

*Some common formulae for determining assay precision (standard deviation, coefficient of variation, etc.) are summarized in Table 16.1.

Table 16.1
Definitions of Common Terms Used to Measure Precision

| Term | Defining equation |
|---|---|
| Mean (or average), C | $C = \frac{\text{(sum of all values } C_i)}{n}$ |
| Standard deviation, SD | $SD = \frac{(\text{sum of all } [C_i - C]^2 \text{ values})^{1/2}}{(n-1)}$ |
| Coefficient of variation CV[b] | CV = 100 SD/C (%) |
| Range of $C_i$ values | Largest $C_i$ minus smallest $C_i$ |

Probability that a value of $C_i$ is "correct"

| Absolute value of $(C_i - C)$/SD | Probability, % |
|---|---|
| 1 | 32 |
| 2 | 5 |
| 3 | 0.2 |

[a]Here $n$ refers to the number of replicate measurements for a given sample; $C_i$ is an individual value of the concentration of compound X in the sample.
[b]When precision is expressed in this chapter as a ±% figure, it refers to the coefficient of variation

might be made up with concentrations of 1, 5, 20, 50, and 100 mg/L. These samples must each be run in replicate over a period of several days (ideally, using two or more LC systems and different operators). The results for each sample are then averaged, and standard deviations are determined. Thus, the precision of the method is known (both within-day and day-to-day) as a function of analyte concentration. Any significant difference between LC assay values (outliers) and known concentrations should be resolved before the assay is approved.

When the method is carried out on a routine basis, it is essential to know that the precision and accuracy originally determined for the method are maintained. This can be checked in various ways:

1. Reproducibility of calibration factors (values of CF)
2. Values for samples, including internal standards
3. Data for daily controls
4. Linearity checks

*Calibration-factor reproducibility.* CF-values usually are determined one or more times during the analysis of a batch of samples on a given day. The frequency of calibration is determined by the tendency of the calibration factor to drift within a day—as determined during method validation. Thus, if replicates that are run after an initial calibration drift to lower or higher values with time, calibration must be repeated often enough so that adequate assay precision is maintained. Repeated calibration during the analysis of a batch of samples can also be used to improve assay precision, by averaging CF-values from replicate calibrations (when these values are stable).

Ideally, CF-values will not drift during the assay of a set of samples, and in this case the reproducibility (standard deviation) of CF-values for a given day is a measure of the precision of the assay on that day. Calibrators should be made up in the same manner as samples are, and a new calibrator should be used for each calibration.

*Sample data.* Analyte concentrations in the samples for assay normally do not provide much information on precision or accuracy, because we usually do not know what concentration to expect. In some cases, however, the analyte concentration is known to fall within certain limits, and results outside these limits are then suspect. Normally such samples will be rerun. In these cases, it usually is wise to make up the sample from scratch, not simply rerun the solution that was just injected (assuming that some sample processing is involved, as opposed to simply injecting the original sample solution).

Some LC procedures call for replicate injections of the sample, usually with averaging of the results for each sample. The precision of replicate determinations provides an additional check on the precision of the LC method. Again, however, it is important that each sample analyzed be processed independently of other replicates, because imprecision often is associated with sample processing rather than LC analysis *per se* (see Section 16.2). Even when replicate assays normally are not run, any sample can be rerun several times; precision can then be calculated for that sample.

If an internal standard is added to each sample at the beginning of the assay procedure (before sample pretreatment, dilution, etc.), then the peak-size measurements for the internal standard can be averaged and a standard deviation determined. The latter can serve as still another measure of assay precision. The internal standard also provides information on the reliability of individual sample results (unlike other system-suitability checks). Thus if an individual sample shows an internal-standard band that is much larger or smaller than values for other samples (deviating by more than three standard deviations), that sample should be re-assayed.

*Daily controls.* Controls are simulated samples that contain known amounts of the analyte. They usually are formulated in large batches, then bottled in separate containers for use on each day that an LC procedure is run. When analyte concentrations can vary over a significant range (e.g., 1–100 mg/L of X), it is useful to use controls at three concentration levels: low, average, and high (e.g., 2, 30, and 80 mg/L for analyte X). A set of controls should be run with each batch of samples. Controls should be made up in the same matrix as "real" samples and processed in the same way during LC analysis.

Results for control samples provide the primary assurance that the assay procedure is giving acceptable precision and accuracy. Values for each control should fall within a ±3 SD range of the known concentration of analyte in the control. When values outside this range are reported, the controls should be rerun to confirm that there really is some problem with the assay procedure. Controls normally are run soon after the initial calibration of the procedure, in order to ensure that valid results will be obtained for subsequent samples. Data on control results can be averaged over time to provide additional information on day-to-day (and system-to-system) repeatability of the assay method.

*Linearity checks.* This usually involves injecting a series of samples of known analyte concentration $C_i$, where $C_i$ varies over the range of reportable values. For example, if the concentration of compound X can vary from 10 to 100 mg/L, the linearity samples might have concentrations of 10, 40, 70, and 100 mg/L. Linearity checks are not run routinely, but in response to a suspected problem with the LC procedure. For example, the daily controls may suggest that values at higher concentration are being under-recovered; an 80-mg/L control may be consistently analyzing at 75 mg/L. A linearity check can then confirm that a calibration-plot nonlinearity exists, in turn suggesting corrective action.

## 16.2. Unacceptable Precision

Problems with the precision of an LC assay procedure may occur during either (a) method development, or (b) the subsequent routine application of the method. Our approach to these two kinds of problems is somewhat different. In the first case, we usually are comparing the precision of the assay with the precision we require in the final results. For example, assay results might be required have a precision of ±1% (1 SD), while our initial method may have a precision of ±2%. In this case we have to improve the assay procedure so as to reduce imprecision to the target value (±1%). This means identifying the various sources of imprecision and reducing the major contributions to imprecision.

In the second case, where a validated method is used for routine analysis, the assay procedure normally will yield the precision required (e.g., ±1% in the preceding example). However, on a given day (because of some unknown problem), the assay results may have a precision of only ±3%. In this case we have to determine what has changed in our LC procedure, and then correct the problem. Troubleshooting really involves only the second situation. However, we will have a better grasp of how to troubleshoot precision problems if we first review the control of assay precision during method development. For a detailed discussion of this, see ref. (2).

### *Minimizing Assay Imprecision During Method Development*

It is useful to characterize imprecision in LC according to whether or not it depends on sample concentration. This is illustrated in the following hypothetical precision data for the determination of compound X:

| Concentration of X, mg/L | Coefficient of variation (CV), % |
|---|---|
| 10 | ±2 |
| 2 | 2 |
| 0.5 | 2.5 |
| 0.1 | 6 |
| 0.02 | 20 |

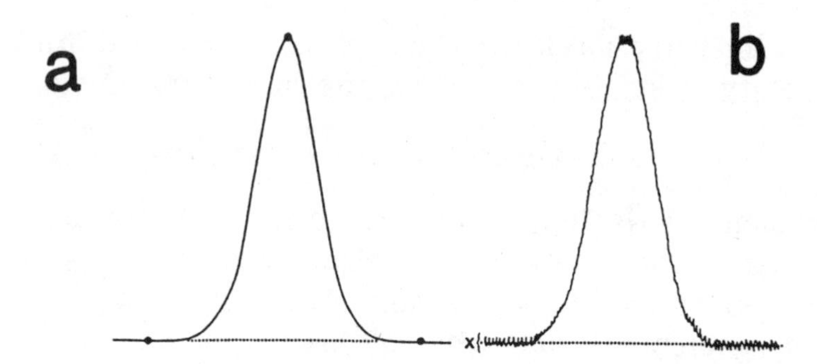


Fig. 16.3. Contribution of baseline noise to imprecision at low analyte concentrations (b) and high concentrations (a).

In general, at higher sample concentrations, precision (CV) does not vary with concentration (e.g., here for concentrations > 0.5 mg/L). However, when lower concentrations are assayed, the coefficient of variation eventually increases (here for concentrations < 0.5 mg/L). A high- vs a low-concentration band are compared in Fig. 16.3. It is seen that baseline noise limits the precision with which band size can be measured for the low-concentration sample.

*Imprecision at low concentration.* Assume that assay precision is observed to vary with sample concentration for concentrations likely to be encountered in "real" samples. This means that our signal-to-noise ratio (peak height vs baseline noise) is too small. There are three approaches to improving precision in this case:

1. Increase the amount of analyte (X) injected onto the column
2. Increase the detector sensitivity for the analyte
3. Decrease the baseline noise
4. Decrease column volume and/or $d_p$

Often it is possible to inject a larger sample volume or to decrease the dilution of the injected sample (case #1). Either approach will result in a larger band for compound X, and should decrease the CV for the assay. Detector sensitivity (case #2) often can be improved by using a different detection-wavelength, or substituting a different type of detector (e.g., fluorimeter replacing a photometric detector). Baseline noise (case #3) can be

Table 16.2
Imprecision in Different Steps of an LC Assay

| Assay step | Usual imprecision | Comments |
|---|---|---|
| Sample processing | ±0.5% | Operator dependent; volumetric dilution imprecise |
| Sample cleanup | ±3–5% | Operator dependent; robotics or column-switching improves |
| Sample injection | < ±1% | Fill loop completely; check autosampler |
| LC separation | ±0–2% | Peak area better; use Rs > 1.5 |
| Detection | < ±1% | Good signal-to-noise (improves); clean flowcell |
| Data processing and calibration | 0–5% | Peak height vs area; internal standard; calibration frequency |

reduced in various ways. Usually baseline noise is greatest when the band elutes early ($k' < 1$), and so it is often unwise to position bands in this part of the chromatogram. Baseline noise often can be reduced by a more careful cleanup of "dirty" samples, so as to reduce the amounts of late-eluters. The later-eluting bands often overlap following chromatograms, and in many cases are the major source of baseline noise. Finally (case #4), for the case of a sample available in limited amounts, detection sensitivity can be increased by decreasing the peak volume (i.e., by decreasing column length and/or diameter, or particle size).

*Imprecision at high concentration.* During method development, generally every effort will be made to minimize imprecision caused by baseline noise. If this is successful, it will be observed that assay precision is the same for all sample concentrations likely to arise (e.g., ±2% for 0.5 to 10 mg/L concentrations in the above example). However, the precision of the assay may still be unacceptable, because of other kinds of imprecision effects. We have summarized these concentration-independent contributions to assay imprecision in Table 16.2.

*Sample Processing* refers to the preparation of the sample for injection, apart from any cleanup procedures. This generally comprises various weighing, filling-to-mark, and dilution steps. If reasonable care is taken, and class-A glassware is used for volumetric flasks and pipets, total imprecision for this step can be limited to ±0.5–1.0%. For simple LC procedures involving only one or a few analytes in high concentration (content assay), and no special problems, most laboratories can achieve an assay precision of about ±1%. In these cases, the imprecision caused by sample processing can be the major source of overall assay error. To lower the imprecision of volumetric dilutions (from flasks or pipets), consider the following recommendations:

1. Use calibrated glassware, or correct volumetric additions by weighing (much more precise)
2. When weighing samples, use a large enough weight to reduce imprecision caused by weighing to < 0.2%
3. Check the precision and technique of each operator (or automatic-dispense module)

*Sample Cleanup* may be required whenever major interferences are present, or if the sample contains components that can damage the column. Typically such operations are carried out manually, and they may involve manipulations of extraction solvents or solid-phase extractions columns (e.g., Sep-Pak from Waters Chromatography Division). The resulting assay procedures tend to be highly operator-dependent, and generally have precisions in the ±3–5% range. Often the major source of imprecision for these assays is the sample cleanup procedure itself [see Sect. 17.2 and discussion of ref. (2)]. Imprecision from the sample-cleanup step can be minimized by:

1. Avoiding evaporation-to-dryness steps
2. Avoiding sample derivatization
3. Maximizing analyte recovery in each step of sample cleanup; extraction efficiencies should be > 95%
4. Using internal standards

*Sample Injection* is a highly precise operation if carried out properly. Generally the imprecision arising from injection should be < 0.2%. For best results:

1. In manual injection, inject a sample-volume that is at least three times the volume of the loop (in order to fill the loop completely; see Chapter 10)

2. If imprecise assay results are suspected to be caused by the injection step, use an internal standard—this corrects for all errors during sample injection

*LC Separation* should not contribute significantly to assay imprecision, if the separation is well designed. Errors from this step can be minimized by:

1. Obtaining good resolution ($Rs > 1.5$) for all band-pairs
2. Controlling separation conditions: mobile-phase composition, temperature, etc. Band-area generally is more precise than peak-height if separation conditions are not well controlled (e.g., column is not thermostatted)
3. Making sure that the LC system hardware is operating properly. Air bubbles, dirty check-valves, worn pump seals, etc. can contribute to flow imprecision and thus degrade the assay precision.

*Detection* should not contribute to assay imprecision, as long as the signal-to-noise ratio is adequate. However, dirty flowcells or improperly adjusted controls can degrade detector performance and indirectly contribute to assay imprecision (by lowering the band-signal). These effects are generally apparent as a major decrease in values of the calibration factor (analyte sensitivity).

*Data Processing* and *Calibration* can have a variable effect on assay precision. A common problem is improper selection of the data-processing parameters: number of data points sampled per unit time, peak detection parameters, etc. This can lead to significant precision problems, but such problems should be less common when all bands are well resolved and the baseline does not drift. Calibration problems lead frequently to large differences in results from different labs (see Section 16.3). Within-lab precision can be affected adversely by unstable calibrators. Therefore, the stability of the calibrator under various storage conditions should be carefully checked for each LC procedure.

The choice of area vs peak-height, or internal- vs external-standardization, is seldom clearcut. That is, often we cannot predict which of these options will lead to the lowest assay CV. The best approach is to try each of these choices during method validation, and then select the option that gives the best overall precision.

### *Minimizing Assay Imprecision During Routine Operation*

Let us assume that an assay procedure has been validated, and the resulting day-to-day precision has been found acceptable. The goal of routine analysis is then to carry out the same procedure and obtain results of comparable precision. If assay precision appears unacceptable on a given day, our job is to find out what has gone wrong—not to change the assay conditions. Troubleshooting the problem in this case is described in Section 16.4. In this section we will briefly review the steps of Table 16.2, from the standpoint of routine analysis as opposed to method development.

1. *Sample processing.* The various flasks, pipets, or balances used may be miscalibrated or out-of-calibration. A new operator may lack the necessary pipeting skills, or may be using a different pipeting technique. The wrong solvent composition for the sample may have been formulated.
2. *Sample cleanup.* If solid-phase extraction is used, the cleanup columns may have changed (e.g., a new batch) so that extraction efficiency is altered. A new operator may be careless or may not be following the proper procedure. Buffers used for extraction may be prepared improperly.
3. *Sample injection.* The volume injected manually (by the syringe) may not be properly controlled; the same approximate volume (±5%) should be injected each time, even when an excess volume is used. The autosampler may be out of adjustment.
4. *LC separation.* Contaminated mobile-phase components can give a higher baseline and poorer signal-to-noise ratios. The resolution of two bands may have degraded; e.g., column deterioration, wrong mobile phase, change in room temperature, etc. Change in experimental conditions also can result in change in retention, so that one or more peaks are misidentified.
5. *Detection.* The wrong detector settings may be used; e.g., band width, time constant, wavelength, etc. The flowcell may be dirty, or a printed-circuit board may be faulty.
6. *Data processing.* The wrong data-system parameters may have been chosen.
7. *Calibration.* The calibrator solution or original compound may have degraded.

## 16.3. Unacceptable Accuracy

Problems with accuracy usually transcend the LC separation. They often are associated with sample-matrix effects, or with inaccurate estimates of the purity of calibration compounds. Sometimes the chemistry involved in sample pretreatment is poorly understood, giving rise to unsuspected bias in the final assay procedure. To do justice to this area we would almost have to write a book on analytical chemistry. For these reasons, our discussion of accuracy problems will be somewhat limited.

Bias in an LC procedure often is associated with one of the following phenomena:

1. Calibration plots that do not go through the origin
2. Matrix effects
3. Interferences

Every effort should be made during method development to eliminate or minimize the above problems. We will discuss each of these situations briefly.

### *Calibration Plots That Do Not Go Through the Origin*

Many laboratories assume that plots such as those of Fig. 16.1 may not necessarily pass through the origin (0,0 point); e.g., the examples of Fig. 16.4 a,b. Usually these labs are influenced by older gas chromatographic assays, where this commonly was the case. In LC procedures, however, plots that pass through the origin should be the general rule. The first step, therefore, should be to make absolutely sure that there is a real problem. It commonly is found for a limited linearity check that the least-squares regression curve gives a small positive or negative intercept on the *y*-axis of a calibration plot; e.g., Fig. 16.4c. However, repetition of the linearity check often shows that the apparent *y*-intercept is an artifact of method imprecision. So a good first step is to run more than one such linearity check and confirm that the intercept is "real"; i.e., constant and nonzero within experimental error.

In the case of "real," nonzero intercepts, it sometimes is found that these are associated with matrix effects (see below). Thus, depending on the condition of the sample when injected (degree of cleanup, kind of solvent or buffer used to dissolve the sample sample, etc.), the size of the intercept varies. For this reason the presence of an intercept in a calibration plot or linearity check is a warning that method accuracy can be compromised by any differences in calibrators vs samples. Under these conditions it is essential that controls for the method be made up in the same kind of matrix that the "real" samples come in. It also is recommended that calibrators likewise duplicate the sample matrix.

### *Matrix Effects*

The sample matrix consists of the sample solution minus all analytes. Ideally the final sample solution will be the same as the mobile phase, but this often is not the case. In favorable situations, the solvent used to dissolve the sample may differ only in minor respects from the mobile phase. However, it is generally advisable to make the solvent for the sample as close in composition to the mobile phase as possible.

Sometimes the original sample contains solids (e.g., "fillers") that do not dissolve in the sample solvent, and must be separated during sample cleanup. Occasionally such solids can selectively adsorb sample components, lowering their recovery from the original sample during cleanup. Usually any analyte losses of this kind vary with analyte concentration, so that recovery is not constant. This can lead to major inaccuracy in the assay of low-concentration samples. A similar situation is encountered in the cleanup of biological samples, in which proteins and other interferences may bind analytes of interest. The cardinal rule here is to make sure that analyte recovery approximates 100% during each step of sample cleanup. Accepting lower recoveries often leads to later accuracy problems.

The method-of-standard-additions is useful to confirm that matrix effects are not important. Here the addition of known amounts of analyte to a sample should result in a predictable increase in analyte concentration as assayed by the LC procedure. It is a good idea to apply this test to several different samples, to make sure that matrix effects are the same (or absent) for a representative group of samples.

### *Interferences*

The appearance of a well-resolved band at the retention time of the analyte (standard) does not guarantee that the band repre-

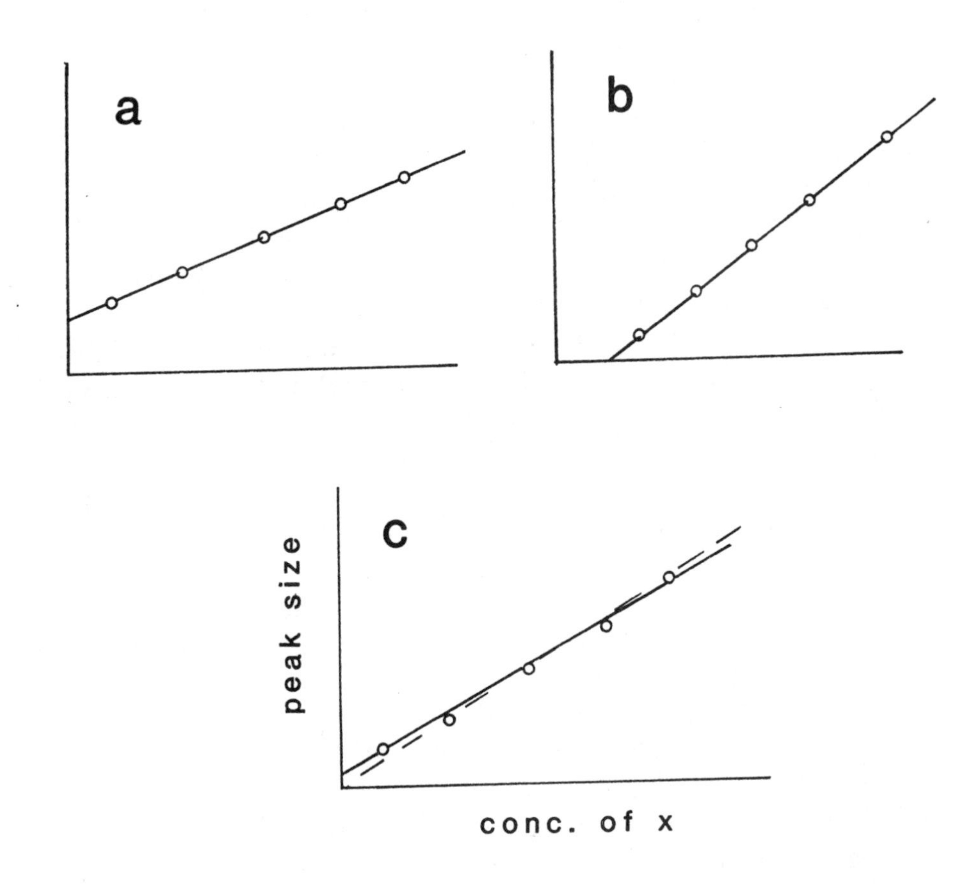


**Fig. 16.4.** Calibration plots that do not pass through the origin.

sents a pure compound. It always is possible for a second compound to have the same retention time as the analyte. Therefore, during method development it is important to check for any interferences that might overlap analyte bands. Commonly used procedures include:

1. Chromatographic cross-checks
2. Wavelength-ratioing and diode-array detectors
3. Other spectroscopic tests, as by LC-mass spectrometry

A chromatographic cross-check refers to the use of a different (reference) assay procedure. Thus, reversed-phase separation might be compared with a TLC analysis. The presence of additional peaks (not observed in the LC procedure) by TLC would indicate the possiblity of "hidden" interferences. Ideally, a second assay procedure can be used to show that the same analyte concentration is obtained by the two different methods.

Wavelength-ratioing [e.g., ref. (3)] refers to the simultaneous use of two wavelengths for detection. The ratio of absorbances at each wavelength is characteristic for a particular compound, so this ratio can be compared between the sample band and standard compound in the calibrator. Identical ratios indicate that there is no interference present in the sample band. The wavelength ratio can also be scanned across the band, with a constant value further confirming an absence of interferences. Diode-array detectors can be used to carry out such checks for band purity at a number of different wavelengths.

The ultimate check on band purity is afforded by a spectroscopic method such as mass spectrometry (MS). The on-line verification of band purity or identity is increasingly being achieved by LC/MS.

Sometimes interferences are absent from the samples used during method development, but show up in later samples. When an interfering compound is suspected to overlap an analyte band, the shape and width of the analyte band (and its retention time) should be compared with that of the standard compound (e.g., in the calibrator). Usually a significant interference will lead either to small shifts in retention, or to significant broadening of the band—especially if bandwidth is measured at 10% of the peak maximum.

## 16.4. Troubleshooting Precision Problems

Locating and solving a precision or accuracy problem begins by discovering that a problem exists. As discussed in Section 16.1, we can monitor precision by keeping track of:

1. Calibration factors
2. Internal-standard peak height or area
3. Controls
4. Replicate sample assays

Table 16.3
A Fast Procedure for Estimating the Precision of a Set of Replicate Values[a]

C = [(largest assay value) + (smallest value)]/2

SD = [(largest value) – (smallest value)]/$x$

where $x$ is given by the number of replicates $n$:

| $n$ | $x$ | $n$ | $x$ |
|---|---|---|---|
| 2 | 1.4 | 20 | 3.9 |
| 3 | 1.9 | 50 | 4.6 |
| 52.6 | 100 | 5.0 | |
| 10 | 3.3 | | |

CV = 100 SD/C

Example for CF values:

CF = 0.105, 0.110, 0.103, 0.106, 0.105

C = (0.110 + 0.103)/2 = 0.1065

$n$ = 5

$x$ = 2.6

SD = (0.110 – 0.103)/2.6 = 0.0027

CV = (0.0027/0.1065) 100 = 2.5%

[a]Symbols defined in Table 16.2.

Assuming that we see indications of a precision problem, how can we confirm that a problem really exists? For example, our assay precision may be ±2% within-day, and the following calibration factors may be reported on a given day: 0.105, 0.110, 0.103, 0.106, and 0.105. A quick check (without the use of a computer or statistical-calculator) can be obtained by the procedure outlined in Table 16.3. In this case, we estimate that CV = 2.6%. This is larger than 2%, but not significantly so for this small number of replicates. With a little practice, workers with a facility for mental arithmetic can readily estimate CV-values without the use of a calculator (using the procedure of Table 16.3; so called "method of the range").

### *Solving Precision Problems*

In addressing precision problems, we have access to two kinds of data: prior results on calibrators, internal standards, etc., and precision experiments that we might carry out to add to our data base. Let's see how we can apply these two kinds of information to diagnosing and resolving precision problems.

The first step is to organize our precision data. Assume that we have been tracking the absolute peak-size of the calibrator or of internal standard peaks, so that at constant time-intervals we obtain an assay result for the same sample. The CV of the assay procedure is ±1% under normal conditions.

Here are different examples of a possible precision problem:

| | Assay value, mg/L | | |
|---|---|---|---|
| Sample | Case 1 | Case 2 | Case 3 |
| 1 | 3.50 | 3.62 | 3.63 |
| 2 | 3.56 | 3.54 | 3.59 |
| 3 | 3.59 | 3.57 | 3.61 |
| 4 | 3.64 | 3.71 | 3.64 |
| 5 | 3.67 | 3.66 | 3.50 |
| 6 | 3.75 | 3.73 | 3.75 |
| CV | 2.5% | 2.5% | 2.5% |

In each case we see that the assay precision has degraded from the normal 1% CV to 2.5% CV. For case #1, we see that the assay values appear to drift steadily upward. That is, our problem appears to be related to *Assay Drift*. In case #2, the assay values bounce around sporadically, without any apparent trend with time. In case #3, the first four assay results are pretty close to each other—for these three values the CV is about 0.6%. However, the next two assay values vary widely. That is, in this case there seems to have been a sudden loss of precision. *It is important in troubleshooting precision problems to recognize when assay drift is involved, or when precision deteriorates at some specific time.*

*Assay Drift.* Assay drift can be caused by:

1. Reaction (and loss) of analyte in either the calibrator or individual samples while awaiting injection (e.g., when loaded onto an autosampler tray)
2. Evaporation of samples while awaiting injection
3. Drift in separation conditions (change in temperature, alteration of the mobile phase, continuous deterioration of the column, etc.)

4. An unstable internal standard (assuming that calibrators and samples were all supplemented with the internal standard at the same time—which is the usual case)
5. Changing performance of the LC hardware; e.g., the flow rate may be changing, the detector sensitivity may have altered, and so on.

Drift caused by sample (or calibrator) reaction or evaporation is not uncommon. This can be checked by preparing a large volume of sample (ready for injection) and aliquotting it into several sample vials. One vial can be capped and stored in the refrigerator, while the remaining 5–10 vials can be injected successively as in normal operation. Just before the last of these vials has been injected, the refrigerated sample should be warmed to ambient temperature and injected right after the last of these samples. If drift in the assay results is observed as previously, but the last (refrigerated) sample is closer in value to the first sample of the set, then either reaction or evaporation of the sample solution is confirmed. Conditions must then be changed to minimize any effect on assay results. For example, calibrators can be run more frequently, or CF values can be plotted vs time, so that a "corrected" CF is used for every assay calculation. Or perhaps better vial-seals should be used.

Drift in separation conditions is more likely to affect assay results when peak heights are used for quantitation. In this case, either retention times or plate numbers will also change during a work-shift; i.e., $t_R$ or $N$-values should correlate with CF values. This can be checked; if confirmed, separation conditions must be controlled more tightly, or peak areas should be used for the assay calculations.

An unstable internal standard can be checked in the same way as sample reaction (see above). The usual indication of this effect is a continued decrease in the peak-size for the internal standard when compared to the peak size of the analyte in the calibrator.

Changing system performance is rare, but can be confirmed by subsituting another module for the pump, detector, etc.

*Sudden change in precision.* A sudden loss in precision during a work-shift, or at the beginning of the shift, is an indication that something has changed. In this case, a history of events

Table 16.4
Isolating Precision Problems

| Problem source (assay step) | Procedure |
|---|---|
| Sample processing | 1. Prepare enough test-sample for 5–10 injections; pool this sample and inject repetitively and determine CV (if sample cleanup is involved, process entire sample in one step) |
| Sample cleanup | 2. If procedure #1 gives good CV, and sample cleanup is involved, the problem probably is in the cleanup step; confirm by processing a large batch of sample, then cleaning up individual portions for separate injection |
| Sample injection | 3. If an autosampler is used, bypass the autosampler and manually inject several replicates |
| LC separation | 4. Compare chromatogram to standard chromatogram |
| Detection | 5. Substitute another detector of the same kind |
| Data processing | 6. Manually measure peak-height and carry out assay calculation |
| Calibration | 7. Recalculate results based on single (initial) calibration, or on average CF for entire data set |

preceding the loss in precision should be organized. Often this will pinpoint the problem source. A different LC system may have been used, a new operator may have worked that shift, etc.

*Isolating precision problems.* As in troubleshooting other problems of the system, problem-isolation is an important tool for diagnosing causes of imprecision. We saw in Section 16.2 that precision problems can arise in different parts of the assay procedure. Table 16.4 outlines how problems of each type can be isolated and confirmed.

Sample Processing and cleanup can be treated as a single operation when diagnosing the source of imprecision. One approach is to prepare a pooled sample that is processed and otherwise treated as a single sample. Alternatively, if this is impractical (e.g., when Sep-Paks are used for individual samples), five to ten cleaned-up samples can be combined just prior to the injection step. Now the sample pool is aliquotted into individual vials, and five to ten replicate assays are carried out. If the imprecision problem disappears with this approach, then either sample processing or cleanup problems are implicated. When a cleanup step is part of the assay procedure, this most often will be the problem area. In either case, consult Table 16.2 and Section 16.2 for corrective action.

Sample Injection as a contributor to imprecision should be considered mainly for the case of external-standard procedures and assays that are normally quite precise (CV < 1%). If an autosampler is used, try bypassing the autosampler and using manual injection. If precision improves for several replicates, then either the autosampler should be adjusted or replaced, or an internal standard should be added to the sample.

If injection imprecision is suspected in the case of manual injection, maybe the injector is leaking (see Chapter 10 for diagnosing injector leaks). Also try overfilling the loop by fivefold to see whether this affects precision. If precision improves for a larger fill-volume, either increase the volume of sample used to flush the loop each time, or control the syringe-injection-volume more precisely. It is possible to obtain good precision with a fill-volume one- to two-times that of the loop volume, but in this case the syringe injection into the loop must be more precise (e.g., for a 50-μL loop, an injected volume of 98–102 μL).

If the LC Separation is affecting precision, this usually will be apparent by comparing current chromatograms for the assay with "good" chromatograms from a time when adequate precision was being obtained. Sometimes the sample itself changes (extra peaks that interfere with data processing). When chromatograms are compared in this way, look for differences in (a) baseline noise and drift, (b) resolution between bands, and (c) peak shape (tailing, broadening).

The Detector seldom interferes with assay precision, but there are exceptions. Commonly it is not appreciated that band-size with photometric detectors (e.g., maximum absorbance of the peak) is not exactly the same from one detector to the next. That is, because of the geometry of LC flowcells, Beer's law is not obeyed exactly. This is illustrated in Table 16.5 for data obtained from the authors' laboratory in the late 1970s. Without going into the history of this study, we observed that two detectors of the same model (A1, A2 in Table 16.5) gave readings for the same

Table 16.5
Variable Response of LC Detectors: A Case Study[a]

| Detector | Absorbance reading, *A* | Action |
|---|---|---|
| A1 | 0.93 | None |
| A2 | 0.63 | None |
| B | 0.14 | Out of adjustment; readjusted reading 0.89 |
| C | 0.04 | Adjustments OK; returned to manufacturing for servicing |

[a]Mobile phase was supplemented with a UV-absorber to give an absorbance reading of 1.00 AU (254 nm) in a Cary spectrophotometer. This mobile phase was then pumped through various detectors in the laboratory (variable and fixed- wavelength units, operated at 254 nm).

sample of 0.93 and 0.63 AU (a Cary spectrophotomer measured 1.00 AU for the same sample). Another LC detector model (B) gave an initial reading of 0.14, but it was found that the wrong detector adjustments had been used. Readjustment gave an absorbance reading of 0.89 AU (closer to the expected value). A fourth detector (C) gave an absorbance reading of only 0.04, and this detector appeared to be malfunctioning.

Many laboratories use LC detectors without verifying that they have the proper sensitivity. As long as the sensitivity for a detector remains unchanged, errors arising from insensitivity should cancel out between calibrators and samples. However, if the sensitivity of the detector is grossly degraded, this may be reflected in a poorer signal-to-noise response. As we saw earlier in this chapter (Fig. 16.3 and related discussion), lower detector sensitivity can translate into higher method imprecision. This can usually be picked up by noting the calibration-factor value. When values of CF change by more than 20–30% from day-to-day or among different detectors, this indicates that detector performance may be contributing to assay imprecision.

Problems in Data Processing can arise for various reasons. The parameters required by the data processor may have been entered incorrectly, or changed accidentally. The chromatogram itself may have changed, making data reduction less precise. When data processing is suspected as the problem, then:

1. Verify that the correct parameters are entered into the data processor
2. Manually measure peak sizes and carry out assay calculations for several samples; compare the results with those from the data processor

Calibration problems are fairly common as a source of assay imprecision or bias. In the case of precision problems, the role of the calibrator is easily checked. Simply compare the precision of the method when run in the normal way (several recalibrations) vs without recalibration. If precision improves using a single calibration, the calibrator is definitely the source of the problem.

## 16.5. References

[1]Snyder, L. R. and Kirkland, J. J. (1979) *Introduction to Modern Liquid Chromatography,* 2nd edn., Wiley-Interscience, New York, Chap. 13.
[2]Snyder, L. R. and van der Wal, Sj. (1981) *Anal. Chem.* **53,** 877.
[3]Webb, P. A., Ball, D., and Thornton, T. (1983) *J. Chromatogr. Sci.* **21**,447.

# Chapter 17

# GRADIENT ELUTION AND SAMPLE PRETREATMENT

## Introduction

Previous chapters discussed problems related to the various LC system modules and the results obtained from LC analysis. This chapter covers two procedures that may or may not be used in your laboratory.

Gradient elution was very popular when it was first introduced, because it increased the separation power of the relatively inefficient columns that were used. With improved column technology, isocratic separation became more popular, and equipment-related problems made gradient elution a less popular technique. With today's high-precision, microprocessor-based LC systems and the increased need for analyzing biological samples, gradient elution is regaining popularity. Problems unique to gradient elution are discussed in Section 17.1.

Sample pretreatment may or may not be a part of routine assays in your lab, but you will use one or more pretreatment steps at some point in your work. Sample pretreatment, though not a part of the LC system, can have a profound impact on the results obtained from an LC assay. Assay imprecision, reduced column lifetime, autosampler problems, and many other problems can be minimized or eliminated by improving the sample pretreatment steps. A brief review of sample pretreatment and related problem areas is given in Section 17.2.

## 17.1. Gradient Elution

Gradient elution was referred to briefly in Chapter 3, and equipment for gradient elution was discussed in Chapter 7. Figures 7.19 and 7.20 show the two general approaches to gradient elution: mixing before the pump, and after the pump. Low-pressure mixing (Fig. 7.19) is more flexible and versatile, and usually is more accurate at the beginning or end of the gradient. The various problems associated with either type of equipment are discussed in Section 7.3.

Problems with gradient elution are of two kinds. First, there are the usual problems associated with equipment failure, experimental technique, deteriorated columns, etc. Second, many problems in gradient elution arise from an imperfect understanding of how this elution mode works, and what to expect when different

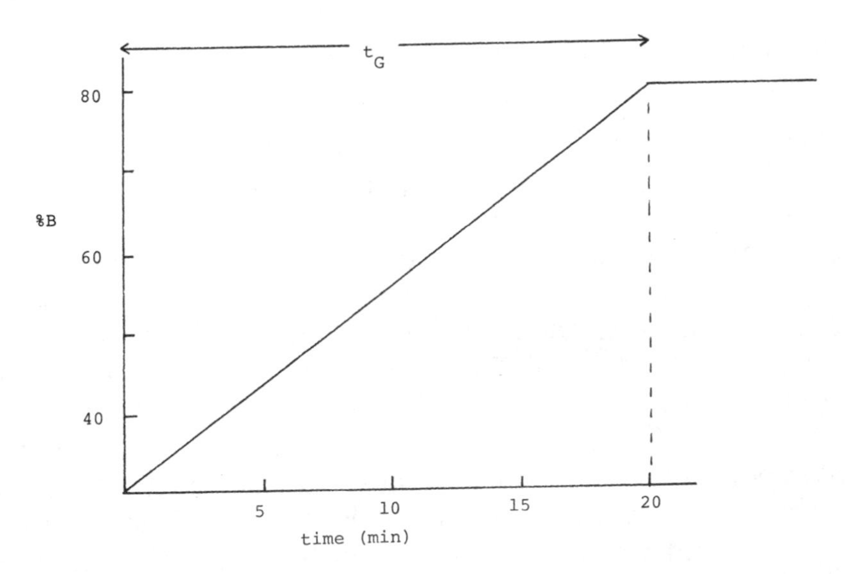


**Fig. 17.1.** Illustration of gradient conditions (see text).

conditions are changed. Generally this results in separation problems: poor resolution, broad bands, and so on. For this reason we will begin with a review of the basics of gradient-elution separation. Table 17.1 (at the end of this section) summarizes various gradient-related problems and the solutions discussed here.

## *Gradient-Elution Basics and Separation Problems*

In gradient elution, the composition of the mobile phase changes during a separation. Typically two solvents are used for the mobile phase, a weak solvent A and a strong solvent B. For reversed-phase separation, A might be water and B might be methanol. In all cases, the strength of the mobile phase should increase during the separation, which means that the concentration of B in the mobile phase also must increase. This is illustrated in Fig. 17.1, for a gradient from 30 to 80% of solvent B (e.g., methanol) in A (e.g., water). The gradient controller is used to determine the composition of the mobile phase during gradient elution, with the operator entering in the various conditions:

1. Starting %B in the mobile phase (30% in Fig. 17.1)
2. Final %B in the mobile phase (80% in Fig. 17.1)
3. Gradient time from beginning to end of the separation (20 min in Fig. 17.1)

4. Gradient shape (linear in Fig. 17.1)
5. Flow rate

Some controllers provide for entering the %-change in B per minute (2.5%/min in Fig. 17.1) instead of gradient time—but this provides an equivalent result.

Gradient elution normally is used for separating wide-polarity-range samples, where the isocratic retention *(k')* of the last band is much greater than for the first band. When the *k'*-ratio for the first and last bands in isocratic separation exceeds about 30, gradient elution is likely to give better results. This is illustrated in Fig. 17.2, for the reversed-phase separation of a mixture of chlorobenzenes. For this sample, isocratic separation (Fig. 17.2b) bunches peaks together at the beginning of the chromatogram, and broadens them excessively at the end of the chromatogram. The corresponding gradient separation (Fig. 17.2a) provides more even spacing of bands and better resolution, as well as narrower (more easily detected) bands at the end of the chromatogram. The problem with separations such as those of Fig. 17.2b is that early bands have very small values of *k'* and late bands have very large values of *k'* (see Section 3.2). Gradient elution works by changing *k'* during the separation, because mobile-phase strength changes.

Thus, weakly retained compounds leave the column first in a weak mobile phase, while strongly retained bands leave last in a strong mobile phase. To a first approximation, all bands in gradient elution will have similar *k'* values and similar band widths. For a detailed discussion of gradient elution and the control of separation, see refs. (1–4).

Early bands can be poorly resolved in gradient elution if the starting %B is too large. In this case the initial mobile-phase strength is too large, and these initial bands elute with *k'*-values that are too small, as is illustrated in the examples of Fig. 17.3 for a starting mobile phase of 46% acetonitrile/water (Fig. 17.3a) and 79% acetonitrile (Fig. 17.3b). Bands 1–5 are too weakly retained in 79% acetonitrile to be resolved in the separation (Fig. 17.3b). When bunching of initial bands is observed during gradient elution, the solution is to change the starting %B to a lower value.

The final %B in the gradient should be adjusted so that the last band leaves the column at about the time the gradient is

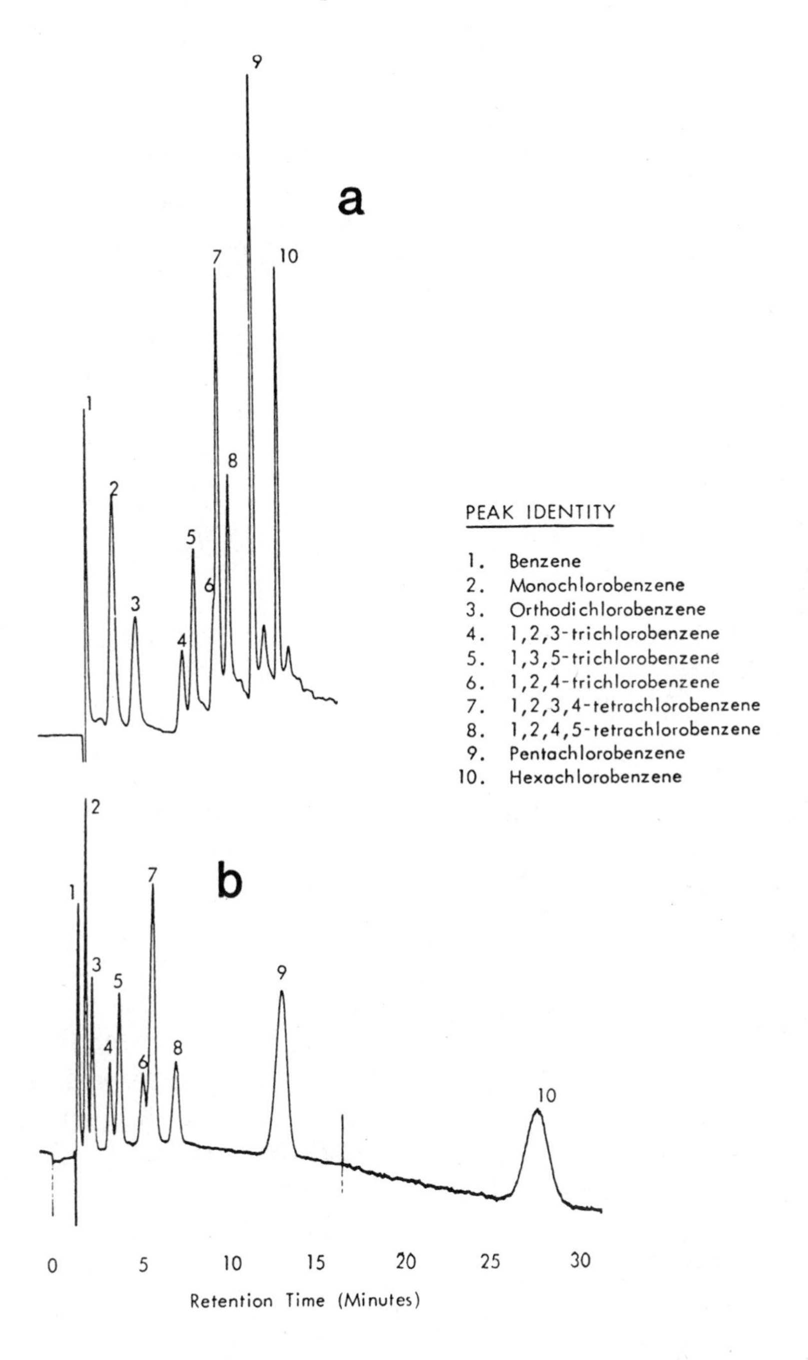


Fig. 17.2. Comparison of gradient and isocratic elution for reversed-phase separation of chlorobenzene sample. (a) Gradient elution from 40 to 100 %v methanol/water; (b) isocratic elution with 50 %v methanol/water. Reprinted from ref. (1) with permission.

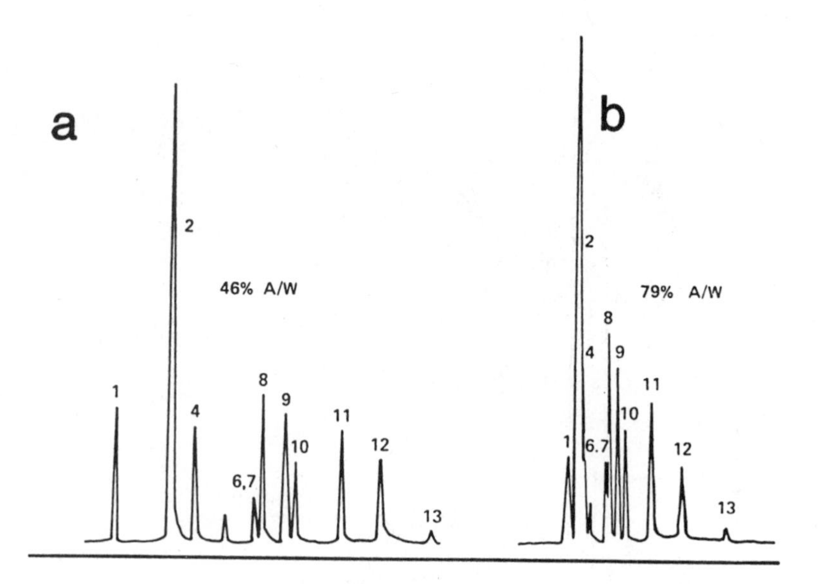


**Fig. 17.3.** Reversed-phase gradient elution with change in %B at start of gradient. (a) Gradient from 46 to 95% acetonitrile/water; (b) gradient from 79 to 95% acetonitrile/water. Reprinted from ref. (3) with permission.

completed. That is, if the gradient time is 20 min (as in Fig. 17.1), then the retention time for the last band should be about 20 min.

Initial gradient separations often show incomplete resolution of one or more band-pairs—just as in isocratic separation. Resolution can be improved in these cases in the same way as for isocratic elution (Chapter 3). That is, we can optimize $k'$, $N$, and/or $\alpha$. Many workers have trouble in using this approach for gradient elution, although this is actually easier than for isocratic separation. The major difficulty in this respect is the fact that $k'$ in gradient elution depends on such conditions as the starting and final %B, the gradient time $t_G$, the column volume $V_m$, and the flow rate $F$. However, there is a simple relationship between $k'$ and these various conditions:

$$k' \cong (\text{constant})\, t_G\, F/(\Delta\%\text{B})\, V_m \qquad (17.1)$$

Here $k'$ is a constant, equal to about 20 for reversed-phase LC, $\Delta$%B is the final %B minus the starting %B, and the column volume $V_m$ equals ($F\,t_0$) (about 1.5 mL for a 15 x 0.46-cm column). According to Eq. (17.1), we can increase $k'$ in gradient elution by

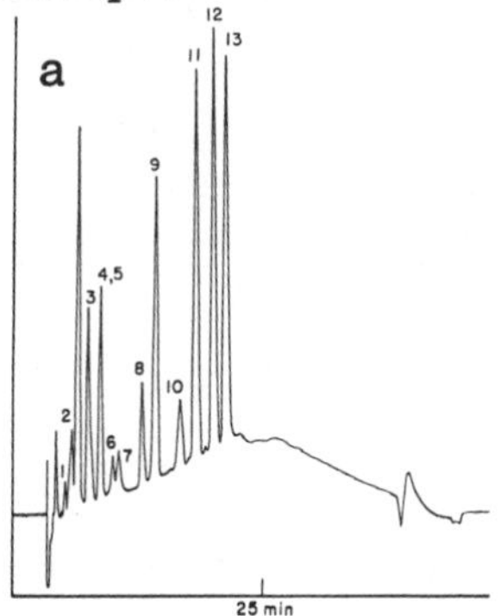


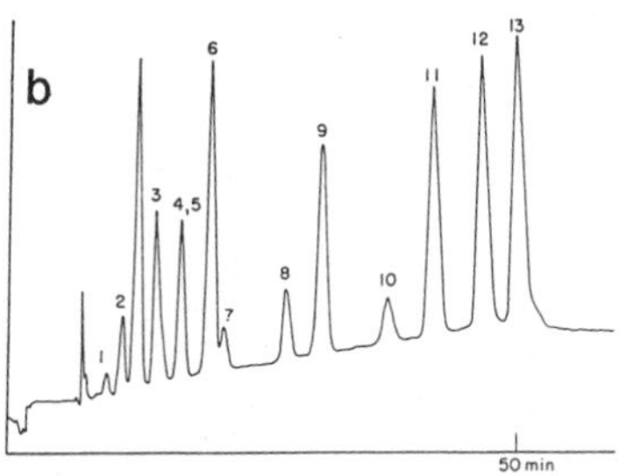


**Fig. 17.4.** Effect of gradient time on ion-exchange separation. (a) 10–100 m*M* salt gradient with 20-min gradient time; (b) same conditions, except 50-min gradient. Reprinted from ref. (1) with permission.

increasing gradient time or flow rate, decreasing column length (which decreases the volume $V_m$), or decreasing the gradient range (change in %B during the gradient). When we increase $k'$ in gradient elution, we obtain the same results as in isocratic elution: wider bands, longer separation time, and better resolution. This is illustrated in Fig. 17.4 for the ion-exchange gradient elution of a mixture of carboxylic acids. In Fig. 17.4a a 20-min gradient time is used, while in 17.4b a 50-min gradient is employed.

We also can increase resolution in gradient elution by increasing $N$ (Section 3.3) We increase $N$ by using a better column (e.g., smaller particle size), a longer column, and/or by decreasing flow rate. However, when we change column length or flow rate, we affect $k'$ [Eq. (17.1)]. That is, conditions for increasing $N$ lead to smaller values of $k'$, which works in the opposite direction so far as resolution is concerned. This generally is confusing to most chromatographers. The solution is to hold $k'$ constant while increasing $N$ in gradient elution. According to Eq. (17.1), if we keep ($t_G F/V_m$) constant while changing $F$ or column length, then $k'$ will also stay constant. For example, if we want to double column length, while reducing flow rate by half, we must increase gradient time fourfold.

These interrelationships between the various separation parameters are easily mastered with the aid of DryLab G software.[2] In this case, optimized gradient conditions can be obtained based on two initial experimental runs. Then the various parameters can be varied by the operator to optimize the separation; the computer makes the appropriate adjustments of related parameters when one or more parameters are changed. For example, the computer makes the appropriate gradient-time adjustments to keep $k'$ constant when the operator desires to change the column length.

Changing $\alpha$ or band-spacing in gradient elution is done in the same way as for isocratic separation: by changing mobile-phase composition, the column type, or the temperature.

*Column Regeneration.* Before the next sample is injected after a gradient run, the column must be re-equilibrated with the starting mobile phase. Generally this can be achieved by flowing 15 column-volumes of the starting mobile phase (30% methanol/water in example of Fig. 17.1) through the column. If the extent of column-flushing between gradient runs is not enough, then problems can arise in run-to-run reproducibility. Generally the retention times for early peaks will vary from one run to the next. When this is observed, it means that more washing of the column (by the starting mobile phase) is required between each run. This can be achieved by increasing the regeneration time or increasing the flow rate during equilibration. Alternatively, if the time between sample injections is kept constant, retention variation will be less, even when the column is not equilibrated completely. In some cases, complete regeneration of the column for gradient elution might require a prohibitive time between runs.

### *Gradient-Elution Problems*

For the most part, problems observed in gradient elution are exactly the same as problems observed in isocratic elution. Therefore if the problems summarized in Table 17.1 (at the end of this section) do not apply to a particular situation, simply treat the problem as if it arises from isocratic separation. Problems specific to gradient elution include the following:

1. Drifting baselines
2. Artifactual peaks
3. Solvent demixing

*Drifting Baseline.* This is fairly common, as illustrated by the run of Fig. 17.4a. In this case, the baseline rises during the gradient because of the greater absorbance of the B-solvent (acetonitrile) vs the A-solvent (water). This problem is magnified by the use of lower UV-wavelengths for detection, and also is more serious for some detectors (which do not filter out second-order wavelengths produced by diffraction gratings). One solution to the problem is to add a UV-absorbing compound to the A- or B-solvent, so as to equalize the absorbance of both solvents. It is essential, however, that the UV-absorbing additive is not retained under the conditions of the gradient (otherwise it will separate during the gradient, and create a major baseline upset). Berry[5] has discussed this problem in detail, and proposed adding nitrous oxide gas to the A-solvent (water) in reversed-phase gradients. A more convenient alternative for 185–200 nm detection is the use of nitric acid or nitrate in small concentrations.[6] At higher detection wavelengths, salts containing elements from the higher rows of the periodic table can be used (e.g., bromate, periodate).

*Artifactual Peaks.* Any UV-absorbing impurities in the mobile-phase components can separate during the gradient, with the appearance of peaks that do not correspond to sample bands. This is illustrated in Fig. 17.5 for a gradient run carried out without injecting sample ("blank" gradient). In the top chromatogram, ordinary distilled water was used in a water (A)/ acetonitrile (B) gradient. This chromatogram is filled with peaks that represent impurities in the water. The lower chromatogram in Fig. 17.5 was run with HPLC-grade water in place of the distilled water. This blank gradient looks much more reasonable; the artifactual peaks from the distilled water have for the most part disappeared.

Artifactual peaks in gradient elution can arise whenever the various mobile-phase components are inadequately purified. Even HPLC-grade solvents can give rise to this problem when detection wavelengths less than 220 nm are used and the detector is

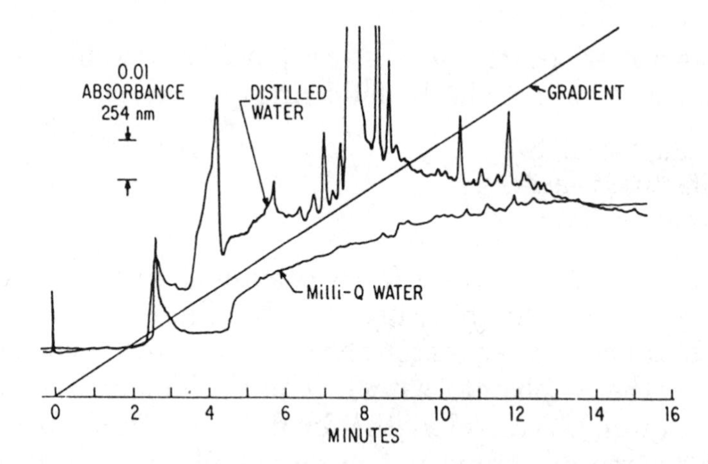


**Fig. 17.5.** Blank gradient runs (no sample) with different solvents. Top: 0–100% acetonitrile/distilled water; bottom; 0–100% acetonitrile/HPLC-grade water (Milli-Q). Reprinted from ref. (1) with permission.

set for maximum sensitivity. A "bad" lot of HPLC-grade solvent can magnify the problem further. For this reason, it usually is a good idea to run a blank gradient before injecting any sample. If artifactual peaks are seen, then different lots of solvent or mobile-phase additives may be tried in order to reduce the problem.

When carrying out reversed-phase gradient elution for samples that require low-UV detection, problems often are encountered with the quality of both the water and the organic solvent. Milli-Q grade water (or equivalent) can be purified further by irradiation with UV light. Acetonitrile is the only organic solvent that can be used for detection at 210 nm or lower, and HPLC-grade acetonitrile forms UV-absorbing impurities on standing.[5] Consequently for low-UV detection (particularly at high sensitivity), the acetonitrile used must be "fresh" or must be purified by using an alumina precolumn.[5]

*Solvent Demixing.* This effect can arise in gradient elution with silica columns. If the B-solvent is very polar (e.g., propanol) and the A-solvent is very nonpolar (e.g., hexane), the B-solvent will be taken up by the column during the initial stage of gradient, followed by a "breakthrough" of the B-solvent during the gradient. The result is a sudden loss of resolution at that point in the chromatogram. This is illustrated by the example of Fig. 17.6, for the conditions just described.

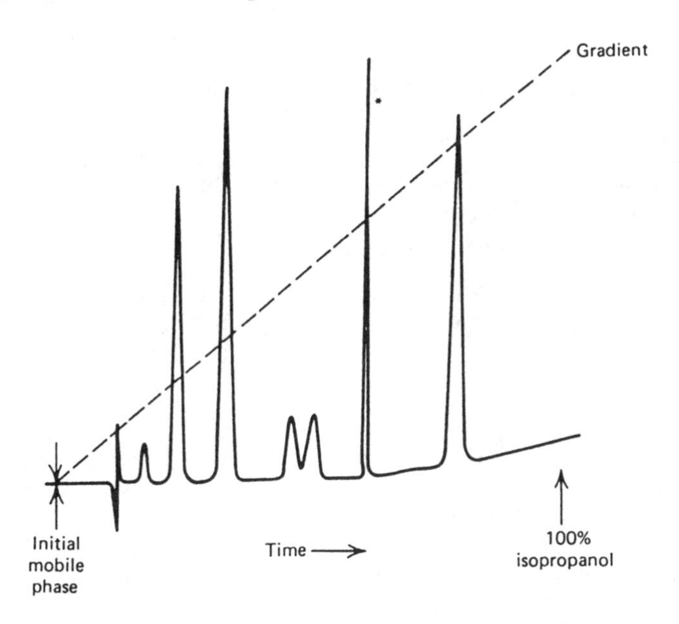


**Fig. 17.6.** Effect of solvent-demixing on gradient separation. Hexane/isopropanol gradient with silica column. See text. Reprinted from ref. (1) with permission.

It is seen that the next-to-last band is much narrower than the bands on either side. This peak has eluted at the point at which the isopropanol B-solvent has broken through (after solvent demixing). The solution to the problem of solvent-demixing is to use a column which has less tendency to sorb the B-solvent. Generally this problem does not occur with bonded-phase columns of various types.

## 17.2. Sample Pretreatment

Sample pretreatment refers to any operation involving the sample prior to its injection, apart from weighing and dilution. Some examples include:

1. Sample filtration
2. Sample extraction
3. Sample derivatization (pre-column derivatization)
4. Liquid chromatography (a second separation of the sample)

These various operations can be carried out either manually or using automation. The goals of sample pretreatment include

Table 17.1
Problems and Solutions for Gradient-Elution

| Cause of problem | Symptom | Solution |
|---|---|---|
| Starting mobile-phase too strong | Bands bunch at front of chromatogram with poor resolution | 1. Decrease %B at beginning of gradient |
| Suboptimum conditions | Poor resolution in middle of chromatogram | 1. Increase $k'$, $N$, and/or $\alpha$; remember that $k'$ varies with gradient time, flow rate, and column length (Eq. 17.1) |
| Poor column-equilibration | Retention times at front of chromatogram not reproducible | 1. Increase regeneration time between gradient runs<br>2. Inject samples at regular intervals |
| A and B solvents of different UV absorbance | Baseline drifts | 1. Add a nonretained UV-absorber to the less-absorbing solvent<br>2. Use different wavelength<br>3. Use different detector |
| Reagent or mobile-phase impurities | Artifactual peaks in in blank gradient | 1. Use HPLC-grade reagents<br>2. Purify mobile-phase components |
| Solvent demixing | Sudden change in resolution part way through chromatogram | 1. Use bonded-phase columns instead of bare silica |

removal of interferences, enhancement of analyte detection, and protection of the column.

Sample pretreatment aims at avoiding problems in LC separation. Therefore a review of when sample pretreatment is appropriate, and how different pretreatment procedures are carried out, will be useful in a troubleshooting context. Sample pretreatment also can be a source of other difficulties, as pointed out in the following discussion. These pretreatment-related problems include sample losses, possible contamination of the sample during pretreatment, incomplete derivatization reactions or multiple reaction products, and so on. The combined effect of these latter possibilities is to reduce assay precision. Consequently, assay procedures that include sample pretreatment are often less precise (Chapter 16). In this section we will address this general problem. A summary of sample-pretreatment problems and solutions can be found in Table 17.2 at the end of this section.

### *Filtration*

The sample solution for LC assay must be homogenous and free of particulates. Any solid particles in the sample can damage the sample injector and block the column inlet. Therefore, it is generally advisable to check the final sample solution by swirling it while holding it up to the light. Any indication of particulates, cloudiness, or opalescence means that the sample should be filtered prior to injection. The filters used should retain 0.5-µm and larger particles and be suitable for LC application. Several companies supply filters of this type (e.g., Gelman, Millipore). The filter must also be matched to the solvent used to dissolve the sample. Guidelines for filter use were summarized in Table 6.2 (Section 6.3).

Sample filtration can result either in (a) contamination of the sample, or (b) loss of analyte by adsorption onto the filter. These problems are not common, but they should be kept in mind if problems arise during the use of filtered samples. Sample contamination can be checked by soaking a filter with the sample solvent, then injecting an aliquot of the solvent. Any extraneous peaks indicate a possible contamination problem. Adsorptive loss of the sample generally will be greatest for low-concentration samples, and this possibility should be suspected whenever recovery decreases at lower analyte concentrations. Problems with filtration-induced sample contamination or loss can usually be controlled by use of the right filter. Manufacturers' guidelines should be consulted in this case. References (6a–6c) contain a discussion of extraction of interferences from membrane filters.

Routine filtration of every sample prior to injection can be costly; it may be unnecessary if only some of the samples actually require filtration. On the other hand, if any sample loss occurs on the filter, every sample should be treated identically (otherwise the filtered samples will show systematically lower recoveries than unfiltered samples).

### *Extraction*

Sample extraction is used to:

1. Separate analytes from insoluble sample matrices
2. Reduce interferences or sample components that can damage the column

*Insoluble Sample-Matrix.* Some samples are either insoluble solids or are mixed with insoluble solids. Examples include dried paints and inks, highly crosslinked polymers, samples of plant or animal origin, and so on. Attempts to dissolve such samples in any reasonable solvent generally are unsuccessful. Usually we are not interested in the solid matrix itself, in which case the analyte of interest can be extracted from the original sample. A simple procedure might involve finely dividing (chopping) the solid sample, weighing an aliquot into a volumetric flask, filling to mark with the extraction solvent, shaking occasionally over a half-hour period, and then aliquotting the supernatant for assay by LC.

In many cases, it will be found that incomplete extraction of the analyte results. This can be checked in two ways: (1) repeat the sampling and analysis as a function of time, to be sure that equilibrium is reached; (2) pour off the extraction solvent, replace it with fresh solvent, and repeat the entire procedure to make sure that at equilibrium there is complete extraction of the solid sample. A reproducible and precise extraction procedure generally requires 95%+ recovery of analyte into the extraction solvent. Extraction can be speeded up by using a laboratory blender for the chopping and extraction processes. Recovery (at equilibrium) can be improved by trying solvents of different polarity (e.g., hexane, ethyl acetate, and methanol)—or mixtures thereof.

*Interferences and Column-Damaging Components in the Sample.* A common use of sample pretreatment is for simplifying complex samples; i.e., cleanup procedures. The aim is a simple separation step that puts the analyte into one fraction, and undesirable sample components into a second fraction. Extraction of solid samples (as described above) is one example. In the case of liquid samples, a liquid–liquid partitioning or extraction can be used. For example, chloroform and water plus sample can be added to a separatory funnel, shaken together, and the phases separated. If the pretreatment is successful, most of the analyte will end up in one phase (e.g., chloroform), and most of the interferences plus compounds that might damage the column would end up in the water phase. The chloroform extract can then be evaporated to dryness, taken up in an aliquot of mobile phase, and analyzed by LC. Different partitioning solvents can be used, the pH of the solvents can be adjusted, and the relative volumes of the two phases can be varied to improve analyte recovery and to minimize interferences.

A detailed discussion of errors or imprecision in solvent extraction procedures has been given.[7] Precision is favored by (a) 95%+ recovery of analyte in its phase, (b) careful control over the time and temperature of the extraction step, (c) use of internal standards, and (d) avoiding emulsions. The precision of the extraction step can be checked independently, as by extracting several samples and then carrying out replicate assays of both the individual and pooled extracts. If the coefficient of variation (CV) for the individual extracts is CV1 and the CV for the pool is CV2, then the extraction-CV is $(CV1^2 - CV2^2)^{0.5}$. When assay imprecision is unacceptably high, and (extraction CV) > CV2, it is necessary to improve the extraction procedure first (to reduce its CV).

Extraction also can be carried out in a solid-phase mode. We have included this option under a following section (Liquid Chromatography).

## *Pre-column Derivatization*

Derivatization generally is used to enhance (a) the detection of trace compounds in "dirty" samples, (b) the determination of pollutants in environmental samples (soil, water or air), and (c) the measurement of low-level products in complex reaction mixtures such as fermentation broths. In each case, the sample is reacted with a derivatizing reagent—a compound that can react with a functional group in the analyte molecule, and whose reaction product is easily detected by UV or fluorescence measurements. The result is an enhanced signal from the derivatized analyte band, but with little increase in background (interferences) signal. Good accounts of precolumn derivatization are given in references (8,9). The general result of precolumn derivatization is increased accuracy and an ability to measure lower concentrations of the analyte. The precision of final results can also be better, although derivatization is itself a significant contribution to overall assay imprecision.

The precision (CV) of the derivatization step can be determined in the same way as for the CV of an extraction step (above), by carrying out replicate derivatizations of the same sample, pooling a portion of each derivatized sample, then comparing the precision of pool replicates vs that of the individual derivatized samples. When assay precision is not acceptable, it is often found that the CV of the derivatization step is the major contribution to overall assay CV. In this case, the derivatization step must be improved. Error in the latter step can arise from: (a) incomplete reaction of the analyte; (b) losses in any cleanup step required following derivatization; (c) degradation of the derivatization reagent. Each of these possibilities should be checked in attempts to improve precision.

### *Liquid Chromatography*

An initial separation of the sample by various liquid chromatography procedures can fractionate analytes of interest from various interferences and other undesirable components of the sample. Low-pressure column procedures, such as those that preceded HPLC, have been used in this connection. Today it is more common (and convenient) to use Solid Phase Extraction. This entails using disposable cartridges or mini-columns, such as Sep-Pak units sold by Waters Chromatography Division. These cartridges are filled with various solid sorbents, such as reversed-phase packings, silica, or ion-exchange resins. Disposable cartridges are used by connecting to them a syringe filled with sample (or various solvents), so that the following operations can be carried out: (a) application of sample; (b) washing off interferences; (c) stripping analyte from the column. The analyte fraction can then be evaporated to dryness, redissolved in mobile phase, and injected into the LC system. In some cases, the analyte fraction can be injected directly (without removal of stripping solvent).

As in the case of solvent-extraction (described above), solid-phase extraction generally introduces significant imprecision into the LC assay. This can arise from poor recovery (< 95%) of analyte from the cartridge, inaccurate measurement of the volume of the final extract solution, losses during the evaporation step, and other causes. Increased precision in solid-phase extraction can be attained by (a) minimizing the number of steps in the cleanup procedure (e.g., eliminating an evaporation-to-dryness step), (b) using an internal standard, and/or (c) carefully controlling the volumes used for removing interferences and analytes from the cartridge. An evaporation step can be avoided in various ways:

1. If the analyte is acidic or basic, use an ion-exchange cartridge, and elute the analyte from the cartridge using an aqueous solution of suitable pH or salt content; the resulting analyte-fraction can usually be injected directly onto a reversed-phase column (in some cases the analyte-fraction must be buffered to a new pH before injection)

2. For acidic or basic analytes, use a reversed-phase cartridge and aqueous eluting solutions, control pH so as to retain the analyte during washing of interferences from the column, and elute the analyte in the analyte-recovery step; the analyte fraction is then handled as in step 1 (low pH favors elution of bases; high pH favors elution of acids)

3. Use a reversed-phase cartridge and elute the analyte with a minimum volume of organic solvent (methanol or acetonitrile); dilute the resulting analyte-fraction 10:1 with water, and inject as large a sample volume as possible; the weak sample-solvent will often allow a relatively large injection volume (e.g., 500–1000 μL).

### *Automation*

Sample pretreatment can be a major source of assay imprecision, and it is often tedious and time-consuming. As a result, more and more laboratories are using automated sample pretreatment. A detailed discussion of this approach and its troubleshooting is beyond the scope of this book. We simply will cite some of the commercial systems available, and make some general comments on column-switching (one way to automate sample pretreatment).

*Varian AASP System.* This unit uses solid-phase extraction in a way that is similar to manual procedures. The individual cartridges are packaged in sets of ten (a cassette), allowing ten samples to be applied to each cassette. The individual cartridges are washed off-line to remove interferences, then the cassettes are placed in the AASP (Automated Analysis with Sample Pretreatment) module. The individual cartridges then are indexed automatically into connections that join the pump and LC column. At the appropriate time, a switching valve diverts flow of mobile phase through the column, washing the analytes onto the LC column for subsequent separation and analysis. This approach eliminates some of the manual steps in sample cleanup, and provides for more reliable and accurate recovery of analyte.

*Du Pont Prep System.* This sample-pretreatment system duplicates most of the steps in manual solid-phase extraction: washing off interferences, recovering the sample in an organic solvent and evaporating the sample to dryness. The various operations are performed by a centrifuge under control of a programmer. With the Prep unit, samples are applied manually to individual cartridges, the cartridges are processed by the system, and samples are redissolved manually and injected.

*Robotics.* Systems sold by Zymark, Perkin-Elmer, and others are designed to allow a robot to replace all of the manual steps involved in sample pretreatment. Solvent extraction, solid-phase extraction, filtration, derivatization, and other operations can now be carried out automatically. Because of the close control over each step, robots can improve precision in LC assays that require sample pretreatment. Troubleshooting these systems requires the development of new skills, beyond those presented in this book. However, much of the LC troubleshooting presented here will still apply for the LC part of a robotics system.

*Column-Switching.* For sample pretreatment, this refers to the use of two (or more) columns in series connected by a switching valve so that some fraction from the first column can be resolved further on the second column. This allows a variety of sample pretreatment procedures to be carried out. Typically column-switching procedures are automated, with injection of samples by an autosampler under the direction of a controller (computer, timer, or microprocessor) that also controls the switching valve. Because of close control over all steps of sample pretreatment plus frequent 95%+ recovery of analyte from the in-

Table 17.2
Sample-Pretreatment Problems and Solutions

| Cause of problem | Symptom | Solution |
|---|---|---|
| Contamination by sample filters | Extraneous peaks in chromatogram | 1. Confirm by soaking filter in sample solvent and injecting aliquot<br>2. Change filter type (see Table 6.2)<br>3. Use alternate cleanup technique |
| Adsorptive loss on sample filter | Smaller than expected peaks for some or all compounds, especially at lower concentrations | 1. Change filter type<br>2. Treat all samples exactly the same<br>3. Use alternate cleanup technique |
| Incomplete extraction | Poor or variable recovery | 1. Increase extraction time, heat solvents, etc.<br>2. Modify cleanup method |
| Interferences and contaminants from sample | Extra bands in chromatogram; shortened column life | 1. Improve cleanup |
| Incomplete recovery | Poor precision | 1. Improve or replace derivatization, separation, extraction, or other cleanup steps<br>2. Automate pretreatment to increase precision |

jected sample, it is possible to attain a CV of the order of ±1% in favorable cases [e.g., ref. (10)]. For a further discussion of column switching for sample pretreatment, see refs. (11–13).

## 17.3. References

[1]Snyder, L. R. and Kirkland, J. J. (1979) *Introduction to Modern Liquid Chromatography,* 2nd edn. Wiley-Interscience, New York, Chap. 16.

[2]Dolan, J. W., Snyder, L. R., and Quarry, M. A. (1987) *Chromatographia* **24,** 261.

[3]Snyder, L. R. (1980) *High-Performance Liquid Chromtography. Advances and Perspectives,* Vol. 1, (Cs. Horvath, ed.) Academic, New York, p. 207.

[4]Snyder, L. R. and Stadalius, M. A. High-Performance Liquid Chromatography. Advances and Perspectives, Vol. 4, Cs. Horvath, ed., Academic Press, New York, 1986, p. 195
[5]Berry, V. V. (1984) *J. Chromatogr.* **290,** 143.
[6]van der Wal, Sj. and Snyder, L. R. (1983) *J. Chromatogr.* **255,** 463.
[6a]Merrill, J. C. (1984) *A Laboratory Comparison of Popular HPLC Filters and Filter Devices for Extractables,* Gelman Sciences, Ann Arbor, MI.
[6b]Merrill, J. C. (1986) *A Laboratory Comparison of Nylon Syringe Filter Units for Extractables in HPLC,* Gelman Sciences, Ann Arbor, MI.
[6c]Merrill, J. C. (October 1987) *Amer. Lab.* **19(10),**74.
[7]Snyder, L. R. and van der Wal, Sj. (1981) *Anal. Chem.* **53,** 877.
[8]Lawrence, J. R. and Frei, R. W. (1976) *Chemical Derivatization in Liquid Chromatography,* Elsevier, Amsterdam.
[9]Sternson, L. (1980) *Chemical Derivatization in Analytical Chemistry,* Plenum, New York, Chap. 3.
[10]Nazareth, A., Jaramillo, L., Giese, R. W., Karger, B. L., and Snyder, L. R. (1984) *J. Chromatogr.* **309,** 357.
[11]Karger, B. L., Giese, R. W., and Snyder, L. R. (1983) *Trends Anal. Chem.* **2,** 1.
[12]Majors, R. (1984) *LC, Liq. Chromatogr. HPLC Mag.* **2,** 358.
[13]Little, C. J., Stahel, O. Lindner, W., and Frei, R. W. (October 1984) *Amer. Lab.,* p. 120.

# Index